Broadband Terahertz Devices and Communication Technologies, 2nd Edition

Broadband Terahertz Devices and Communication Technologies, 2nd Edition

Guest Editors

Jiao Zhang

Lu Zhang

Kaihui Wang

Xiongbin Yu

MDPI

Basel • Beijing • Wuhan • Barcelona • Belgrade • Novi Sad • Cluj • Manchester

Guest Editors
Jiao Zhang
Pervasive Communication
Research Center
Purple Mountain
Laboratories
Nanjing
China

Lu Zhang
College of Information
Science and Electronic
Engineering
Zhejiang University
Hangzhou
China

Kaihui Wang
School of Information Science
and Technology
Fudan University
Shanghai
China

Xiongbin Yu
Pengcheng Laboratory
Shenzhen
China

Editorial Office
MDPI AG
Grosspeteranlage 5
4052 Basel, Switzerland

This is a reprint of the Special Issue, published open access by the journal *Micromachines* (ISSN 2072-666X), freely accessible at: https://www.mdpi.com/journal/micromachines/special_issues/9TTIFXUQH8.

For citation purposes, cite each article independently as indicated on the article page online and as indicated below:

Lastname, A.A.; Lastname, B.B. Article Title. *Journal Name* **Year**, *Volume Number*, Page Range.

ISBN 978-3-7258-8309-7 (Hbk)
ISBN 978-3-7258-8310-3 (PDF)
https://doi.org/10.3390/books978-3-7258-8310-3

Contents

About the Editors

Jiao Zhang

Jiao Zhang is currently an Associate Researcher and Assistant Dean at Purple Mountain Laboratories, as well as an engineering doctoral supervisor at Southeast University. He received a Ph.D. degree from the Fudan University, Shanghai, China, in 2020. Between 2021 and 2023, he held the position of Post-Doctoral Research Fellow at Southeast University and Purple Mountain Laboratories, China. He focuses on photonics-aided millimeter wave/terahertz communication for 5G-A/6G networks, broadband optical fiber access, and high-speed optical fiber communication. He has authored and co-authored more than 230 peer-reviewed papers in top IEEE and OPTICA journals and conferences, and published 1 academic monograph. He is also a Senior Member of IEEE, and included in the "World's Top 2% Scientists" list for consecutive years. Dr. Zhang has been honored with prestigious recognitions including the "First Prize of Shanghai Natural Science Award", Jiangsu Provincial "333 High-Level Talent Cultivation Project" award, "National Microwave Photonics Academic Rising Star", and "2025 6G Rising Star"' Young Scholar.

Lu Zhang

Lu Zhang is a tenured Associate Professor and Doctoral Supervisor at the School of Information Science and Electronic Engineering, Zhejiang University, China. His research interests include photonic millimeter wave/terahertz communication and sensing systems, high-speed optical fiber transmission systems, and intelligent photonic signal processing. He received his Bachelor's degree from Chien-Shiung Wu College, Southeast University, and his Ph.D. from Shanghai Jiao Tong University under the supervision of Professor Shilin Xiao. During his doctoral studies, he participated in a CSC-sponsored joint Ph.D. program with the research groups of Professor Jiajia Chen and Professor Lena Wosinska at the Royal Institute of Technology (KTH), Sweden, where he also worked as a research engineer. Dr. Zhang currently serves as the deputy director of the Institute of Radio-frequency and Photonic Information Processing (RPIP) and is a Chief Young Scientist of the National Key R&D Program of China. He is also a Senior Member of IEEE.

Kaihui Wang

Kaihui Wang is an associate researcher at Fudan University. He obtained his Bachelor's degree in 2016 and Doctoral degree in 2021, both from Fudan University. He has been engaged in research on high-speed optical fiber communications, radio-over-fiber convergence systems, and low-complexity digital signal processing circuit design for many years, accumulating solid theoretical foundations and extensive practical experience. He has published more than 150 high-quality academic papers to date and has been included in Elsevier's "World's Top 2% Scientists" list for consecutive years. He is also the recipient of awards including the IEEE Photonics Society Graduate Student Scholarships.

Xiongbin Yu

Xiongbin Yu received his Ph.D. degree in the division of advanced electronics and optical science from Osaka University, Osaka, Japan, in 2020. Subsequently, he was specially appointed as an Assistant Professor with the Laboratory for Future Interdisciplinary Research of Science and Technology (FIRST), Tokyo Institute of Technology, Tokyo, Japan. Between 2021 and 2023, he was part of the Institute of Strategic Research, Huawei Company, Ltd., Shenzhen, China. He is currently a Researcher affiliated with the Department of Broadband Communication, Pengcheng Laboratory, Shenzhen, China. His research focuses on exploring terahertz electronic and photonic devices in relation to their communications and sensing applications.

micromachines

Editorial

Editorial for the Special Issue on Broadband Terahertz Devices and Communication Technologies, 2nd Edition

Jiao Zhang [1,*], Lu Zhang [2,*], Kaihui Wang [3,*] and Xiongbin Yu [4,*]

1 Purple Mountain Laboratories, Nanjing 211111, China
2 College of Information Science and Electronic Engineering, Zhejiang University, Hangzhou 310027, China
3 School of Information Science and Technology, Fudan University, Shanghai 200433, China
4 Pengcheng Laboratory, Shenzhen 518000, China
* Correspondence: zhangjiao1@pmlabs.com.cn (J.Z.); zhanglu1993@zju.edu.cn (L.Z.); khwang@fudan.edu.cn (K.W.); yuxb@pcl.ac.cn (X.Y.)

The exponential growth of user data traffic imposes increasingly stringent requirements on communication networks regarding bandwidth, data rates, and diversified capabilities to support future social and industrial information exchange. Consequently, academic and industrial communities have actively pursued research and development (R&D) efforts to explore the feasibility of mobile broadband systems operating in frequency bands above 100 GHz. The terahertz band, spanning 0.1–10 THz and situated between microwave and optical frequencies, is regarded as a critical breakthrough for revolutionizing communication technologies, owing to its abundant spectrum resources [1–3]. As such, it has been widely recognized as a promising candidate for future bandwidth-intensive applications, particularly sixth-generation (6G) communications. In 2024, the ITU Radiocommunication Sector (ITU-R) approved a new report entitled "Technical Feasibility of IMT in Bands Above 100 GHz", providing comprehensive technical information on frequency bands between 100 GHz and 400 GHz [4]. The terahertz band is capable of satisfying the ultra-high bandwidth and ultra-large capacity demands of emerging scenarios, thereby facilitating advanced services such as virtual reality, augmented reality, and satellite–terrestrial communications [5]. Furthermore, terahertz technology has also found promising applications in data center interconnection and integrated space–ground networks, reflecting its growing role in diverse high-performance communication infrastructures [6].

Building upon the success of the first edition of the Broadband Terahertz Devices and Communication Technologies Special Issue [7], this second edition continues to explore the state-of-the-art advances in the terahertz field. This Special Issue assembles 15 peer-reviewed articles, including 2 review papers and 13 original research articles. The covered research scope involves key broadband terahertz devices, system-level verification of terahertz communication systems, and innovative technologies at other low-frequency bands, which can also provide valuable insights and inspiration for subsequent terahertz research.

Five studies (Contribution 1–5) present power amplifiers, low-noise amplifiers and traveling wave tube amplifiers at different frequency bands, which could effectively expand the coverage range. Hu et al. (Contribution 1) and Shen et al. (Contribution 2) proposed a V-band high-gain wideband power amplifier (PA) and a compact V-band low-noise amplifier (LNA), respectively. The former PA prototype achieved a peak gain of 27.3 dB at 64 GHz, with a 3 dB bandwidth exceeding 13 GHz and a maximum saturated output power of 19.7 dBm using a 130 nm SiGe BiCMOS process. The latter LNA prototype exhibits a gain variation of less than 1.5 dB, a noise figure variation of less than 1.2 dB at test frequencies from 40 GHz to 65 GHz, and also shows excellent temperature robustness over a broad range from −55 °C to 85 °C. Yuan et al. (Contribution 3) designed two wideband high-efficiency power amplifiers operating in the X and Ku bands with peak saturated output powers of 31.2 dBm and 30.8 dBm, respectively, and fabricated using a 0.25 mm GaAs pseudomorphic high electron mobility transistor (pHEMT) process. Liu et al. (Contribution 4) proposed a dual-side hybrid embedding network to mitigate gain degradation in terahertz amplifiers at frequencies near $f_{\max}$; a four-stage amplifier is designed with a power gain of 19.3 dB at 280 GHz, corresponding to an 11.5 dB improvement over a conventional unboosted amplifier. Furthermore, Jamil et al. (Contribution 5) presented the design and optimization of a periodic cusped magnet-focused sheet beam optical system for application in a 340 GHz traveling wave tube amplifier.

Wang et al. (Contribution 6) presented a wideband D-band frequency sextupler chain in a 100 nm GaAs pHEMT process, comprising an input-stage tripler, an inter-stage harmonic-rejection power amplifier, and an output-stage doubler. The balanced tripler suppresses the fundamental, while the inter-stage PA enhances harmonic rejection and provides sufficient drive. The single-balanced doubler extracts the second harmonic while suppressing odd-order components. The prototype achieves 2.33 dBm peak output power with 0.33 dB conversion gain and an 18.9% fractional bandwidth (126.3–152.7 GHz). The 5th and 7th harmonics are suppressed by over 20 dBc, demonstrating suitability for wideband D-band signal generation.

Liu et al. (Contribution 7) proposed a dual-band bandpass filter with a wide upper stopband, integrating stepped-impedance resonators and a low-pass filter. The filter operates at 2.5 and 5.35 GHz with insertion losses of 0.12 dB and 0.6 dB, respectively, and achieves a wide upper stopband extending from 6.1 to 25 GHz to suppress high-frequency interference. Simulated and measured results show good agreement.

Electromagnetic metamaterials, engineered to manipulate wave properties including amplitude, phase, and wavefront, hold a prominent position in the field. Xu et al. (Contribution 8) developed a Metamaterial Incident Photon Reconstruction Theory (MIPRT), which reveals that metamaterials modulate incident photons through coherent destructive interference rather than absorption and re-emission. Validated in a single-layer metamaterial, MIPRT derives a unique resonant phase-amplitude relationship and offers a novel perspective on electromagnetic response mechanisms. Through scattering theory analysis of representative structures, Xu et al. (Contribution 9) derived and verified two explicit corollaries on transmission-reflection coupling, offering new insight into fundamental modulation mechanisms.

Terahertz detectors hold a considerable promise for applications spanning communications, imaging, spectroscopy, and sensing. Wang et al. (Contribution 10) designed a photonic terahertz antenna for the 275–296 GHz band, integrating modified uni-traveling carrier photodiodes (MUTC-PDs), impedance matching networks, a Wilkinson power combiner, and a Vivaldi antenna. Simulations show 1.58 dBm saturated output power at 280 GHz, a 4.13-fold enhancement over standalone devices via optimized impedance matching and power combining. The antenna achieves 7.93 dBi peak gain with reflections below −10 dB. Combining efficiency exceeds 95% under 20° phase imbalance, while the Wilkinson combiner shows only 0.76 dB insertion loss at 285 GHz. Tang et al. (Contribution 11) presented a 380 GHz zero-bias detector based on an ACST Schottky diode for terahertz radiometer receivers. A high-impedance grounding topology suppresses parasitic resonances and enhances stability, while a compact U-shaped waveguide transition realizes an inline configuration for simplified integration. Measurements demonstrate a peak voltage responsivity of 2318 V/W and a linearity of 0.9996 at 380 GHz, validating the proposed design for high-frequency radiometer and sensing applications.

Benefiting from broadband terahertz devices, terahertz communication systems have undergone rapid evolution in recent years. Gao et al. (Contribution 12) compared all-

https://doi.org/10.3390/mi17060672

optical and all-electric receivers over 15 m and demonstrated HD video transmission over 200 m and 2 km using a 100 GHz all-optical transceiver. The all-optical transceiver achieves 11.318 Gbps error-free transmission over 200 m without clock recovery, whereas the all-electric receiver is limited to 3.125 Gbps over 15 m. These results validate the all-optical receiver for ultra-wideband, high-capacity 6G wireless communications.

Ensuring transmission security for high-speed terahertz wireless links remains a paramount concern in the field of terahertz research. Wang et al. (Contribution 13) proposed a secure, spectrum-efficient radio-over-fiber (RoF) system using compressive sensing (CS) and chaotic encryption. An 8 Gbit/s system is demonstrated over 10 km fiber and 20 GHz wireless links. Spectrum efficiency is enhanced by compressing data and reducing quantization bits. The system recovers signals with 6-bit fronthaul quantization, yielding SSIMs of 0.952 and 0.073 for legitimate (Bob) and unauthorized (Eve) receivers, respectively, at a 0.75 compression ratio, validating the proposed security.

Terahertz communication is pivotal for 6G, addressing data demands beyond current microwave and mmWave capabilities. However, Tbps kilometer-range transmission faces severe free-space path loss (>120 dB/km) and atmospheric absorption—the "dual attenuation dilemma." Yu et al. (Contribution 14) summarized their advancements in long-distance terahertz communication through photonic–electronic co-design, including photonically assisted generation, polarization-multiplexed MIMO with MRC, high-gain antenna-lens arrays, and InP amplifiers. Experimental milestones include 1.0488 Tbps PS-64QAM (330–500 GHz), 30.2 km D-band transmission, 3 km fog-penetrating link at 312 GHz, and 100 Gbps satellite–terrestrial links beyond 36,000 km, validating THz viability for extreme-capacity 6G backhaul.

Finally, the multi-scenario applications of terahertz technology with free-space optics are reviewed. Liu et al. (Contribution 15) reviewed integrated terahertz and free-space optical (FSO) systems for future 6G and space–air–ground integrated networks (SAGIN). It examines advances in channel modeling, transmission performance, and integrated architectures, analyzing attenuation, turbulence, pointing errors, and noise under diverse atmospheric conditions. Terahertz achieves multi-Tbps rates over kilometers but suffers from molecular absorption and weather attenuation, whereas FSO offers superior bandwidth-distance products yet is vulnerable to turbulence-induced fading. Hybrid terahertz and FSO systems using adaptive switching enhance reliability and spectral efficiency, demonstrating Tbps-class data rates. Challenges remain in hardware integration and resource allocation, while future directions include unified channel models and intelligent switching for robust infrastructures.

The contributions presented in this Special Issue collectively advance terahertz devices and communication technologies, offering both profound theoretical insights and substantial engineering value. As Guest Editors, we extend our sincere gratitude to the authors for their outstanding contributions, to the reviewers for their rigorous and constructive feedback, and to the editorial team for their steadfast dedication throughout the publication process.

Conflicts of Interest: The authors declare no conflicts of interest.

List of Contributions:

1. Hu, J.; Wan, J.; Shen, Y.; Zhao, W.; Luo, J. A V-Band Wideband Power Amplifier with High Gain in a 130 nm SiGe BiCMOS Process. *Micromachines* **2024**, *15*, 1077. https://doi.org/10.3390/mi15091077.
2. Shen, Y.; Luo, J.; Zhao, W.; Dai, J.-Y.; Cheng, Q. A Compact V-Band Temperature Compensation Low-Noise Amplifier in a 130 nm SiGe BiCMOS Process. *Micromachines* **2024**, *15*, 1248. https://doi.org/10.3390/mi15101248.

3. Yuan, Y.; Zhao, X.; Fan, J.; Yu, Z. Design of Dual Continuous-Mode Class-J Power Amplifiers with Harmonic Matching Networks for X and Ku Bands. *Micromachines* **2025**, *16*, 1362. https://doi.org/10.3390/mi16121362.
4. Liu, X.; Yu, J.; Wang, Y.; Huang, Y.; Zhang, F.; Wang, Z.; Cheng, Y. Dual-Side Hybrid Embedding Network for Gain Enhancement of Terahertz Amplifiers at Frequencies Near fmax. *Micromachines* **2026**, *17*, 432. https://doi.org/10.3390/mi17040432.
5. Jamil, M.H.; Lin, Z.; Sharif, H.; Abbas, N.S.; He, W. Design and Simulation of an Electron Optical System for Terahertz Vacuum Devices. *Micromachines* **2025**, *16*, 928. https://doi.org/10.3390/mi16080928.
6. Wang, P.; Chen, Z.; Guo, Y.; Qi, Y.; Yang, P. A Wideband D-Band Frequency Sextupler Chain with High Harmonic Rejection in 100 nm GaAs pHEMT Technology. *Micromachines* **2025**, *16*, 984. https://doi.org/10.3390/mi16090984.
7. Liu, L.; Lin, Y.; Chen, Q.; Zhang, L.; Weng, M. A Dual-Band Bandpass Filter with Wide Upper Stopband Using Stepped-Impedance Resonators and an Integrated Low-Pass Filter. *Micromachines* **2026**, *17*, 75. https://doi.org/10.3390/mi17010075.
8. Xu, B.; Zhong, R. Metamaterial Incident Photon Reconstruction Theory Based on Resonant Dipole Phase. *Micromachines* **2026**, *17*, 130. https://doi.org/10.3390/mi17010130.
9. Xu, B.; Zhong, R. Intrinsic Relations Between Transmission and Reflection in Metamaterials. *Micromachines* **2026**, *17*, 493. https://doi.org/10.3390/mi17040493.
10. Wang, Y.; Liu, X.; Yu, J. MUTC-PD-Based High-Efficiency Photonic Terahertz Generation and Radiation in the 275–296 GHz Band. *Micromachines* **2026**, *17*, 196. https://doi.org/10.3390/mi17020196.
11. Tang, H.; Deng, Y.; Zhang, D. A Compact 380 GHz Zero-Bias Schottky Diode Detector for High-Sensitivity Radiometer Applications. *Micromachines* **2026**, *17*, 352. https://doi.org/10.3390/mi17030352.
12. Gao, S.; Jiang, Y.; Li, Z.; Zhong, Q.; Zhu, M.; Zhang, J. 2 km Uncompressed HD Video Wireless Transmission at 100 GHz Based on All-Optical Frequency Up- and Down-Conversion. *Micromachines* **2024**, *15*, 1488. https://doi.org/10.3390/mi15121488.
13. Wang, Z.; Zhang, L.; Zhang, J.; Ozolins, O.; Pang, X.; Yu, X. High Spectrum Efficiency and High Security Radio-Over-Fiber Systems with Compressive-Sensing-Based Chaotic Encryption. *Micromachines* **2026**, *17*, 80. https://doi.org/10.3390/mi17010080.
14. Yu, J.; Chen, J. Research on Tbps and Kilometer-Range Transmission of Terahertz Signals. *Micromachines* **2025**, *16*, 828. https://doi.org/10.3390/mi16070828.
15. Liu, J.; Yang, X.; Wei, Y.; Zhao, F. Integrated THz/FSO Communications: A Review of Practical Constraints, Applications and Challenges. *Micromachines* **2025**, *16*, 1297. https://doi.org/10.3390/mi16111297.

References

1. Zhu, M.; Zhang, J.; Hua, B.; Lei, M.; Cai, Y.; Tian, L. Ultra-wideband fiber-THz-fiber seamless integration communication system toward 6G: Architecture, key techniques, and testbed implementation. *Sci. China Inf. Sci.* **2023**, *66*, 113301. [CrossRef]
2. Yu, J. Photonics-assisted millimeter-wave wireless communication. *IEEE J. Quantum Electron.* **2017**, *53*, 8000517. [CrossRef]
3. Tripathi, S.; Sabu, N.V.; Gupta, A.K.; Dhillon, H.S. Millimeter-wave and terahertz spectrum for 6G wireless. In *6G Mobile Wireless Networks*; Springer: Berlin, Germany, 2021; pp. 83–121.
4. *ITU-R M.2541-0*; Technical Feasibility of IMT in Bandsabove 100 GHz. ITU: Geneva, Switzerland, 2024.
5. Guan, S.; Cheng, J.; Chang, S. Recent Progress of Terahertz Spatial Light Modulators: Materials, Principles and Applications. *Micromachines* **2022**, *13*, 1637. [CrossRef] [PubMed]
6. Ahmad, I.; Tan, W.; Ali, Q.; Sun, H. Latest Performance Improvement Strategies and Techniques Used in 5G Antenna Designing Technology, a Comprehensive Study. *Micromachines* **2022**, *13*, 717. [CrossRef] [PubMed]
7. Zhang, L.; Pang, X.; Pitchappa, P. Editorial for the Special Issue on Broadband Terahertz Devices and Communication Technologies. *Micromachines* **2023**, *14*, 1044. [CrossRef] [PubMed]

https://doi.org/10.3390/mi17060672

micromachines

MDPI

Article

A V-Band Wideband Power Amplifier with High Gain in a 130 nm SiGe BiCMOS Process

Jianing Hu [1], Jialong Wan [2], Yi Shen [1], Wei Zhao [1] and Jiang Luo [1,3,*]

[1] School of Electronics and Information, Hangzhou Dianzi University, Hangzhou 310018, China
[2] School of Microelectronics, South China University of Technology, Guangzhou 511442, China
[3] State Key Laboratory of Millimeter Waves, Southeast University, Nanjing 210096, China
* Correspondence: luojiang@hdu.edu.cn

Abstract: This paper introduces a high-gain wideband power amplifier (PA) designed for V-band applications, operating across 52 to 65 GHz. The proposed PA design employs a combination of techniques, including pole-gain distribution, base-capacitive peaking, and the parallel configuration of multiple small-sized transistors. These strategies enable significant bandwidth extension while maintaining high gain, substantial output power, and a compact footprint. A two-stage PA using the combination technique was developed and fabricated in a 130 nm SiGe BiCMOS process. The PA prototype achieved a peak gain of 27.3 dB at 64 GHz, with a 3 dB bandwidth exceeding 13 GHz and a fractional bandwidth greater than 22.2%. It delivered a maximum saturated output power of 19.7 dBm and an output 1 dB compression point of 18 dBm. Moreover, the PA chip occupied a total silicon area of 0.57 mm^2, including all testing pads with a compact core size of 0.198 mm^2.

Keywords: millimeter-wave (mm-wave); SiGe BiCMOS; bandwidth extension; power amplifier

1. Introduction

The V-band encompasses a rich spectrum of 40 to 75 GHz and has been widely utilized in high-speed wireless communications, high-resolution radar, and radio telescopes in recent years [1–4]. These applications have driven an increasing demand for multi-band, wideband millimeter-wave (mm-wave) front ends. Undoubtedly, the power amplifier (PA) plays a crucial role as an indispensable component in a mm-wave transmitter. It is expected to provide wide bandwidth, high gain, high output power (P_{out}), and excellent linearity while maintaining a flat frequency response within the target band, ensuring optimal spectrum efficiency. Due to the effective scaling of device sizes, SiGe BiCMOS has proven to be an efficient technology for mm-wave systems [5–7]. Moreover, its compatibility with bulk silicon CMOS for biasing circuits and digital logic makes it well suited for cost-sensitive mass production. Consequently, there is a significant demand for a wideband high-performance BiCMOS PA tailored for advanced mm-wave front ends.

Considerable work has been focused on increasing the operational bandwidth of mm-wave PAs [8–16]. Several mm-wave wideband PAs developed in SiGe BiCMOS technologies were reported in [10–13]. A 60 GHz single-stage, single-ended cascode PA has been designed and fabricated using a 0.13 μm SiGe BiCMOS process, with a focus on maximizing output power while minimizing chip area [11]. The PA features a 12 GHz bandwidth spanning from 54 to 66 GHz, a power gain of 18 dB, and a maximum output power of 14.7 dBm, all within a compact chip area of only 0.3 mm^2. In reference [10], a transformer-based two-way combining common emitter PA employs a capacitor coupling technique to significantly suppress phase imbalance. The prototype SiGe PA operates under 1.5 V low-voltage conditions, delivering a bandwidth of 42–48.5 GHz, a peak gain of 29.5 dB, and a very flat output 1 dB compression point (OP_{1dB}) ranging from 14.5 to 14.8 dBm. However, the saturated output power (P_{sat}) is only 16.2 dBm, indicating considerable room for improvement. Distributed architectures offer significant advantages in bandwidth expansion. Lee et al. reported an 18–50 GHz SiGe HBT common-base non-uniform distributed PA [12]. By utilizing a multi-section lumped element artificial transmission line, they achieved optimal load conditions for each active stage across the entire bandwidth. The proposed mm-wave PA reaches a peak P_{sat} of 19.0 dBm and a peak PAE of 19.1% at 32.0 GHz. Nevertheless, this PA occupies a large silicon area of 1.83 mm^2. To enhance efficiency and output power over a wide frequency range, a traditional transmission line-based impedance inversion technique was applied to a 79 GHz differential Doherty PA in a 130 nm SiGe BiCMOS technology [13]. The proposed amplifier achieves a 1 dB flat bandwidth of 12 GHz, with a P_{sat} of 17.0 dBm and a peak power-PAE of 11.6%. However, achieving this excellent wideband response resulted in a noticeable compromise in the output return loss. Zhao et al. introduced a design approach for interstage and output matching networks in wideband power amplifiers, utilizing wideband inductively coupled resonators and Norton transformations [8]. The prototype PA was fabricated using a 28 nm CMOS process, delivering a P_{sat} of 13 dBm over a 40–67 GHz bandwidth and an OP_{1dB} of 12 dBm without the need for power combining. However, the peak gain is constrained to 13 dB. Additionally, several other wideband PAs based on CMOS technology have also been reported [14–18]. The primary challenge in broadband PA design is achieving a wideband response while balancing key parameters like gain, P_{out}, OP_{1dB}, and physical size. Although the aforementioned techniques can somewhat broaden the bandwidth, there is still an urgent need for solutions that can significantly extend the bandwidth while maintaining high gain, good gain flatness, high P_{out} and OP_{1dB}, as well as a compact footprint.

In this paper, we propose an alternative V-band wideband PA with high gain and high P_{out}. The design strategy utilizes a pole-gain tuning approach with dual transmission lines and a common-base (CB) capacitive peaking technique to achieve a broadband gain frequency response. Additionally, it employs a parallel configuration of multiple small-sized transistors to enhance P_{out}. The proposed techniques are demonstrated in a two-stage cascode PA using 130 nm SiGe BiCMOS technology. This article is organized as follows. Section 2 discusses the design of the proposed wideband PA, including the overall scheme selection and circuit schematic design. The post-simulation and measurement results that validate the design technique are presented in Section 3. Finally, the paper concludes in Section 4.

2. Circuit Analysis and Design

Figure 1 illustrates the simplified schematic of the proposed two-stage common-emitter-common-base (CE-CB) V-band PA. The CE-CB configuration is chosen for its superior voltage gain, higher breakdown voltage, and improved isolation compared to standalone common-emitter (CE) or CB topologies. Two pole-gain tuning networks (PGTN) with dual transmission lines are applied to both the first and second stages to adjust the dominant poles at different frequencies, thereby achieving a wideband frequency response. The base capacitors C_{b1} and C_{b2} are connected to the base terminals of transistors Q_2 and Q_4, respectively, to reduce input parasitic capacitance and enhance power gain, particularly the maximum power gain (MPG) at high operating frequencies. Transmission lines are utilized as inductive passive components due to their smaller dimensions and higher quality factor compared to traditional bulky spiral inductors.

2.1. Bandwidth Extension with Pole-Gain Tuning

Given the significant parasitic parameters at mm-wave frequencies, device models have become exceedingly complex, making the optimization of mm-wave amplifiers a considerably challenging task. However, the pole-gain tuning technique has been demonstrated to be an effective approach for enhancing amplifier bandwidth, as it provides great advantages in graphical analysis for bandwidth expansion [19,20]. Generally, the gain-frequency response of an amplifier can be expressed as a function of a complex variable.

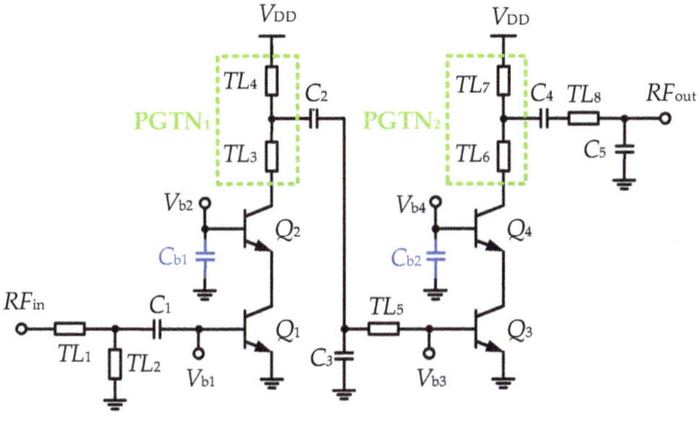

Figure 1. Simplified schematic of the proposed V-band wideband PA.

This function includes the coefficient K, zeros z_i, and poles p_j. Equations (1) and (2), respectively, provide the magnitude and phase responses of the function. As the value of s approaches zero, the function value gradually decreases to 0. Conversely, as s nears a pole, the function value experiences a substantial increase. The behavior of the gain-frequency function is governed by its zeros and poles. Therefore, the wideband characteristics of the amplifier can be obtained by tuning different dominant poles.

$$|H(s)| = K\frac{\prod_{i=1}^{m}|s - z_i|}{\prod_{j=1}^{n}|s - p_j|} \tag{1}$$

$$\angle H(s) = \sum_{i=1}^{m}\angle(s - z_i) - \sum_{j=1}^{n}\angle(s - p_j) \tag{2}$$

The dominant poles of the proposed V-band PA with PGTN are plotted in Figure 2, where $\rho_j = |s - p_j|$ means the distance from s to pole j in the coordinate chart, and $\theta_j = \angle(s - p_j)$ means the angle from the horizontal. Note that the gain-frequency response function is mainly controlled by poles close to the imaginary axis; poles far away from the imaginary are ignored. Two pairs of complex conjugate poles, p_1 and p_2, and p_3 and p_4, are introduced through PGTNs (PGTN$_1$ and PGTN$_2$) in the first and second stages, respectively. The p_1 (p_2) govern the low-frequency gain characteristics and are referred to as the low-frequency dominant poles P_L. The pole p_3 (p_4) dominates the high-frequency gain behavior and is designated as the high-frequency dominant pole P_H.

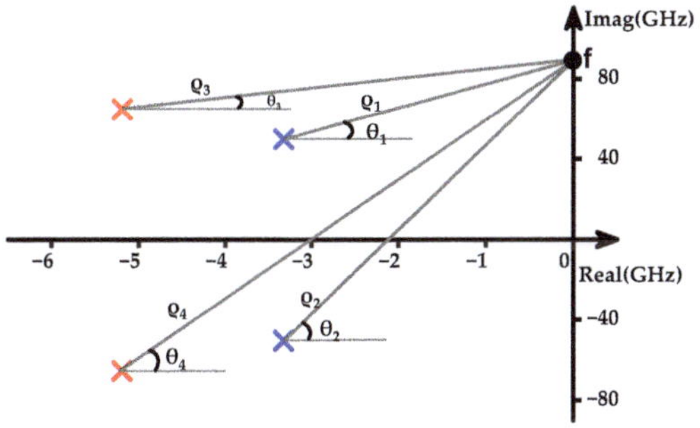

Figure 2. Dominant poles plot for the proposed V-band amplifier.

The $PGTN_1$ in the first stage is primarily composed of transmission line inductors TL_3 and TL_4, which are used to generate the high-frequency dominant poles p_3 and p_4. TL_3 is directly connected to the collector of transistor Q_2, isolating the load of the amplifier from the output capacitor. This configuration introduces an initial delay in current reaching other parts of the network, thereby delaying the rise time of the collector current in exchange for improved bandwidth. Figure 3a illustrates the locus of the dominant poles p_1 and p_2, and p_3 and p_4 as TL_3 increases from 55 pH to 85 pH, while Figure 3b shows the gain-frequency response of the proposed V-band PA under different conditions of TL_3. It can be observed that the high-frequency dominant poles p_3 and p_4 are more sensitive to variations in TL_3 than the low-frequency dominant poles p_1 and p_2. The dominant poles, p_3 and p_4, quickly move toward the imaginary axis and gradually shift to lower frequencies, while p_1 and p_2 approach the imaginary axis much more slowly. This results in a significant reduction in the minimum vector of poles p_3 and p_4, thereby enhancing the amplitude of the high-frequency dominant poles p_3 and p_4. Figure 3b provides a more intuitive physical insight into how the gain-frequency response of the amplifier is influenced by the high-frequency dominant poles.

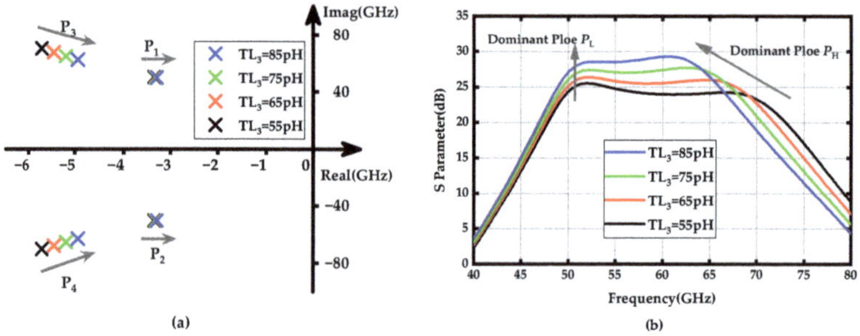

Figure 3. (**a**) Locus of poles p_1 (p_2) and p_3 (p_4) when the TL_3 increases from 55 to 85 pH; (**b**) Gain-frequency response of the proposed amplifier when the TL_3 values vary from 55 to 85 pH.

Similarly, the parallel peaking transmission line inductor TL_4 delays the current flow into its own branch, allowing more of the initial charging current to flow toward the capacitive load. This reduces the rise time and can also extend the bandwidth. Figure 4a depicts the locus of the dominant poles p_1 and p_2, and p_3 and p_4 as TL_4 increases from 35 pH to 65 pH and illustrates the locus of the dominant poles p_1 and p_2, and p_3 and p_4 as TL_4 increases from 55 pH to 85 pH, while Figure 4b displays the PA's gain-frequency response for different TL_4 values. It can be seen that TL_4 not only adjusts the gain-frequency response of the high-frequency dominant pole but also significantly impacts the amplitude of the low-frequency dominant pole, offering greater flexibility in tuning the amplifier's frequency response.

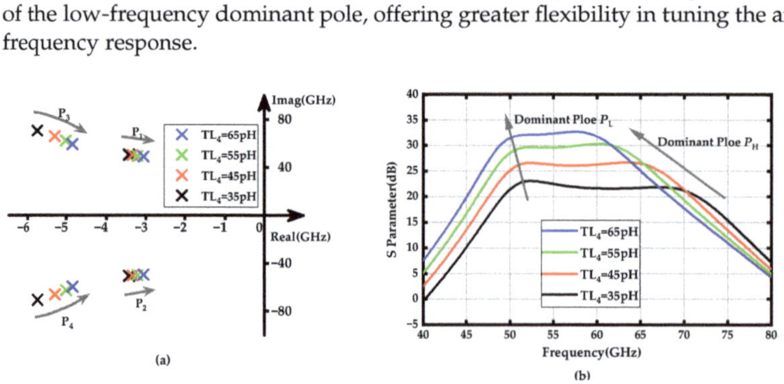

Figure 4. (**a**) Locus of poles p_1 (p_2) and p_3 (p_4) when the TL_4 increases from 35 to 65 pH; (**b**) Gain-frequency response of the proposed amplifier when the TL_4 values vary from 35 to 65 pH.

To achieve further bandwidth expansion, the low-frequency dominant poles p_1 and p_2 should be more flexibly tunable. These poles are established by the $PGTN_2$, which consists of the transmission line inductors TL_6 and TL_7, located in the second stage of the amplifier. Figure 5a shows the locus of p_1 and p_2, as well as p_3 and p_4, as TL_6 increases from 40 to 100 pH. It can be observed that the low-frequency dominant poles p_1 and p_2 are significantly affected by changes in TL_6; as the value of TL_6 increases, they rapidly shift away from the imaginary axis and towards lower frequencies. In contrast, although the high-frequency dominant poles p_3 and p_4 also move away from the imaginary axis, they do so at a much slower and more parallel rate, resulting in relatively minor degradation of their amplitude. Figure 5b demonstrates the gain-frequency response verification results for TL_6 variations affecting the low and high-frequency dominant poles. Furthermore, the impact of the transmission line inductor TL_7 on the amplifier's dominant poles was analyzed, as depicted in Figure 6. These results indicate that both dominant poles shift closer to the imaginary axis, with the lower-frequency poles p_1 and p_2 moving more rapidly. This shift helps mitigate the drawback of enhanced bandwidth at the expense of gain degradation caused by the increase in TL_6.

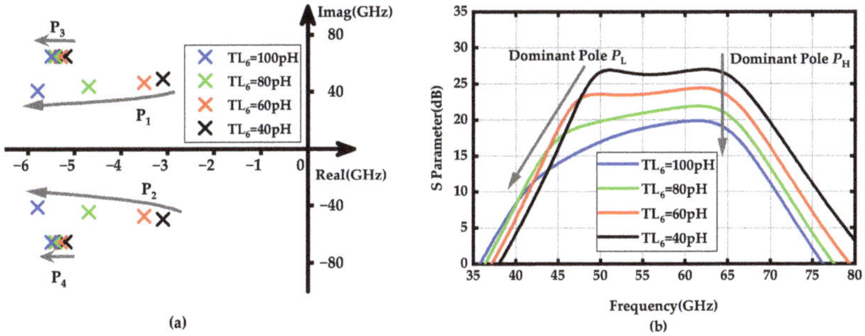

Figure 5. (**a**) Locus of poles p_1 (p_2) and p_3 (p_4) when the TL_6 increases from 40 to 100 pH; (**b**) Gain-frequency response of the proposed amplifier when the TL_6 values vary from 40 to 100 pH.

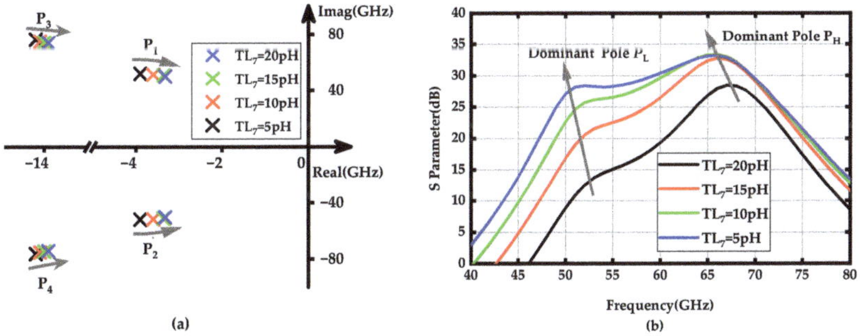

Figure 6. (**a**) Locus of poles p_1 (p_2) and p_3 (p_4) when the TL_7 increases from 40 to 100 pH; (**b**) Gain-frequency response of the proposed amplifier when the TL_7 values vary from 40 to 100 pH.

In summary, the proposed PGTN technique provides an effective approach to extending amplifier bandwidth. Firstly, the generated dominant pole pairs, p_1 (p_2) and p_3 (p_4), remain in the left-half plane, thus ensuring amplifier stability. Secondly, the high-frequency dominant poles p_3 and p_4 generated by $PGTN_1$ always move closer to the imaginary axis, significantly enhancing the high-frequency response. Meanwhile, the movement direction and magnitude of the low-frequency dominant poles, p_1 (p_2), can be precisely controlled, providing considerable flexibility in adjusting the amplifier's overall gain-frequency response.

2.2. Common-Base Capacitive Peaking Technique

At millimeter-wave frequencies, significant parasitic effects can degrade the amplifying capability of active transistors. In this work, the capacitive peaking technique is applied to the bases of the CB transistors Q_2 and Q_4 to reduce input parasitic capacitance, thereby enhancing the amplifier's gain, particularly in the high-frequency band. The simplified small signal equivalent circuit of the one-stage CE-CB amplifier is plotted in Figure 7. Disregarding the emitter junction resistance of the transistor, the gain-frequency response can be approximately calculated as follows:

$$A_v = G_m \cdot Z_{out} \approx g_{m1} \cdot (Z_x||Z_L) \tag{3}$$

$$Z_{out} = Z_L|| \left[r_{o2} + (r_{o2} \cdot g_{m2} + 1) \cdot \left(\frac{1}{s \cdot C_s} || r_{o1} \right) \right] \tag{4}$$

$$C_s = \frac{C_b \cdot C_{\pi 2}}{C_b + C_{\pi 2}} \tag{5}$$

where G_m and Z_{out} are the total equivalent transconductance and output impedance of the CE-CB amplifier, respectively, Z_X denotes the impedance observed at node X, g_{m1} and g_{m2} are the transconductances of transistors Q_1 and Q_2, while $C_{\pi 1}$, $C_{\pi 2}$, r_{o1} and r_{o2} are the equivalent input parasitic capacitors and the equivalent output resistors of the transistors Q_1 and Q_2, respectively. As inferred from Equations (4) and (5), an increase in C_b will result in a rise in the equivalent output impedance Z_{out}, thereby enhancing the gain Av.

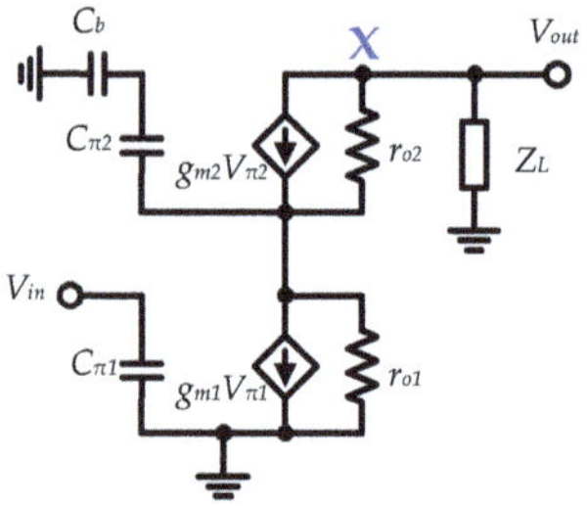

Figure 7. Simplified small-signal equivalent circuit for one-stage CE-CB amplifier.

Figure 8 demonstrates the simulated MPG versus variations of the C_b. As the values of C_b increase from 0 to 400 fF, the simulated MPG shows a significant improvement at high frequencies. This enhancement contributes to achieving a flat wideband gain-frequency response, as the transistor's MPG naturally deteriorates with increasing frequency.

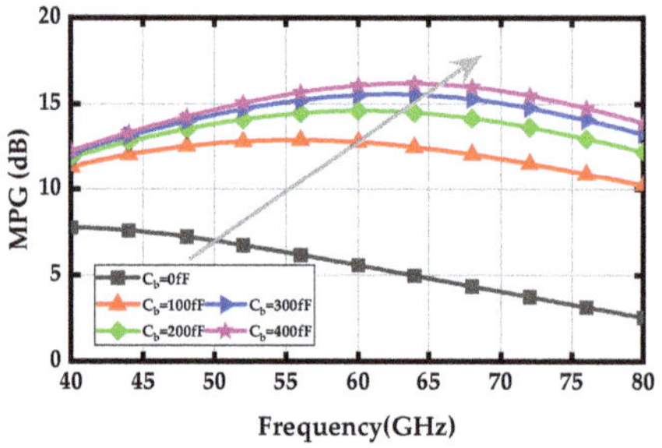

Figure 8. Simulated MPG with variations of base capacitor C_b.

2.3. Layout Configuration of Power Stage Transistors

To achieve high P_{out} within the target mm-wave frequencies, large-size transistors are typically used in the power stage. However, this approach can introduce significant parasitic effects, potentially diminishing the output power. These parasitic effects arise not only from the transistors themselves but also from the metal interconnects and vias that connect them. Therefore, optimizing the size and layout of the transistors in the power stage is crucial to minimizing parasitic effects and improving the overall performance of the PA. To construct large-sized power transistor cells with minimal parasitic parameters, a configuration with multiple smaller transistors in parallel is commonly used. As shown in Figure 9a, the power stage employs a parallel and staggered layout of four transistor cells, each with an emitter length of 14 μm. To minimize parasitic capacitance and losses, the interconnects for the emitters and collectors are designed using top-layer metal in an interleaved pattern. Additionally, since the peaking capacitor C_b is directly connected to the CB transistor, it has a significant impact on the overall output impedance of the power stage at mm-wave frequencies. Therefore, C_b is also modeled together with the transistors as an integrated whole.

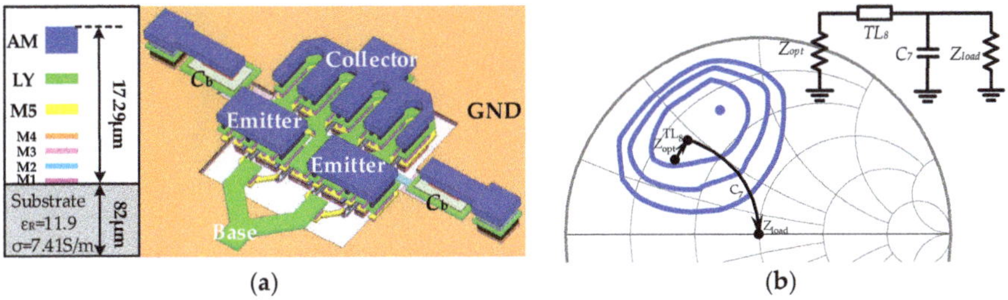

Figure 9. (**a**) The 3D model of power stage transistors with peaking capacitor C_b; (**b**) Simulated power contours of the power stage.

On the other hand, load-pull simulations were conducted on the output stage at various frequencies to determine the optimal power load impedance. As illustrated in Figure 9b, impedance contours for P_{out} were plotted. Note that the optimal power load impedance often differs significantly from the output matching point. In this design, we selected an impedance of Z_{opt} = 15.1 + 17.3j, which is close to the optimal power load impedance, as the initial value for matching. This choice aimed to strike a balance between P_{out}, bandwidth, and output matching. Figure 9b also depicts the impedance transformation path, where the L-type network formed by TL_8 and C_7 transforms the 50 Ω load impedance to Z_{opt}. Furthermore, the source-pull simulation was applied to optimize the output impedance of the first stage, ensuring it delivers sufficient power to drive the power stage.

3. Experimental Results

The proposed wideband BiCMOS PA has been designed and fabricated using a 0.13-μm SiGe BiCMOS process, which is a high-performance technology with the compatible 0.13 μm CMOS process. The process provided five thin metal layers and two thick top metal layers labeled AM and LY, respectively. To enhance the quality factor (Q), all transmission lines used for impedance matching in the circuit were designed using the thick top metal layer AM. The bottom metal layer M1 served as a ground shield to reduce the parasitic capacitance generated by the transmission lines. Additionally, a series of MIM capacitors were employed in the bypass network to ensure effective power decoupling. Figure 10 shows the die micrograph of the proposed PA. To enhance the amplifier's performance accuracy, passive components containing transmission lines, capacitors, testing pads, interconnections, vias, and more were meticulously simulated and optimized as a whole using a full-wave 3D electromagnetic field simulator. The whole chip occupied a silicon area of 0.75 mm × 0.76 mm, including all testing pads. The amplifier drew a DC current of 35 mA from a 3.3 V supply voltage, resulting in a power consumption of 115.5 mW.

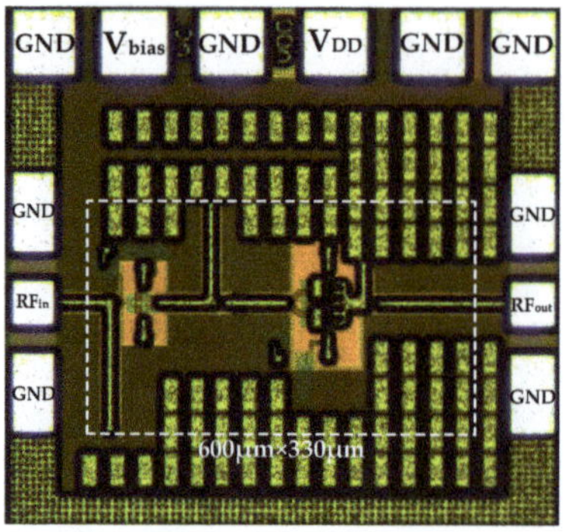

Figure 10. Microphotograph of the proposed wideband PA.

On-wafer *S*-parameter measurements were carried out over the 45 to 65 GHz frequency range using an MPI TS200-SE probe station with ground-signal-ground (G-S-G) probes having a 150 μm pitch. The measurement setup featured a Keysight PNA-X-N5247B network analyzer with a frequency range of 10 MHz to 67 GHz and a Keysight N6705C DC power analyzer, recognized for its high-resolution capabilities. Calibration was conducted using the standard short-open-load-through (SOLT) method.

Figure 11 shows the simulated and measured *S*-parameters of the proposed wideband PA, which exhibit a consistent trend. The PA demonstrated an excellent wideband gain response ($|S_{21}|$) with a peak value of 27.3 dB at 64 GHz. The measured −3 dB bandwidth spanned from 52 GHz to 65 GHz. It is worth noting that the high corner frequency of the −3 dB gain bandwidth exceeded 65 GHz, but this could not be captured due to limitations in our testing equipment. Therefore, the PA achieved a −3 dB gain bandwidth of over 13 GHz, with a fractional bandwidth (FBW) exceeding 22.2%. The input return loss ($|S_{11}|$) and output return loss ($|S_{22}|$) are also illustrated in Figure 9, and they were compromised to some extent due to the desired wideband gain response and high output power. Please note that the discrepancies between the simulation and measurement results were primarily due to the foundry adding numerous randomized dummy metal layers during the manufacturing process to pass the Design Rule Check (DRC). These dummy metal layers introduced additional parasitic elements, which degraded the quality factor (Q) and effective inductance of the T-matching network.

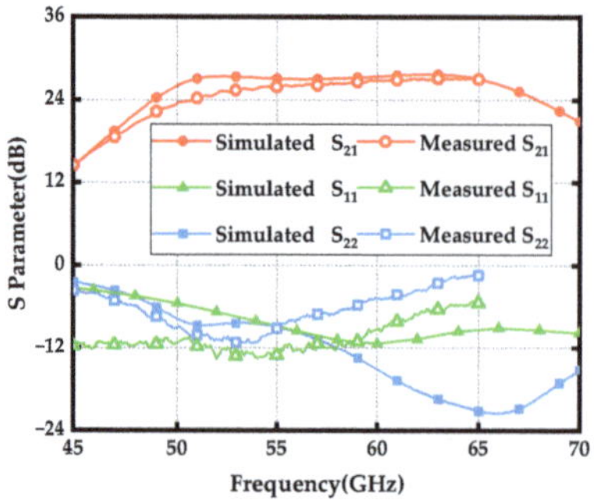

Figure 11. Simulated (solid symbol line) and Measured (hollow symbol line) S-parameters for the wideband PA.

Figure 12 presents the calculated stability factor *K* and delta Δ derived from the measured *S*-parameters. Across the measured frequency range of 45 to 65 GHz, the stability factor *K* consistently exceeded 3.2, while the delta Δ remained below 1. These results confirm that the proposed wideband amplifier maintained stability under all operating conditions.

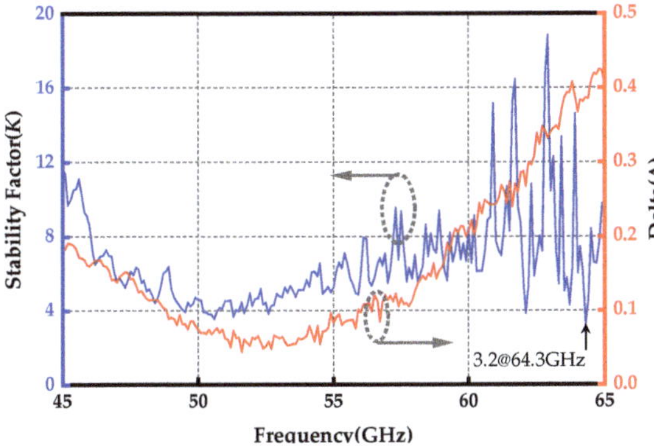

Figure 12. Measured stability factor *K* and delta △.

The power-handling capability of the wideband amplifier was determined by evaluating the output 1 dB compression point (OP_{1dB}) and the saturated output power within the target −3 dB bandwidth. As illustrated in Figure 13, the measured OP_{1dB} and saturated output power varied with operating frequency. Within the specified −3 dB bandwidth frequency range, the OP_{1dB} was observed to range from 15.4 to 18 dBm. In the frequency range of 52 to 65 GHz, the saturated output power varied from 17.4 to 19.7 dBm, demonstrating excellent power flatness with a deviation of less than ±1.15 dBm.

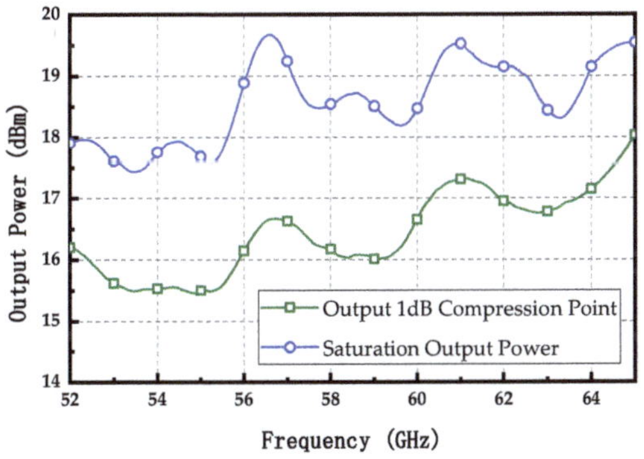

Figure 13. Measured output 1 dB compression point and saturation output power.

In Table 1, the measured performance of the proposed BiCMOS PA is summarized and compared with the recently demonstrated silicon-based PAs. The proposed PA demonstrated a competitive overall performance, offering a significant bandwidth in extension while maintaining good results in gain, OP_{1dB}, and saturated output power.

Table 1. Performance summary and comparison with recently reported PAs.

Ref.	Technology	Bandwidth (GHz)	FBW (%)	Gain (dB)	P_{1dB} (dBm)	P_{sat} (dBm)	P_{dc} (mw)	Size (mm^2)
[17]	22 nm SOI CMOS	56.5–63.5 (7)	11.7	22@60 GHz	15.7	18.6	351	0.43
[18]	65 nm CMOS	50.2–59.4 (9.2)	16.78	20.8@54.4 GHz	5.9	13.0	-	0.3
[8]	28 nm CMOS	40–67 (27)	51	13@50 GHz	12	13	-	0.33
[11]	130 nm SiGe BiCMOS	54–66 (12)	20.0	18@61.5 GHz	12	14.7	-	0.30
[10]	130 nm SiGe BiCMOS	42–48.5 (6.5)	14.36	29.5@44 GHz *	14.8	16.2 *	-	1.00
[9]	130 nm SiGe BiCMOS	52–83.4 * (31.4)	46.38 *	17.5@77 GHz	>16	19.1	-	0.68
This work	130 nm SiGe BiCMOS	52–65 (>13)	>22.2	27.3@64 GHz	18	19.7	115.5	0.57

FBW = bandwidth/f_c, where f_c is the center frequency of −3 dB bandwidth. * estimated from the figure.

4. Conclusions

A combination of techniques, including pole-gain distribution with PGTN, CB capacitive peaking, and parallel configuration of multiple small-sized transistors, was developed and successfully implemented in a *V*-band PA fabricated in a 130 nm SiGe BiCMOS process. These techniques enable a wideband flat gain-frequency response without compromising other key performance metrics. The PA achieved a peak gain of 27.3 dB, a −3 dB bandwidth exceeding 13 GHz, a maximum P_{sat} of 19.7 dBm, and a maximum OP_{1dB} of 18 dBm while occupying a small footprint of 0.57 mm^2. This design approach shows significant potential as an effective method for designing wideband amplifiers.

Author Contributions: Conceptualization, J.H. and J.L.; Methodology, J.H., J.W., Y.S. and W.Z.; Writing—original draft, J.H. and Y.S.; Data curation, J.W., Y.S. and W.Z.; Formal analysis, J.H., J.W. and W.Z.; Writing—review and editing, J.W. and J.L.; Supervision, J.L.; Project administration, J.L. All authors have read and agreed to the published version of the manuscript.

Funding: This research was supported in part by the National Key Research and Development Program of China under Grant 2023YFB3811503, in part by the Zhejiang Provincial Natural Science Foundation of China under Grant LQ23F040009, in part by the State Key Laboratory of Millimeter Waves, Southeast University under Grant K202316.

Data Availability Statement: The data presented in this work are available within the article.

Conflicts of Interest: The authors declare no conflicts of interest.

References

1. So, C.; Sung, E.-T.; Hong, S. A V-Band Four-Channel Phased Array Transmitter Front-End With 0.7 Phase Step and 20 dB Gain Dynamic Range. *IEEE Trans. Microw. Theory Tech.* **2024**, *72*, 2799–2808. [CrossRef]
2. Sutbas, B.; Ng, H.J.; Eissa, M.H.; Kahmen, G. A low-power V-band radar transceiver front-end chip using 1.5 V supply in 130-nm SiGe BiCMOS. *IEEE Trans. Microw. Theory Tech.* **2023**, *71*, 4855–4868. [CrossRef]
3. Gadallah, A.; Franzese, A.; Eissa, M.H.; Drenkhahn, K.E.; Kissinger, D.; Malignaggi, A. A 4-channel V-band beamformer featuring a switchless PALNA for scalable phased array systems. In Proceedings of the 2021 IEEE MTT-S International Microwave Symposium (IMS), Atlanta, GA, USA, 7–25 June 2021; pp. 839–841.

4. Gadallah, A.; Eissa, M.H.; Kissinger, D.; Malignaggi, A. A V-band miniaturized bidirectional switchless PALNA in SiGe: C BiCMOS technology. *IEEE Microw. Wirel. Compon. Lett.* **2020**, *30*, 786–789. [CrossRef]
5. Sarkar, A.; Aryanfar, F.; Floyd, B.A. A 28-GHz SiGe BiCMOS PA with 32% efficiency and 23-dBm output power. *IEEE J. Solid-State Circuits* **2017**, *52*, 1680–1686. [CrossRef]
6. Yin, T.; Pappu, A.M.; Apsel, A.B. Low-cost, high-efficiency, and high-speed SiGe phototransistors in commercial BiCMOS. *IEEE Photonics Technol. Lett.* **2005**, *18*, 55–57. [CrossRef]
7. Ding, R.; Xuan, Z.; Yao, P.; Baehr-Jones, T.; Prather, D.; Hochberg, M. Power-efficient low-noise 86 GHz broadband Amplifier in 130 nm SiGe BiCMOS. *Electron. Lett.* **2014**, *50*, 741–743. [CrossRef]
8. Zhao, J.; Bassi, M.; Bevilacqua, A.; Ghilioni, A.; Mazzanti, A.; Svelto, F. A 40–67GHz power amplifier with 13 dBm PSAT and 16% PAE in 28 nm CMOS LP. In Proceedings of the ESSCIRC 2014-40th European Solid State Circuits Conference (ESSCIRC), Venice Lido, Italy, 22–26 September 2014; pp. 179–182.
9. Zhou, P.; Yan, P.; Chen, J.; Chen, Z.; Hong, W. A 77 GHz Power Amplifier with 19.1 dBm Peak Output Power in 130 nm SiGe Process. *Micromachines* **2023**, *14*, 2238. [CrossRef] [PubMed]
10. Wang, L.; Chen, J.; Hou, D.; Hong, W. A 45 GHz Low-Voltage Cascode Power Amplifier Based on Capacitor Coupling Technology. In Proceedings of the 2021 IEEE MTT-S International Wireless Symposium (IWS), Nanjing, China, 23–26 May 2021; pp. 1–3.
11. Ferchichi, A.; Rehman, S.U.; Carta, C.; Ellinger, F. 60-GHz SiGe-BiCMOS power amplifier with 14.7 dBm output power and 18 dB power gain. In Proceedings of the 2019 12th German Microwave Conference (GeMiC), Stuttgart, Germany, 25–27 March 2019; pp. 229–231.
12. Lee, S.; Ju, I.; Gong, Y.; Cardoso, A.S.; Connor, J.D.; Cho, M.-K.; Cressler, J.D. Design of an 18–50 GHz SiGe HBT cascode non-uniform distributed power amplifier. In Proceedings of the 2020 IEEE BiCMOS and Compound Semiconductor Integrated Circuits and Technology Symposium (BCICTS), Monterey, CA, USA, 16–19 November 2020; pp. 1–4.
13. Schoepfel, J.; Rücker, H.; Pohl, N. A Differential SiGe HBT Doherty Power Amplifier for Automotive Radar at 79 GHz. In Proceedings of the 2023 IEEE 23rd Topical Meeting on Silicon Monolithic Integrated Circuits in RF Systems, Las Vegas, NV, USA, 22–25 January 2023; pp. 44–46.
14. Chen, J.-C.; Chang, T.-Y.; Chiang, Y.-C. A V-band power amplifier using Marchand balun for power combining in 90n-nm CMOS process. In Proceedings of the 2017 IEEE Wireless Power Transfer Conference (WPTC), Taipei, Taiwan, 10–12 May 2017; pp. 1–3.
15. Chou, C.-F.; Hsiao, Y.-H.; Wu, Y.-C.; Lin, Y.-H.; Wu, C.-W.; Wang, H. Design of a $ V $-Band 20-dBm wideband power amplifier using transformer-based radial power combining in 90-nm CMOS. *IEEE Trans. Microw. Theory Tech.* **2016**, *64*, 4545–4560. [CrossRef]
16. Liu, J.Y.-C.; Tang, A.; Wang, N.-Y.; Gu, Q.J.; Berenguer, R.; Hsieh, H.-H.; Wu, P.-Y.; Jou, C.; Chang, M.-C.F. A V-band self-healing power amplifier with adaptive feedback bias control in 65 nm CMOS. In Proceedings of the 2011 IEEE Radio Frequency Integrated Circuits Symposium, Baltimore, MD, USA, 5–7 June 2011; pp. 1–4.
17. Cui, M.; Tibenszky, Z.; Carta, C.; Ellinger, F. Design of a Compact Power Amplifier with 18.6 dBm 60 GHz 20.5% PAE in 22 nm FD-SOI. In Proceedings of the 2020 15th European Microwave Integrated Circuits Conference (EuMIC), Utrecht, The Netherlands, 10–15 January 2021; pp. 141–144.
18. Wang, Y.; Chen, T.; Wu, L. A 60-GHz Two-Way Transformer-Coupled Power Amplifier in 65-nm CMOS. In Proceedings of the 2022 IEEE MTT-S International Microwave Workshop Series on Advanced Materials and Processes for RF and THz Applications (IMWS-AMP), Guangzhou, China, 27–29 November 2022; pp. 1–3.
19. Feng, G.; Boon, C.C.; Meng, F.; Yi, X.; Yang, K.; Li, C.; Luong, H.C. Pole-converging intrastage bandwidth extension technique for wideband amplifiers. *IEEE J. Solid-State Circuits* **2017**, *52*, 769–780. [CrossRef]
20. Luo, J.; He, J.; Feng, G.; Apriyana, A.; Fang, Y.; Xue, Z.; Huang, Q.; Yu, H. A D-band amplifier in 65 nm bulk CMOS for short-distance data center communication. *IEEE Access* **2018**, *6*, 53191–53200. [CrossRef]

micromachines

MDPI

Article

A Compact V-Band Temperature Compensation Low-Noise Amplifier in a 130 nm SiGe BiCMOS Process

Yi Shen [1], Jiang Luo [1,*], Wei Zhao [1], Jun-Yan Dai [2] and Qiang Cheng [2,*]

[1] School of Electronics and Information, Hangzhou Dianzi University, Hangzhou 310018, China
[2] State Key Laboratory of Millimeter Waves, Southeast University, Nanjing 210096, China
* Correspondence: luojiang@hdu.edu.cn (J.L.); qiangcheng@seu.edu.cn (Q.C.)

Abstract: This paper presents a compact V-band low-noise amplifier (LNA) featuring temperature compensation, implemented in a 130 nm SiGe BiCMOS process. A negative temperature coefficient bias circuit generates an adaptive current for temperature compensation, enhancing the LNA's temperature robustness. A T-type inductive network is employed to establish two dominant poles at different frequencies, significantly broadening the amplifier's bandwidth. Over the wide temperature range of −55 °C to 85 °C, the LNA prototype exhibits a gain variation of less than 1.5 dB at test frequencies from 40 GHz to 65 GHz, corresponding to a temperature coefficient of 0.01 dB/°C. At −55 °C, 25 °C, and 85 °C, the measured peak gains are 25.5 dB, 25 dB, and 24.4 dB, respectively, with minimum noise figures (NF) of 3.0 dB, 3.5 dB, and 4.2 dB, and DC power consumptions of 22.3 mW, 27.6 mW, and 34.4 mW. Moreover, the total silicon area of the LNA chip is 0.37 mm^2, including all test pads, while the core area is only 0.09 mm^2.

Keywords: millimeter-wave (mm-wave); V-band; SiGe BiCMOS; temperature compensation; low-noise amplifier

1. Introduction

The V-band spans a wide frequency spectrum from 40 GHz to 75 GHz, making it an ideal candidate for short-range communications and satellite detection due to its short wavelength and sensitivity to atmospheric conditions [1]. These applications have driven the increasing demand for V-band front-end modules, particularly in the 60 GHz band, which has been standardized by IEEE 802.15.3c [2,3]. Given the complex environments encountered in satellite detection and communications, minimizing receiver performance fluctuations across varying ambient temperatures is crucial to ensuring stable and reliable millimeter-wave (mm-wave) front ends. As the first active component in a millimeter-wave receiver, the low-noise amplifier (LNA) amplifies weak signals received by the antenna while suppressing noise in subsequent stages, ensuring a good signal-to-noise ratio. Consequently, high-performance LNAs with robust temperature tolerance have garnered significant attention.

III-V compound processes, such as indium phosphide (InP) and gallium arsenide(GaAs), are known for their high electron mobility, low-noise figure (NF), and high-power density, making them perfect for mm-wave applications, especially in the design of LNAs and power amplifiers (PAs), where they offer exceptional performance. For example, a 50–75 GHz broadband LNA using the 35 nm InP HEMT process was developed, providing a gain of 15–25 dB and an optimal NF of up to 2.2 dB [4]. Using self-biasing and negative feedback techniques, Macro et al. implemented two broadband LNAs in the V-band and Q-band with a 70 nm GaAs process [5,6]. The V-band LNA achieved an NF of less than 1.7 dB and a gain of 15 dB, while the Q-band LNA achieved an NF of less than 1.55 dB and a peak gain of 19 dB. While compound-based LNAs offer superior performance, they generally come with higher power consumption and manufacturing costs, and they lag behind CMOS/SiGe BiCMOS processes in terms of integration and compatibility [7].

Due to aggressive device scaling, CMOS/SiGe BiCMOS technology has become a strong competitor to III-V group semiconductors for high-performance, cost-effective mm-wave integrated circuits. Silicon-based LNAs in V-band are published in [8–14], with most of the literature providing measured results at room temperature (25 °C) [8–11], leaving their temperature tolerance unknown. Xu et al. implemented a 60 GHz wideband LNA using an advanced 22 nm FD-SOI process. Benefiting from gain distribution techniques, this LNA achieved a 3 dB bandwidth of 18.5 GHz and a peak gain of 20.2 dB, with a minimum noise figure (NF) of 3.3 dB and power consumption of 8.1 mW. Through innovations in topology and circuit bias point design, the work in [9] reported an ultra-low-power wideband CMOS LNA in the V-band, consuming only 8.6 mW at 60 GHz with a low supply voltage of 0.8 V. It achieved a peak gain of 20.4 dB, though the NF of 5 dB leaves room for improvement. Leveraging digitally controlled current reuse, Chang et al. [10] developed a low-power CMOS digitally controlled variable-gain LNA, achieving a gain control range of 6.6–19.8 dB with a 3 dB bandwidth of over 10 GHz. However, while optimizing for low power and wideband performance, compromises were made in NF and input return loss. References [12–14] propose LNAs with temperature tolerance characteristics. By utilizing a bias circuit based on current mirrors and operational amplifiers (OPAMPs), these designs effectively mitigate the gain reduction of the LNA caused by rising temperatures. The prototype of the 60 GHz LNA shows a gain variation of less than 1.7 dB over the temperature range of −20 °C to 100 °C, with a 3 dB bandwidth exceeding 11.5 GHz, although the peak gain is only 9.9 dB. Reference [13] describes a G-band SiGe BiCMOS LNA that employs a bias network based on positive temperature coefficient (PTAT) and negative temperature coefficient (CTAT) resistors. This prototype exhibits a gain variation of less than 2.2 dB in the frequency range of 180–210 GHz over temperatures from −20 °C to 80 °C, with a total power consumption of 35 mW. A standardized temperature compensation design method for mm-wave amplifier design is presented in reference [14], which adjusts the bias circuit by optimizing the reference voltage. This method has been validated in a Ka-band amplifier, although the temperature compensation bias circuit is relatively complex and occupies a significant area.

In this paper, a SiGe BiCMOS V-band wideband LNA with excellent temperature tolerance is proposed, featuring high gain, low noise, and a compact core area. The design employs a current mirror-based adaptive temperature compensation circuit that is simple, easy to implement, and has a small footprint, effectively counteracting temperature-induced variations in gain and NF. Furthermore, a T-type pole-tuning network significantly extends the LNA's bandwidth. The paper is organized as follows: Section 2 details the circuit design of the proposed LNA; Section 3 presents experimental results to verify the design methodology; and the conclusions are summarized in Section 4.

2. Circuit Analysis and Design

Figure 1 illustrates the circuit diagram of the proposed V-band temperature-compensated LNA, which consists of two cascaded common emitter–common base (CE-CB) amplifier stages, a T-type pole tuning network, and a temperature-compensated bias circuit. The CE-CB topology was chosen over common emitter (CE) and common base (CB) configurations because it offers higher voltage gain, better reverse isolation, and the ability to simultaneously optimize noise figure and power matching. The T-type pole tuning network, made up of inductors L_3, L_4, and L_5, is used for interstage matching in the LNA, allowing for adjustment of the high- and low-frequency dominant poles to achieve a flat wideband gain–frequency response. To counteract performance degradation due to temperature variations, a negative temperature coefficient bias circuit generates an adaptive bias current for the base terminals of heterojunction bipolar transistors (HBTs) Q_4 and Q_6, significantly improving the LNA's temperature robustness.

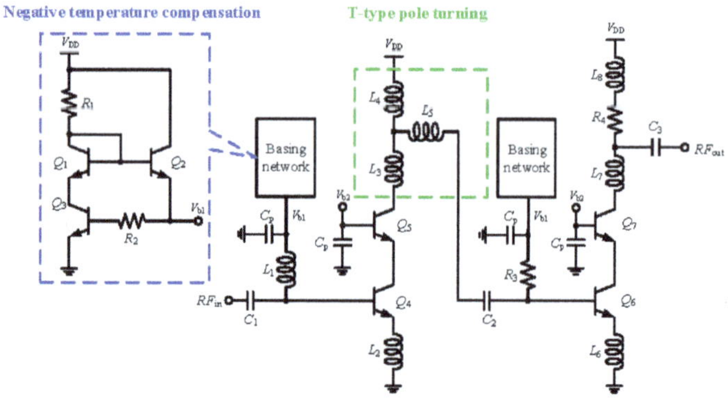

Figure 1. Schematic of the proposed LNA.

2.1. Size of Transistor and Bias Conditions

It is crucial to determine the sizes and biasing conditions of the active devices before designing the passive components, which ultimately influence the overall performance, such as the minimum noise figure (NF_{min}), maximum gain (G_{max}), and input 1 dB compression point (IP_{1dB}). For LNA designs based on current-restricted HBTs, the current density (J_C) comprehensively reflects the size and biasing state. According to [15], at the same frequency, transistors of different sizes can achieve NF_{min} or G_{max} at the same current density. J_C of HBTs can be expressed by emitter current per emitter length, written as follows:

$$J_C = \frac{I_C}{L_E} \tag{1}$$

The emitter width of HBTs is typically selected to be a minimum of 120 nm for optimal high-frequency performance. Noting that in addition to G_{max} and NF_{min}, the power consumed (P_{DC}) should also be considered. The variation of G_{max}, NF_{min}, and P_{DC} verse current density when L_E = 8 um at a frequency of 60 GHz is shown in Figure 2. To comprehensively evaluate the effect of J_C on the overall performance of the LNA, a figure of merit (*FoM*) is introduced, which can be expressed as follows:

$$FoM = \frac{G_{max}[lin]}{F_{min} \cdot P_{DC}} \tag{2}$$

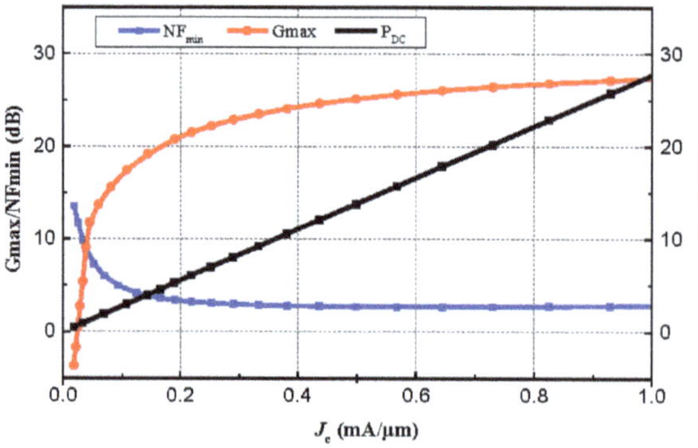

Figure 2. Simulated G_{max}, NF_{min}, and P_{DC} with variations of J_C at 60 GHz and 25 °C when L_E = 8 μm.

By analyzing the J_C and *FoM* values at different emitter lengths, the optimal device size and biasing state for the desired operating frequencies can be determined. As shown in Figure 3, the HBT device with an emitter length of L_E = 8 µm achieves a better *FoM* at an approximate current density of 0.24 mA/µm compared to sizes of L_E = 4 µm and L_E = 12 µm. Furthermore, in the V-band frequencies, the parasitic parameters caused by the metal interconnections and vias of the HBT devices cannot be overlooked. Therefore, it is necessary to model the passive interconnects surrounding the HBT devices and perform full-wave electromagnetic simulations. Considering the current handling capacity of a single HBT device, a parallel configuration of multiple small-sized transistors is preferred while maintaining the same emitter length. The CB-CE stage employs a parallel and staggered layout of four HBTs with an emitter length of 4 µm, which is also modeled in 3D using full-wave electromagnetic simulation software, as shown in Figure 4a. Figure 4b compares the modeled FoM1 of the interconnect lines around the HBT device with the unmodeled FoM2. It is evident that significant parasitic effects degrade the *FoM*, with the optimal current density increasing from 0.24 mA/µm to 0.38 mA/µm. Therefore, after considering the design requirements, the current density for the first phase is ultimately set at 0.32 mA/µm.

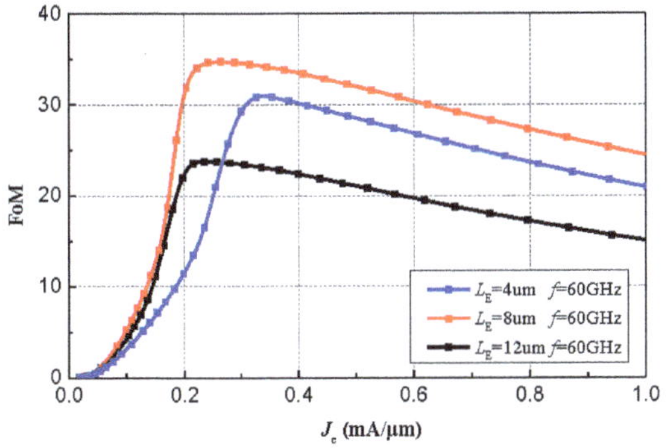

Figure 3. Simulated *FoM* with variations of J_C at 60 GHz and 25 °C under different L_E.

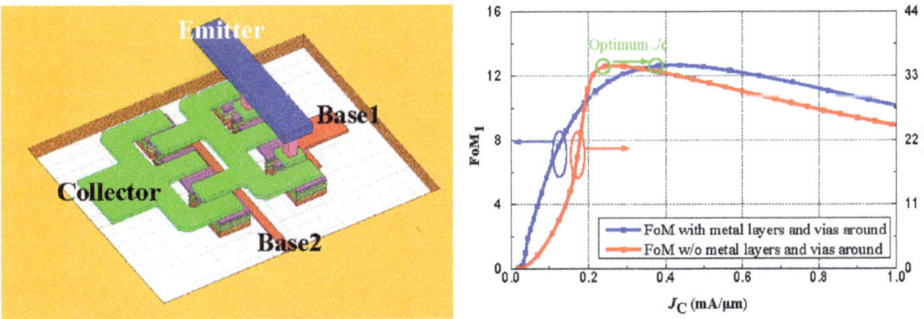

(a) (b)

Figure 4. (**a**) The 3D model of first stage of LNA; (**b**) Comparation between FoM1 and FoM2 with variations of J_C at 60 GHz and 25 °C when L_E = 8 µm.

2.2. Interstage Bandwidth Extension

The L-shaped and Π-shaped pole tuning networks have been proven to be effective means of bandwidth extension and are widely used in broadband LNA designs [16,17]. In this design, a T-type inductive pole tuning network is employed for interstage matching to achieve bandwidth extension.

Figure 5 presents the schematic of the inter-stage T-type pole turning network. Z_{q1} denotes the output impedance of the first stage. To illustrate the effects of L_3 and L_4 on the two main poles: Pole_L and Pole_H, combining L_5 of the T-type matching network with the input impedance of the second-stage cascode into Z_{q2}. Z_{q1} and Z_{q2} can be expressed as follows:

$$Z_{q1} = r_{o5} + \frac{g_{m5}r_{o4}r_{o5} + r_{o5}}{1 + sC_{be5}} \quad (3)$$

$$Z_{q2} = sL_5 + R_{b6} + \frac{1}{sC_{be6}} + sL_6(C_{be6}g_{m6} + 1) \quad (4)$$

where $r_{o4/5}$ and $C_{be5/6}$ are the equivalent output resistors of $Q_{4/5}$ and the equivalent input parasitic capacitors of $Q_{5/6}$, respectively, while $g_{m5/6}$ is the transconductance of $Q_{5/6}$. Noting that, the Z_{q1} features capacitive characteristics, while the Z_{q2} features inductive characteristics, and we define $Z_{q1} = R_{q1} + 1/sC_{q1}$ and $Z_{q2} = R_{q2} + sL_{q2}$ for analytical convenience. Then the simplified $Av(s)$ of the two-stage cascode can be expressed as follows:

$$Av(s) \approx \frac{-g_{m4}(sC_{q1}R_{q1} + 1)[sL_4(R_{q2} + sL_{q2}) + sL_3(sL_4 + R_{q2} + sL_{q2})]}{(sL_4 + sL_{q2} + R_{q2})(sC_{q1}L_3 + C_{q1}R_{q1}) + sC_{q1}L_4(R_{q2} + sL_{q2})} \quad (5)$$

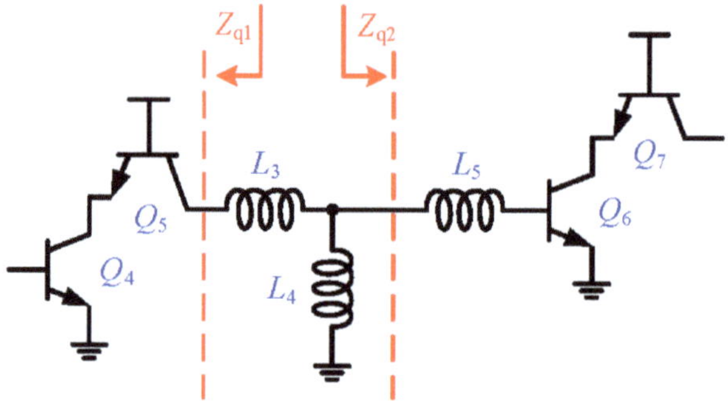

Figure 5. The schematic of the inter-stage T-type network.

The denominator of Figure 5 features two poles that are mainly dominated by L_3 and L_4. Figure 6 show the graphical analysis through S-parameter simulations that demonstrate the changes of the dominate Pole$_L$ and Pole$_H$ with the variations of L_3 and L_4. The increase of L_3 will move Pole$_H$ to lower frequency with minimal impact on Pole$_L$. In contrast, the decrease of L_4 will move Pole$_L$ to a higher frequency with minimal impact on Pole$_H$. Thus, a wideband gain–frequency response can be realized by carefully designing the parameters of the T-type inductive network to tune the positions of the high-frequency and low-frequency dominant poles.

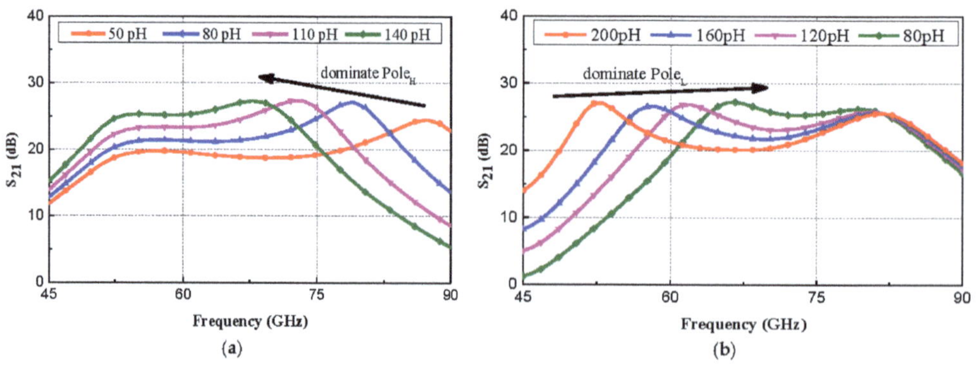

Figure 6. (**a**) S_{21} of the proposed LNA when L_3 vary from 50 to 140 pH; (**b**) S_{21} of the proposed LNA when L_4 vary from 50 to 140 pH.

2.3. Negative Temperature Coefficient Compensation Circuit

Due to the inherent factors of the silicon-based process, thermal variation will inevitably cause unexpected changes in the electron mobility and threshold voltage of the transistors, which will also affect the LNA performance. HBT devices are current-controlled devices, and the variations in collector current caused by temperature change are more pronounced. Therefore, a robust design that stabilizes gain over a wider temperature range is desirable.

Figure 7 is the schematic of the biasing circuit and the parameters of the HBT and resistors in the proposed LNA. The collector current of Q_4 can be expressed as follows:

$$I_T = I_{S4}\exp(\frac{V_{BE4}}{V_T}) \tag{6}$$

where $V_T = KT/q$, and the saturation current $I_{S4} \propto \mu KTni^2$. For convenience, q and k can be regarded as fixed coefficients, T is the absolute temperature, μ represents the carrier mobility, and ni denotes the intrinsic carrier concentration of silicon. Noting that $\mu \propto \mu_0 T^{1+m}$, $ni^2 \propto T^3\exp(-Eg/KT)$ [18]. Then I_{S4} can be expressed as follows:

$$I_{S4} = bT^{4+m}\exp(\frac{-E_g}{KT}) \tag{7}$$

where b is regarded as a proportion factor; and E_g and m are assumed to be constant and the values of these in HBT are approximately 1 eV and −1.5, respectively. Substituting (7) into (6) we obtain the following:

$$I_T = bT^{4+m}\exp(\frac{-E_g + qV_{BE4}}{KT}) \tag{8}$$

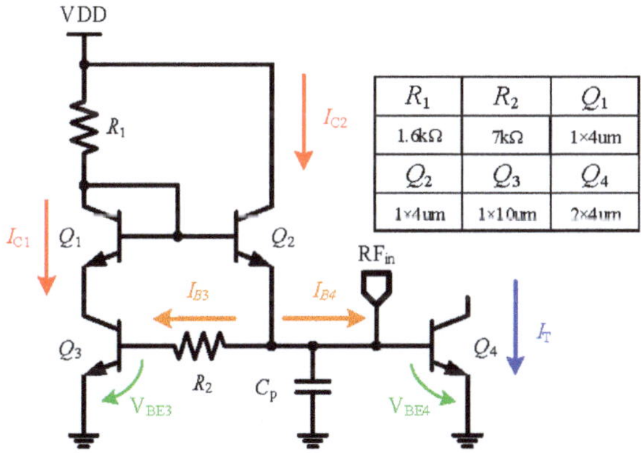

Figure 7. The schematic of the NTC biasing circuit in the proposed LNA.

From (7) and (8), high sensitivity of I_{S4} to temperature greatly affects robustness of HBT to temperature. It is essential to introduce a compensation biasing circuit with a negative temperature coefficient (NTC) to ensure Q_4 operates at an optimal J_C as the temperature changes and, referring to the method in [14], the optimal J_C at each temperature should be obtained first. As shown in Figure 8, the blue curve represents the ideal J_C that obtained through simulations to ensure the same S-parameter at each temperature. The black curve illustrates the J_C with temperature without the temperature compensation circuit. Although the optimal J_C is achieved at 25 °C, there are significant variations in J_C with temperature fluctuations, leading to considerable deviations from the ideal curve. The red curve represents the J_C after implementing the NTC circuit. While there are slight deviations from the ideal curve at −60 °C and 90 °C, the fitting is ideal within the range of −55 to 85 °C, which indicates that the S-parameters will not experience significant variations.

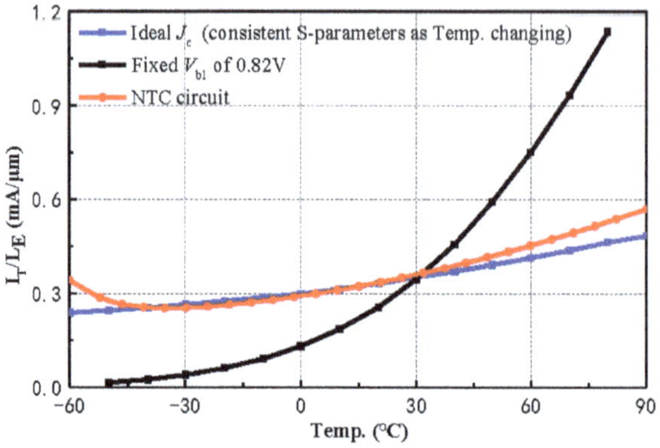

Figure 8. Comparison of J_C in basing situations of fixed Vb1, NTC circuit, and ideal J_C.

In Figure 7, R_2 is a resistor that exhibits good temperature characteristics, and C_p is bypass capacitance to minimize the influence of the biasing circuit. R_1, Q_1, and Q_2 form a conventional current mirror structure. So, assuming $I_{E2} = nI_{E1}$ and n is a constant that shows slight sensitive to temperature. Q_4 relies on the base current I_{B4} to control the collector current I_T, with a current gain of β_4. Therefore, Equation (6) can be derived.

$$I_T = \beta_4(n - \frac{1}{\beta_3})I_{E1} \tag{9}$$

The KVL equation between Q_3 and Q_4 can be derived as follows:

$$V_{BE4} - V_{BE3} = R_2\frac{I_{E1}}{\beta_3} \tag{10}$$

I_T can be derived by substituting (6) and (9) into (10), as follows:

$$I_T = \beta_4(n - \frac{1}{\beta_3})\frac{V_T\beta_3}{R_2}\ln\frac{I_{E4}I_{S3}}{I_{E3}I_{S4}} \tag{11}$$

The temperature-sensitive parameters, I_E and I_S, cancel the effect of T in $I_{E4}I_{S3}/I_{E3}I_{S4}$. Additionally, the slow variation of logarithmic $\ln(I_{E4}I_{S3}/I_{E3}I_{S4})$ with T also reduces the impact of the temperature.

To intuitively demonstrate the compensatory effects of the proposed NTC on the LNA performance, Figure 9 presents simulated results of small-signal gain (S_{21}) and NF as a function of temperature at a fixed bias voltage under NTC conditions at an operating frequency of 55 GHz. In the uncompensated scenario, the gain and noise figure of the LNA sharply degrade as the temperature drops to around −15 °C, primarily due to the pronounced temperature sensitivity of the HBT devices. In contrast, with the implementation of the proposed NTC, both gain and NF remain stable over a broad temperature range from −60 °C to 90 °C, demonstrating remarkable robustness.

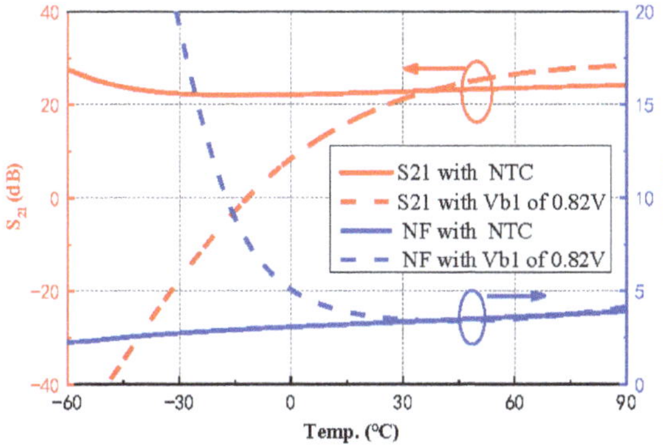

Figure 9. Comparison between simulated S_{21} and NF in basing situations of fixed Vb1 and NTC at 55 GHz.

3. Experimental Validation and Results

The proposed temperature compensation LNA is fabricated using a 0.13 μm SiGe BiCMOS process. Figure 10 shows the die micrograph of the proposed LNA. The whole chip occupies a silicon area of 0.49 mm × 0.76 mm including all testing pads and a core area of 0.3 mm × 0.32 mm.

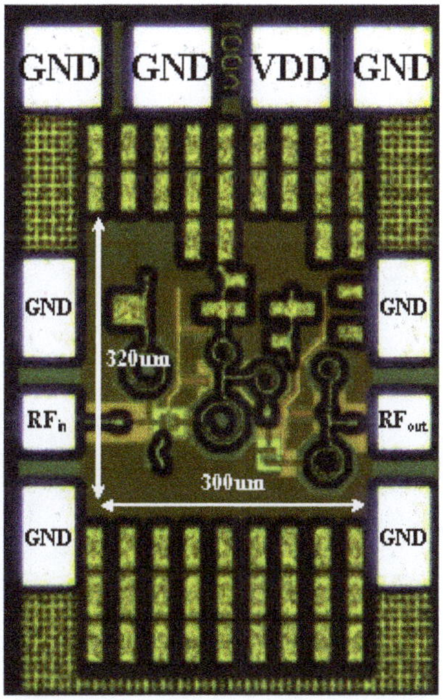

Figure 10. Microphotograph of the LNA chip.

An MPI TS200-SE probe station equipped with ground–signal–ground (G-S-G) probes with a 150 μm pitch was used for on-wafer testing and the calibration was performed using the standard short-open-load-through (SOLT) method. The S-parameters of LNA were measured using a Keysight PNA-X N5247B network analyzer(Keysight Technologies, Inc., Santa Rosa, CA, USA) with a maximum frequency range of 10 MHz to 67 GHz. Due to the limited calibration accuracy of the PNA-X N5247B network analyzer near its upper frequency limit of 67 GHz, the measurements in this work were conducted between the range of 40 GHz and 65 GHz.

Figure 11a presents the comparison between simulated and measured S-parameters at 25 °C, which exhibits similar trends. From measured results, the LNA achieves a peak gain of 25 dB, a 3 dB bandwidth exceeding 11 GHz, and both S_{11} and S_{22} are better than −10 dB within 40~61 GHz. The measured S-parameters of the proposed LNA at −55 °C, 25 °C, and 85 °C are demonstrated in Figure 11b. Owing to the NTC basing network, the LNA's measured S-parameters exhibit excellent robustness to temperature, with ΔGain less than 1.5 dB over 40~65 GHz.

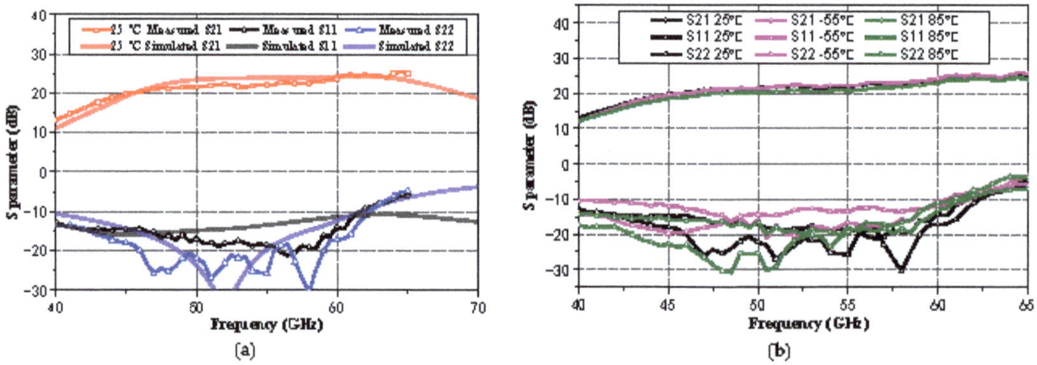

Figure 11. (**a**) Comparison between simulated and measured S-parameters at 25 °C; (**b**) Measured S-parameters from the proposed LNA at −55 °C, 25 °C, and 85 °C.

Figure 12 shows the measured and simulated NF of the proposed LNA at −55 °C, 25 °C, and 85 °C. At 25 °C and 85 °C, the measured results match well with the simulation. However, at −55 °C, the measured NF exhibits a deviation of 0.7 dB from the simulation. One factor for the discrepancy could be the decreasing of measured S_{21} at −55 °C compared to the simulation one, which leads to inadequate noise suppression.

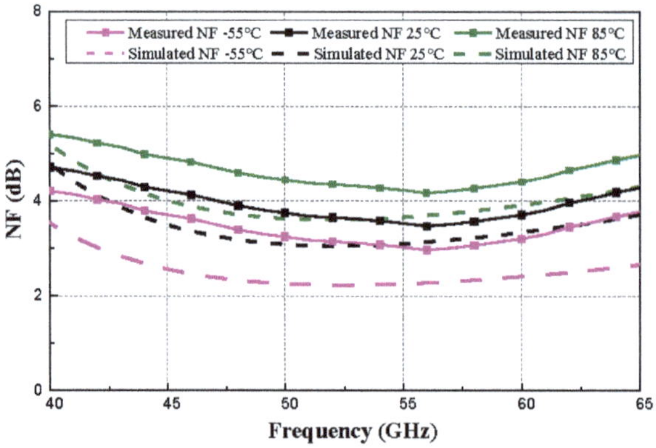

Figure 12. Simulated and measured NF at −55 °C, 25 °C, and 85 °C.

Figure 13a displays the measured input 1 dB compression point (IP_{1dB}), which indicates the amplifier's linearity. Unlike the measured S-parameters, the IP_{1dB} is optimal at 25 °C. This could be attributed to two factors: (1) A high gain at −55 °C will sacrifice linearity. (2) The use of an NTC basing network, which adjusts J_C to ensure good consistency in the S-parameters across −55~85 °C, may not be optimal for linearity. Within the measured frequency range of 40 to 65 GHz, the stability factor *Kf* was consistently greater than 1.2, while the delta Δ remained under 1. These results demonstrate that the proposed wideband LNA achieved stability in all operating conditions.

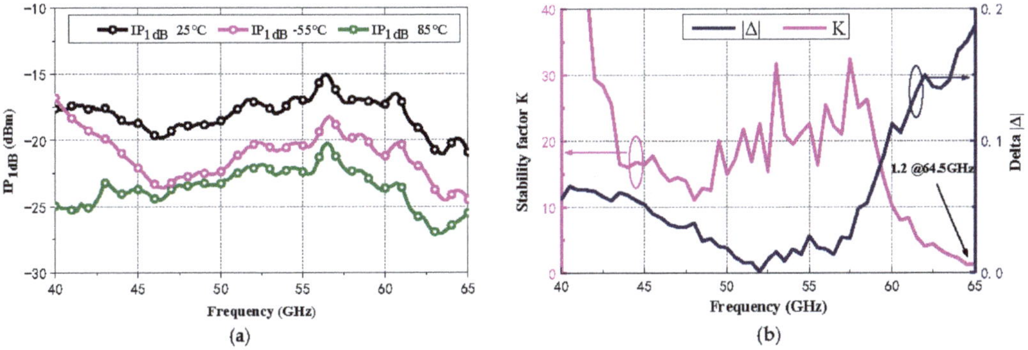

Figure 13. (**a**) Measured IP_{1dB} at −55 °C, 25 °C, and 85 °C; (**b**) Measured stability factor K and delta |△|.

Table 1 summarizes the performance of the proposed LNA and compares it with other recently reported LNAs. The temperature compensation circuits proposed in references [12,14] effectively improve the LNA performance, but their designs are relatively complex and occupy a large area, while also significantly degrading other LNA parameters. The 60 GHz LNA in reference [12] experiences a 2.1 dB degradation in NF at high temperatures. The size of the temperature compensation circuit in reference [14] is 0.168 mm^2, accounting for one-third of the total chip area. The LNA reported in [19] experienced a 5 dB gain degradation at temperatures ranging from −5 to 85 °C, indicating significant room for improvement. The proposed LNA exhibits good stability in key performance metrics, such as gain, NF, and bandwidth, when facing variations in environmental temperature, with a compact core area of only 0.09 mm^2. Compared to LNAs without temperature compensation, such as [9,10,20], the proposed LNA shows superior gain and bandwidth at room temperature. Furthermore, it maintains a high gain of over 20 dB with a less than 1.5 dB variation across a temperature range of −55 to 85 °C. Due to system design specifications requiring a 3.3 V operating voltage, the proposed LNA exhibits relatively high-power consumption.

Table 1. Performance summary and comparison with reported V-band LNAs and temperature compensation amplifier.

Ref.	Process	Temp. (°C)	Peak Gain (dB)	ΔGain (dB)	Temp. Coefficient $\Delta Gain_{max}/S_{21\,RT}/\Delta T$ (dB/°C) [12]	BW_{3dB} (GHz)	BW_{5dB} (GHz)	IP_{1dB} (dBm)	NF_{min} (dB)	P_{DC} (mW)	Core Area (mm^2)
[12]	65 nm CMOS	−20	10.4	<1.7	1.4×10^{-3}	>11 (56~67) *	>13.5 (53.5~67) *	\	\	\	0.72 &
		25	9.9			>11.5 (55.5~67)	>13.5 (53.5~67)*	\	3.9	17.2	
		100	8.8			>11.5 (55.5~67)	>13.5 (53.5~67) *	\	6.0	23.4	

Table 1. *Cont.*

Ref.	Process	Temp. (°C)	Peak Gain (dB)	ΔGain (dB)	Temp. Coefficient $\Delta Gain_{max}/S_{21\,RT}/\Delta T$ (dB/°C) [12]	BW_{3dB} (GHz)	BW_{5dB} (GHz)	IP_{1dB} (dBm)	NF_{min} (dB)	P_{DC} (mW)	Core Area (mm^2)
		−45	23.6			N/A	N/A	\	\	\	
[14]	90 nm CMOS	25	23.8	<1.2	0.29×10^{-3}	4 (24.5~28.5)	5 (24~29) *	\	\	25.2	0.5 &
		125	23			N/A	N/A	\	\	\	
		−5	22.5			10 (55~65) *	>11 (54~65) *	\	\	\	
[19]	90 nm CMOS	25	21	<5	2.6×10^{-3}	8 (55~63) *	>11 (54~65) *	−20 *	6.5	49	1.06 &
		85	17.5			11 (54~65) *	>12.5 (52.5~65) *	\	\	\	
[9]	45 nm RF-SOI	25	20.4	\	\	13.6 (54.4~68)	18 (53~71)	−23.6	5	8.6	0.138
[10]	40 nm CMOS	25	19.8	\	\	10 (55~65)	13 (54~67)	−29.5	6	18	0.22 &
[20]	65 nm CMOS	25	12.8	\	\	10 (55~65)	>12 (54~66)	\	3.6	8.8	0.33
		−55	**25.5**			**>13 (52~65)**	**>18.5 (46.5~65)**	**−20**	**3.0**	**22.3**	
This work	**130nm BiCMOS**	**25**	**25**	**<1.5**	$\mathbf{0.4 \times 10^{-3}}$	**>11 (54~65)**	**>19.5 (45.5~65)**	**−17.5**	**3.5**	**27.6**	**0.09**
		85	**24.4**			**>10 (55~65)**	**>18 (47~65)**	**−22.5**	**4.2**	**34.4**	

* Estimated values from papers; & DC pads are included.

4. Conclusions

An adaptive temperature compensation technology has been successfully implemented in a compact V-band wideband LNA fabricated using a 130 nm SiGe BiCMOS process. This innovation enables the LNA to exhibit excellent temperature robustness over a broad range from −55 °C to 85 °C. The V-band LNA demonstrates a gain variation of less than 1.5 dB, an NF variation of less than 1.2 dB, and an IP_{1dB} variation of less than 5 dBm over measured frequencies from 40 GHz to 65 GHz. The measured DC power consumptions at −55 °C, 25 °C, and 85 °C are 22.3 mW, 27.6 mW, and 34.4 mW, respectively, with a supply voltage of 3.3 V. At room temperature, the LNA demonstrates a gain of 25 dB, an NF of 3.5 dB, and an IP_{1dB} of −17.5 dBm. The LNA chip has a total silicon area of 0.37 mm^2, including all test pads, with the core area being only 0.09 mm^2.

Author Contributions: Conceptualization, Y.S. and J.L.; methodology, Y.S. and W.Z.; writing—original draft, Y.S.; data curation, W.Z. and J.-Y.D.; formal analysis, J.-Y.D.; writing—review and editing, Y.S., J.L., and Q.C.; supervision, J.L. and Q.C.; project administration, J.L. and Q.C. All authors have read and agreed to the published version of the manuscript.

Funding: This research was supported in part by the National Key Research and Development Program of China under grant 2023YFB3811503, in part by the Zhejiang Provincial Natural Science Foundation of China under grant LQ23F040009, and in part by the State Key Laboratory of Millimeter Waves, Southeast University under grant K202316.

Data Availability Statement: The data presented in this work are available within the article.

Conflicts of Interest: The authors declare no conflicts of interest.

References

1. Frounchi, M.; Coen, C.; Cheon, C.D.; Lourenco, N.; Williams, W.; Cressler, J.D. A V-Band SiGe image-reject receiver front-end for atmospheric remote sensing. In Proceedings of the 2018 IEEE BiCMOS Compound Semiconductor Integrated Circuits and Technology Symposium, San Diego, CA, USA, 15–17 October 2018; pp. 223–226.
2. Hansen, J. A novel stochastic millimeter-wave indoor radio channel model. *IEEE J. Sel. Areas Commun.* **2002**, *20*, 1240–1246. [CrossRef]
3. *IEEE Standard 802.15.3c-2009*; IEEE Standard for Information Technology—Local and Metropolitan Area Networks—Specific Requirements—Part 15.3: Amendment 2: Millimeter-Wave-Based Alternative Physical Layer Extension. IEEE Standards Association: Piscataway, NJ, USA, 2009.
4. Samoska, L.A.; Varonen, M.; Kangaslahti, P.; Fung, A.; Gawande, R.; Soria, M.; Lai, R.; Sarkozy, S. V-band MMIC LNAs and mixers for observing the early universe. In Proceedings of the 2016 41st International Conference on Infrared, Millimeter, and Terahertz waves (IRMMW-THz), Copenhagen, Denmark, 25–30 September 2016; pp. 1–2.
5. Polli, G.; Vittori, M.; Ciccognani, W.; Colangeli, S.; Costanzo, F.; Salvucci, A.; Limiti, E. Ka-/V-band self-biased LNAs in 70 nm GaAs/InGaAs Technology. In Proceedings of the 2018 14th Conference on Ph.D. Research in Microelectronics and Electronics (PRIME), Prague, Czech Republic, 2–5 July 2018; pp. 197–200.
6. Vittori, M.; Polli, G.; Ciccognani, W.; Colangeli, S.; Salvucci, A.; Limiti, E. Q-band self-biased MMIC LNAs using a 70 nm InGaAs/AlGaAs process. In Proceedings of the 2017 IEEE Asia Pacific Microwave Conference (APMC), Kuala Lumpur, Malaysia, 13–16 November 2017; pp. 630–633.
7. Cressler, J.D.; Niu, G. *Silicon-Germanium Heterojunction Bipolar Transistors*; Artech House: Norwood, MA, USA, 2003.
8. Xu, X.; Wagner, J.; Carta, C.; Ellinger, F. A 60 GHz Broadband LNA With Joined Variable Gain Control and Switching in 22 nm FD-SOI. *IEEE Access* **2024**, *12*, 111627–111637. [CrossRef]
9. Moody, J.; Lepkowski, S.; Forbes, T.M. A Low Power V-Band LNA with Wide Supply Voltage Range Exploiting Complementary Current Reuse and Power Efficient Bias Point. In Proceedings of the IEEE MTT-S International Microwave Symposium, San Diego, CA, USA, 11–16 June 2023; pp. 135–138.
10. Chang, Y.-T.; Lu, H.-C. A *V*-Band Low-Power Digital Variable-Gain Low-Noise Amplifier Using Current-Reused Technique with Stable Matching and Maintained OP1dB. *IEEE Trans. Microw. Theory Tech.* **2019**, *67*, 4404–4417. [CrossRef]
11. Frounchi, M.; Cressler, J.D. A SiGe Millimeter-Wave Front-End for Remote Sensing and Imaging. In Proceedings of the IEEE Radio Frequency Integrated Circuits Symposium (RFIC), Los Angeles, CA, USA, 4–6 August 2020; pp. 227–230.
12. Kawai, S.; Wang, T.; Mitomo, T. A Temperature Variation Tolerant 60 GHz Low Noise Amplifier with Current Compensated Bias Circuit. In Proceedings of the IEEE Asian Solid-State Circuits Conference (A-SSCC), Singapore, 11–13 November 2013; pp. 429–432.
13. Urain, A.; Del Río, D.; Torres, R.; Berenguer, R. A G-Band SiGe BiCMOS LNA With an On-Chip and Compact Temperature Compensation Biasing Circuit. In Proceedings of the IEEE/MTT-S International Microwave Symposium, San Diego, CA, USA, 11–16 June 2023; pp. 1132–1135.
14. Qi, Q.; Chen, Z. A K-Band CMOS Amplifier with Temperature Compensation for Gain Variation Reduction. *IEEE Microw. Wirel. Compon. Lett.* **2018**, *28*, 150–152. [CrossRef]
15. Dickson, T.O.; Yau, K.H.K.; Chalvatzis, T.; Mangan, A.M.; Laskin, E.; Beerkens, R. The invariance of characteristic current densities in nanoscale MOSFETs and its impact on algorithmic design methodologies and design porting of Si(Ge) (Bi)CMOS high-speed building blocks. *IEEE J. Solid-State Circuits* **2006**, *41*, 1830–1845. [CrossRef]
16. Khanpour, M.; Tang, K.W.; Garcia, P.; Voinigescu, S.P. A wideband W-band receiver front-end in 65-nm CMOS. *IEEE J. Solid-state Circuits* **2008**, *43*, 1717–1730. [CrossRef]
17. Tomkins, A.; Garcia, P.; Voinigescu, S.P. A passive W-band imaging receiver in 65-nm bulk CMOS. *IEEE J. Solid-state Circuits* **2010**, *45*, 1981–1991. [CrossRef]
18. Wang, R.; Zhu, W.; Wang, Y. An Adaptive Analog Temperature Compensated W-Band Front-End with ±0.0033 dB/°C Gain Variation Across −30 °C to 120 °C. *IEEE Trans. Circuits Syst. II Express Briefs* **2024**, *71*, 542–546. [CrossRef]
19. Shin, S.; Leung, M.; Hsiao, S. A Temperature Variation Compensated 60-GHz Low-Noise Amplifier in 90-nm CMOS technology. In Proceedings of the Asia-Pacific Microwave Conference, Melbourne, VIC, Australia, 5–8 December 2011; pp. 211–214.
20. Yaghoobi, M.; Yavari, M.; Kashani, M.H.; Ghafoorifard, H.; Mirabbasi, S. A 55–64-GHz Low-Power Small-Area LNA in 65-nm CMOS With 3.8-dB Average NF and ∼12.8 dB Power Gain. *IEEE Microw. Wirel. Compon. Lett.* **2019**, *29*, 128–130. [CrossRef]

micromachines

MDPI

Article

Design of Dual Continuous-Mode Class-J Power Amplifiers with Harmonic Matching Networks for X and Ku Bands

Yang Yuan [1], Xuesong Zhao [1], Jingxin Fan [1,2] and Zhongjun Yu [1,2,*]

[1] Aerospace Information Research Institute, Chinese Academy of Sciences, Beijing 100190, China
[2] School of Electronic, Electrical and Communication Engineering, University of Chinese Academy of Sciences, Beijing 101408, China
* Correspondence: yuzj@ucas.ac.cn

Abstract

In this article, two wideband high-efficiency Class-J power amplifiers operating in X and Ku bands, respectively, are designed based on continuous mode. The optimal impedance regions of the transistors are determined using harmonic load-pull techniques. An on-chip output matching network with second harmonic control functionality is designed to achieve Class-J operation. To verify the feasibility of designed circuits, both power amplifiers are designed and fabricated using a 0.25 mm GaAs pseudomorphic high electron mobility transistor (pHEMT) process. The power amplifiers are both biased at 6 V/−1 V. The measured results show the X-band and Ku-band power amplifiers achieve peak saturated output powers of 31.2 dBm and 30.8 dBm, respectively. The power-added efficiencies (PAEs) of the two amplifiers within their operating bands reach up to 48% and 45.3%, respectively. Compact size and high efficiency make them suitable for integration into phased array transmit/receiver (T/R) modules.

Keywords: Class-J; continuous mode; harmonic control; load-pull; MMIC; power amplifier

1. Introduction

Modern active phased arrays demand multifunctionality and lightweight design [1]. Accordingly, the widebanding and miniaturization of transmit/receiver (T/R) modules become imperative. Among these requirements, thermal design is a core factor restricting the miniaturization of T/R modules, while the efficiency of T/R modules is primarily determined by their internal power amplifiers [2]. However, the widebanding of power amplifiers leads to efficiency degradation. Therefore, designing wideband high-efficiency power amplifiers is key to realizing the multifunctionality and lightweight design of phased arrays.

Based on waveform engineering analysis, continuous-mode power amplifiers maintain high efficiency over a wide bandwidth, making them ideal for designing saturated power amplifiers in T/R modules [3]. A waveform-engineered sequential load-modulated balanced amplifier, which employs continuous Class-F^{-1} and Class-J operations, achieves a saturated drain efficiency of 60.2–68.3% across 1.80–2.75 GHz [4]. An ultrawideband power amplifier utilizing a novel multi-branch matching network and extended continuous Class-B/J modes delivers a drain efficiency of 56–70% over 0.4–3.8 GHz, with a relative bandwidth of 162% [5]. The power amplifiers reported in [4,5] employ off-chip high-Q output matching networks (OMNs), achieving high efficiency over a wide bandwidth. However, their large overall footprint renders them incompatible with the integration into transmit/receive (T/R) modules. In [6], an X-band Class-J power amplifier based on active load modulation is implemented in a 0.25 mm GaAs pseudomorphic high electron mobility transistor (pHEMT) process, achieving a peak power-added efficiency (PAE) of 50% over 9.1~10.8 GHz with a die area of 1.89×1.69 mm^2. Ref. [7] demonstrates a broadband power amplifier fabricated in a 0.25 μm GaN HEMT process, realizing a PAE of 36.3–47% across 2–6 GHz.

In this article, two wideband high-efficiency Class-J power amplifiers operating in the X band and Ku band, respectively, are designed and fabricated using a 0.25 mm GaAs pHEMT process. On-chip second harmonic control network is integrated into the OMN to realize the Class-J operation. The measured results show that both power amplifiers exhibit high PAE within their respective bandwidths. Furthermore, their compact footprint renders them suitable for integration with miniaturized T/R modules.

2. Continuous-Mode Class-J Power Amplifier

Continuous-mode power amplifiers expand the solution space for the optimal load impedance, thereby enhancing the design flexibility of the output matching network and allowing the power amplifier to maintain high drain efficiency over a broad frequency band [8].

Taking the continuous Class-J power amplifier as an example, the drain-source current i_{d_J} and voltage v_{d_J} of the transistor can be given by Equations (1) and (2), respectively [9].

$$i_{d\_J} = \frac{I_m}{\pi} + \frac{I_m}{2}\cos\theta + \frac{2I_m}{3\pi}\cos 2\theta - \frac{2I_m}{15\pi}\cos 4\theta + \ldots \tag{1}$$

$$\begin{aligned} v_{d\_J} &= V_D(1-\cos\theta)(1-\alpha\sin\theta) \\ &= V_D\left[1-\sqrt{\alpha^2+1}\sin\left(\theta+\arctan\tfrac{1}{\alpha}\right)+\tfrac{\alpha}{2}\sin 2\theta\right] \end{aligned} \tag{2}$$

Here, $\alpha \in [-1, 1]$, I_m and V_D are the peak current and the bias voltage, respectively. The power amplifier operates in Class-J mode for all values of a, with the exception of $a = 0$ (Class-B). Based on Equations (1) and (2), the impedance conditions for continuous Class-J power amplifiers are

$$Z_{n\_J} = \begin{cases} \frac{2V_D}{I_m}(1+j\alpha), n = 1 \\ -j\frac{3\pi V_D}{4I_m}\alpha, n = 2 \\ 0, \quad n > 2 \end{cases} \tag{3}$$

where n is the harmonic order. Equation (3) demonstrates that different values of a produce distinct impedance traces on the Smith chart, as shown in Figure 1, which facilitates broadband matching of the amplifier. Furthermore, as the additional impedance introduced at the second-harmonic frequency is purely reactive, it theoretically imposes no degradation on the amplifier efficiency. Figure 2 shows the waveforms of v_{d_J} and i_{d_J} for different values of a. For the Class-J amplifiers, it can be observed that affected by the second harmonic, the waveform of v_{d_J} shifts left or right under different values of a (relative to $a = 0$) [10].

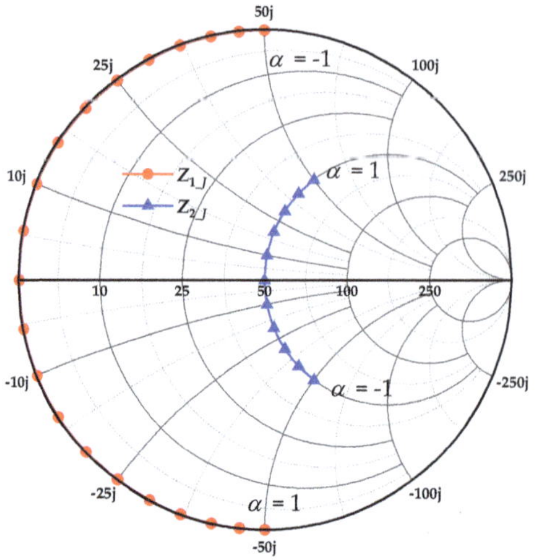

Figure 1. The normalized impedance traces of Z_{n_J} with different values of *a*.

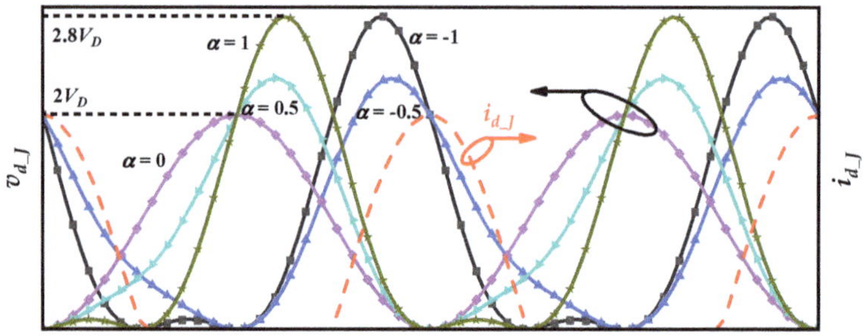

Figure 2. The waveforms of v_{d_J} and i_{d_J} for different values of *a*.

3. Harmonic Load Pull

To achieve the design of a continuous-mode broadband high-efficiency power amplifier, in addition to performing fundamental load pull, it is also necessary to determine the range of the second harmonic impedance. The selected transistor based on a 0.25 mm GaAs pHEMT process features a maximum drain-source voltage of 18 V. As can be seen from Figure 2, in continuous-mode operation, the peak voltage of v_{ds} exceeds the drain bias voltage V_D by approximately 2.8 times. To avoid transistor breakdown caused by excessive peak voltage, V_D is ultimately set to 6 V.

Under a 6 V/−1 V bias condition, fundamental load pull is conducted on the output-stage transistors of the X-band and Ku-band power amplifiers. The resulting optimum impedances serve as the initial values for subsequent iterations. To simulate practical losses, the real part of the second harmonic impedance is fixed at 5 W, while a parameter sweep is performed over its imaginary part ($Imload_2$). Figure 3 plots the saturated output power (P_{sat}) and PAE of the X-band transistor versus $Imload_2$. The results reveal a sharp degradation in both P_{sat} and PAE within the $Imload_2$ range of $-j24$ W to $j27$ W.

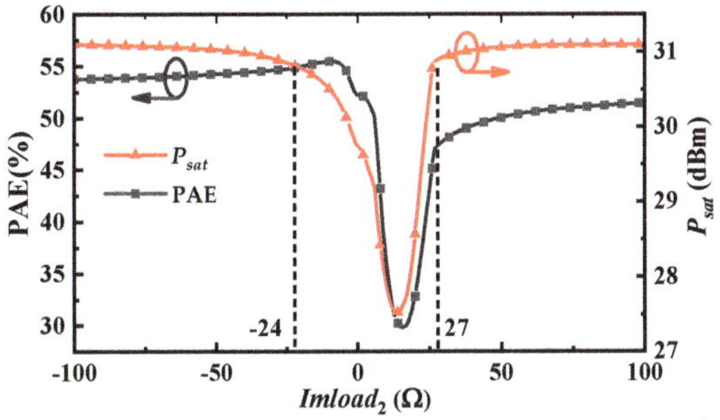

Figure 3. The simulated results of P_{sat} and PAE at different values of $Imload_2$.

Based on the second harmonic parameter sweep, appropriate harmonic load values are selected. After multiple iterations of fundamental load-pull and harmonic impedance sweeps, the final fundamental and harmonic impedance values are obtained. Table 1 presents the results of source pull and harmonic load pull for the output-stage transistors of the X-band and Ku-band power amplifiers.

Table 1. The results of source pull and harmonic load pull for the output-stage transistors of the X-band and Ku-band power amplifiers.

Frequency	8 GHz	10 GHz	12 GHz
Z_s	(11.9 + j21.5) Ω	(5.2 + j15) Ω	(5.8 + j14.3) Ω
Z_{f0}	(18.5 + j11.6) Ω	(14.6 + j9.1) Ω	(10.1 + j10.4) Ω
Z_{2f0}	>j30 W	>j30 W	>j30 W
Frequency	**13 GHz**	**15 GHz**	**17 GHz**
Z_s	(7.3 + j10.8) Ω	(6.3 + j2.4) Ω	(3.1 + j1.7) Ω
Z_{f0}	(8.9 + j9.5) Ω	(7.8 + j8.1) Ω	(6.1 + j8.5) Ω
Z_{2f0}	>j30 W	>j30 W	>j30 W

4. Design of the Harmonic Matching Network

As shown in Figure 1 and Table 1, the second harmonic load is purely reactive. Consequently, the output matching network must achieve a dual function: transforming the 50 Ω source impedance to the fundamental impedance Z_{f0}, while concurrently providing the required harmonic impedance Z_{2f0}. Although the off-chip branch-line structure can readily implement such harmonic matching, its large circuit area makes it unsuitable for integration in phased array T/R modules [11]. Fortunately, advances in semiconductor technology now permit the on-chip integration of harmonic control networks. While this approach may incur a slight efficiency penalty compared to off-chip, high-Q matching circuits, the significantly reduced amplifier area facilitates integration within T/R modules. Figure 4 shows the schematic diagram of the designed impedance matching network with a second harmonic control function.

In Figure 4, L_p and C_p resonate at $2f_0$ to realize second harmonic short-circuiting, whereas at the fundamental frequency, they are equivalent to a shunt grounded capacitor C_{eq}. Equation (4) presents the relationship between C_p and C_{eq}. L_s is employed to tune $Imload_2$ to meet the requirements of Z_{2f0} specified in Table 1. L_1, C_1, C_2, C_{eq}, and L_s form a fifth-order matching network, achieving fundamental impedance matching for both the X and Ku bands. The drain bias inductor L_{d1} is placed before the series resonant network; to avoid affecting the second harmonic, it has a large inductance value and does not participate in impedance matching.

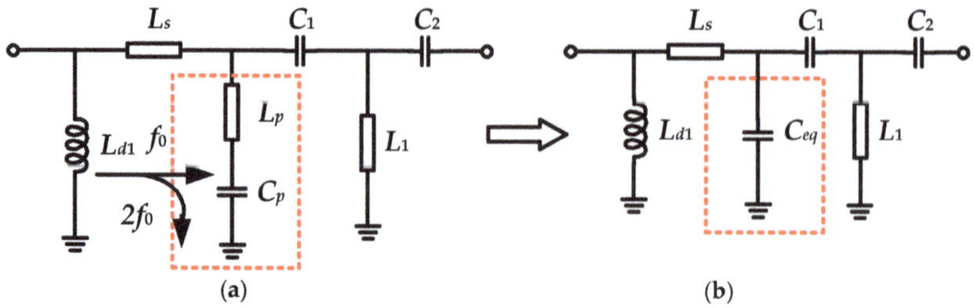

Figure 4. The schematic diagram of the (**a**) designed harmonic matching network and (**b**) its fundamental equivalent circuit.

$$C_p = \frac{3}{4} C_{eq} \tag{4}$$

Figure 5 presents the simulated impedance traces of the output matching networks (OMNs) for the X band and Ku band. The highest efficiency and output power of the output-stage transistor are achieved when its second harmonic impedance is tuned to lie within the green region. Combined with Table 1, it can be seen that at both the frequency bands, the designed harmonic matching network can not only meet the second harmonic impedance control requirements but also satisfy the fundamental load matching requirements.

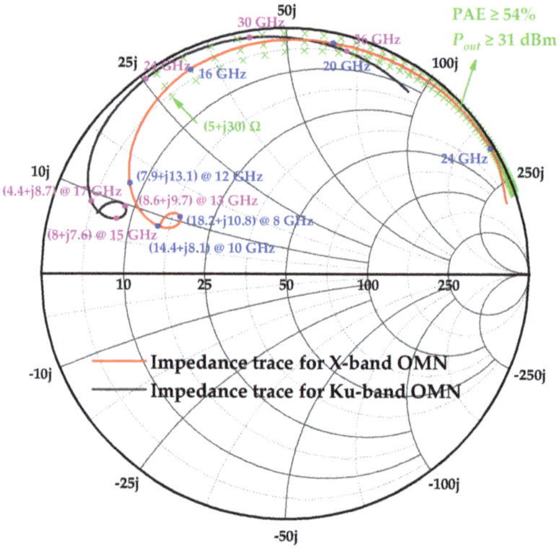

Figure 5. The simulated impedance traces of the OMNs for the X band and Ku band.

5. Design of the X and Ku Bands Class-J Power Amplifiers

After completing the output impedance matching, the driver stages of the X-band and Ku-band amplifiers are designed. Figure 6 presents the schematic diagrams of the interstage and input matching networks, where L_d and L_g serve as the drain and gate bias lines, respectively. Table 2 shows the source-pull and load-pull results of the driver transistor. Combined with Z_s of the output transistors in Table 1, the component values are optimized.

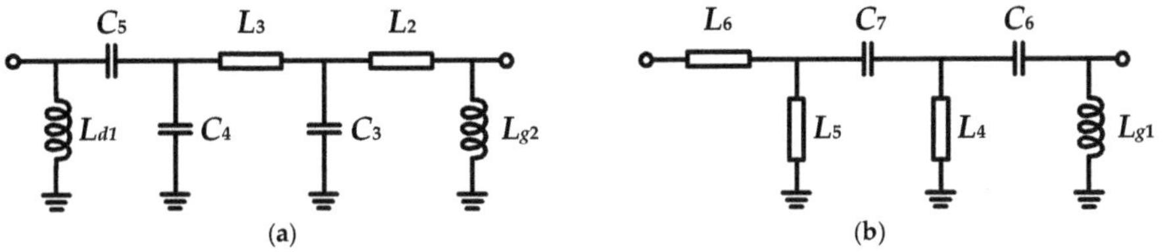

Figure 6. The schematic diagrams of the (**a**) inter-stage and (**b**) input matching networks.

Table 2. The results of source pull and load pull for the driver transistors of the X band and Ku band.

Frequency	**8 GHz**	**10 GHz**	**12 GHz**
Z_S	$(17.4 + j16)\ \Omega$	$(15.2 + j14.5)\ \Omega$	$(13.9 + j12.1)\ \Omega$
Z_L	$(31.3 + j28.5)\ \Omega$	$(29.5 + j26)\ \Omega$	$(28.6 + j23)\ \Omega$
Frequency	**13 GHz**	**15 GHz**	**17 GHz**
Z_S	$(2.4 + j4.6)\ \Omega$	$(2 + j5.2)\ \Omega$	$(2 + j4.3)\ \Omega$
Z_L	$(9.3 + j14.6)\ \Omega$	$(7.1 + j12)\ \Omega$	$(3.9 + j13)\ \Omega$

Figure 7 shows the optimized schematic diagrams of the proposed X-band and Ku-band power amplifiers. The optimized component values are provided in Table 3. Among them, some components are omitted as their minimal contribution to matching. All inductors are implemented using microstrip lines. Figure 8 presents the output-stage drain-source voltage and current waveforms of the amplifiers operating in the X and Ku bands. It can be observed that affected by the second harmonic, the peaks of the drain voltage all shift to the left, which is consistent with the characteristics of Class-J power amplifiers.

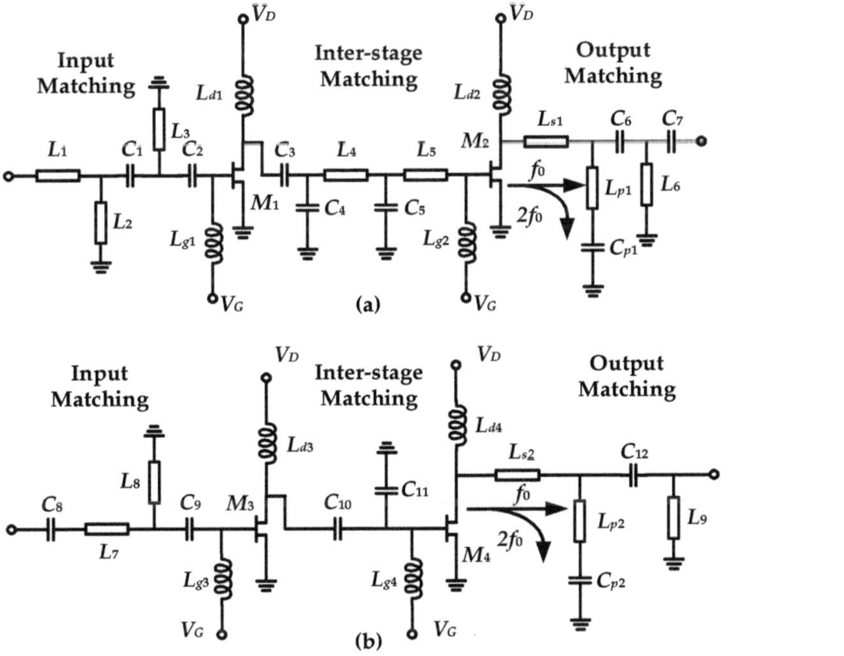

Figure 7. The schematic diagrams of the designed power amplifiers at (**a**) X band and (**b**) Ku band.

Table 3. The optimized component values shown in Figure 6.

Components	Values	Components	Values
L_1	10 mm × 1500 µm	C_{12}	0.52 pF
L_2	10 mm × 340 µm	L_{g1}	10 mm × 1550 µm
L_3	10 mm × 300 µm	L_{g2}	10 mm × 1400 µm
L_4	15 mm × 400 µm	L_{g3}	10 mm × 900 µm
L_5	15 mm × 300 µm	L_{g4}	10 mm × 1000 µm
L_6	10 mm × 1800 µm	L_{d1}	15 mm × 1350 µm
L_7	10 mm × 1050 µm	L_{d2}	30 mm × 1600 µm
L_8	10 mm × 250 µm	L_{d3}	15 mm × 650 µm
L_9	10 mm × 800 µm	L_{d4}	30 mm × 1100 µm
C_1	0.67 pF	L_{s1}	20 mm × 450 µm
C_2	6.10 pF	L_{s2}	30 mm × 400 µm
C_3	0.52 pF	L_{p1}	10 mm × 160 µm
C_4	0.62 pF	L_{p2}	30 mm × 120 µm
C_5	1.01 pF	C_{p1}	0.39 pF
C_6	0.68 pF	C_{p2}	0.35 pF
C_7	2.30 pF	M_1	6 × 100 µm
C_8	0.19 pF	M_2	10 × 150 µm
C_9	0.52 pF	M_3	8 × 200 µm
C_{10}	0.56 pF	M_4	12 × 200 µm
C_{11}	0.83 pF		

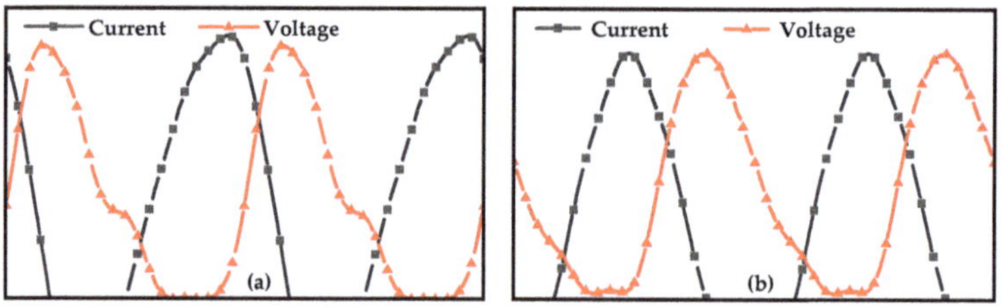

Figure 8. The output-stage voltage and current waveforms of the amplifiers operating at (**a**) 10 GHz and (**b**) 14 GHz.

6. Simulated and Measured Results

The two proposed continuous-mode power amplifiers are fabricated using a 0.25 mm GaAs pHEMT process. Figure 9 shows the microscope photographs of the two power amplifiers, with die areas of 2.75 mm × 1.2 mm and 2.2 mm × 1.2 mm, respectively. The S-parameters of the chips are measured using a Cascade Summit 1200 M probe station and a Keysight N5245A microwave network analyzer. Both chips undergo large-signal measurement after being eutectically bonded to a package with molybdenum-copper carriers, as shown in Figure 10. During the measurement, a heat sink is placed at the bottom to ensure efficient heat dissipation of the power amplifier chips.

Figure 11a presents the S-parameter simulated and measured results of proposed X-band power amplifier. It can be observed that the simulated results are in good agreement with the measured results. Within the frequency range of 8 GHz to 12 GHz, the power amplifier achieves an average small-signal gain of approximately 20 dB, and the measured input reflection coefficient is better than −8 dB. Figure 11b shows the large-signal characteristic simulated and measured results. Within the operating frequency band, the power amplifier delivers a saturated output power exceeding 30 dBm, with a PAE higher than 39% across the band and a maximum of 48.5%.

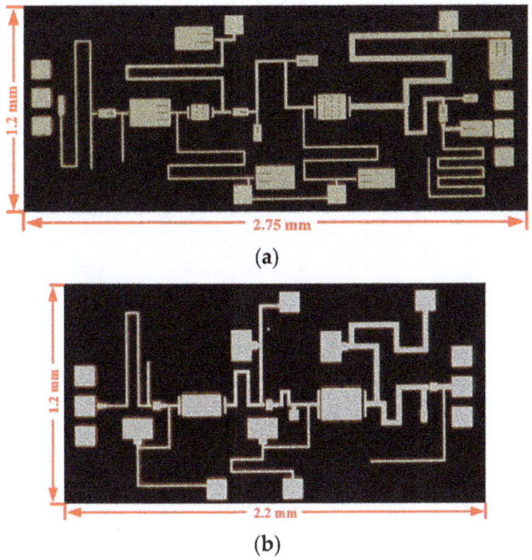

Figure 9. The microscope photograph of the designed continuous-mode power amplifiers for (**a**) the X band and (**b**) the Ku band.

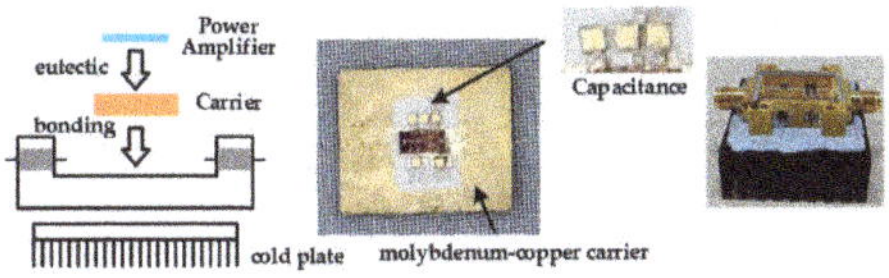

Figure 10. The mearsured carrier for the amplifier chips.

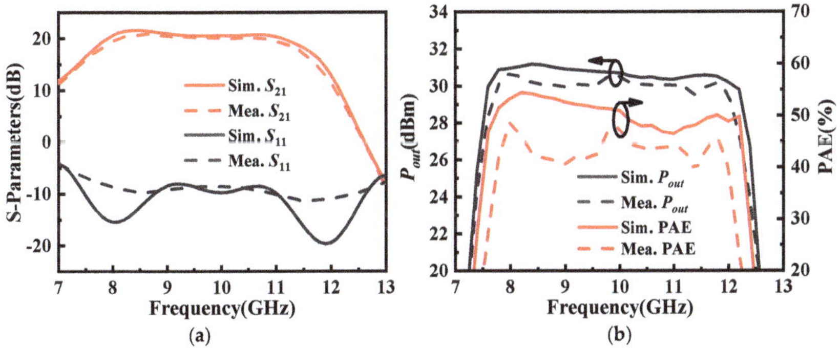

Figure 11. The simulated and measured results of the designed X-band power amplifier: (**a**) S-parameters and (**b**) large-signal characteristics.

The simulated and measured results of the designed Ku-band power amplifier are shown in Figure 12. Over the frequency range of 12 GHz to 18 GHz, the power amplifier achieves an average small-signal gain of approximately 16.5 dB, and the measured input reflection coefficient is better than −6.3 dB. The measured saturated output power achieves an average of 30.8 dBm across the band and 29 dBm at the band edges. Meanwhile, the measured PAE within the band reaches a maximum of 45.3% and an average of 40.32%.

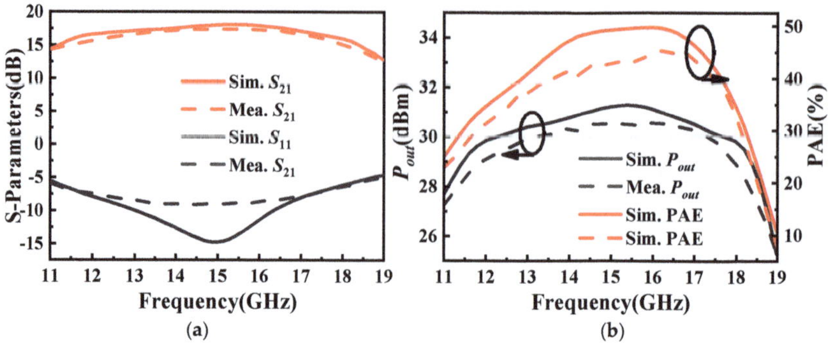

Figure 12. The simulated and measured results of the designed Ku-band power amplifier: (**a**) S-parameters and (**b**) large-signal characteristics.

The detailed performance metrics of two fabricated power amplifiers are listed in Table 4 and compared with several similar GaAs-process-based power amplifier studies in the literature [12–17]. Compared with the power amplifiers in [12,13,16], the proposed design demonstrates a higher PAE under a comparable bandwidth. Although [14,15] achieve high efficiency by employing only fundamental load-pull techniques, they are narrowband. The well-engineered matching network in [17] enables excellent bandwidth but sacrifices PAE.

Table 4. Comparison with other similar research on power amplifiers in GaAs process.

Ref.	Technology	Frequency (GHz)	BW (%)	P_{sat} * (dBm)	PAE * (%)	Area (mm^2)
[12]	0.15-μm GaAs	7~11.5	48.7	33.8	43	3.6 × 1.8
[13]	0.25-μm GaAs	8~13	47.6	31	35	1 × 3.5
[14]	0.25-μm GaAs	8.5~11.5	30	35	49	3 × 1.8
[15]	0.25-μm GaAs	8.5~9.5	11.1	29.6	44.4	2 × 2
		15.5~16.5	6.3	30.1	38.4	
[16]	0.25-μm GaAs	11.7~17.6	40.3	31	38.8	3.8 × 3.3
[17]	0.25-μm GaAs	6~18	100	40.5	29	5 × 3.6
This work	0.25-μm GaAs	8~12	40	31.2	48	2.75 × 1.2
		12~18	40	30.8	45.3	2.2 × 1.2

* maximum values.

7. Conclusions

In this article, two wideband high-efficiency power amplifiers operating in the X and Ku bands, respectively, are designed and fabricated using a 0.25 μm-GaAs pHEMT process. A harmonic matching network is introduced in both power amplifiers, which extends the operating bandwidth without sacrificing efficiency. The proposed X-band power amplifier achieves a peak saturated output power of 31.2 dBm and a maximum PAE of 48%. The proposed Ku-band amplifier delivers a peak saturated output power of 30.8 dBm and a maximum PAE of 45.3%. Both chips feature a compact footprint, rendering them suitable for integration into miniaturized phased array T/R modules.

Author Contributions: Conceptualization, Y.Y. and Z.Y.; data curation, Y.Y., X.Z. and J.F.; funding acquisition, Z.Y.; methodology, Y.Y.; validation, Z.Y.; visualization, Z.Y.; writing—original draft, Y.Y. and X.Z.; writing—review and editing, Y.Y. All authors have read and agreed to the published version of the manuscript.

Funding: This work was supported by the Science and Disruptive Technology Research Fund Project of Aerospace Information Research Institute, Chinese Academy of Sciences (2024-AIRCAS-SDTP-02).

Data Availability Statement: The data presented in this study are available on request from the corresponding author due to (specify the reason for the restriction).

Conflicts of Interest: The authors declare no conflicts of interest.

References

1. Maini, A.K. *Handbook of Defence Electronics and Optronics: Fundamentals, Technologies and System*; John Wiley & Sons: Hoboken, NJ, USA, 2018.
2. Batra, J.; Balakrishnan, G.; Aiello, R.; Foerster, J.R.; Dabak, A. Design of a multiband OFDM system for realistic UWB channel environments. *IEEE Trans. Microw. Theory Techn.* **2004**, *52*, 2123–2138. [CrossRef]
3. Cripps, S.C.; Tasker, P.J.; Clarke, A.L.; Lees, J.; Benedikt, J. On the continuity of high efficiency modes in linear RF power amplifiers. *IEEE Microw. Wirel. Compon. Lett.* **2009**, *19*, 665–667. [CrossRef]
4. Chu, C.; Sharma, T.; Dhar, S.K.; Darraji, R.; Wang, X.; Pang, J.; Zhu, A. Waveform engineered sequential load modulated balanced amplifier with continuous class-F^{-1} and class-J operation. *IEEE Trans. Microw. Theory Tech.* **2021**, *70*, 1269–1283. [CrossRef]
5. Xuan, X.; Cheng, Z.; Zhang, Z.; Le, C. Highly efficient ultrawideband power amplifier based on a novel multi-branch matching network. *IEEE Microw. Wirel. Technol. Lett.* **2023**, *33*, 1325–1328. [CrossRef]
6. Alizadeh, A.; Hassanzadehyamchi, S.; Medi, A.; Kiaei, S. An X-band class-J power amplifier with active load modulation to boost drain efficiency. *IEEE Trans. Circuits Syst. I Regul. Pap.* **2020**, *67*, 3364–3377. [CrossRef]
7. Gong, T.; Cheng, Z.; Zheng, B.; Xuan, X.; Le, C.; Fan, W.; Zhang, Z. SC band wideband GaN power amplifier MMIC for radar application. *IEICE Electron. Express* **2024**, *21*, 20240596. [CrossRef]
8. Wright, P.; Lees, J.; Tasker, P.J.; Benedikt, J.; Cripps, S.C. An Efficient, Linear, Broadband Class-J-Mode PA Realised Using RF Waveform Engineering. In Proceedings of the 2009 IEEE MTT-S International Microwave Symposium Digest, Boston, MA, USA, 7–12 June 2009; IEEE: New York, NY, USA, 2009; pp. 653–656.
9. Alizadeh, A.; Hassanzadehyamchi, S.; Medi, A. Integrated output matching networks for class–J/J− 1 power amplifiers. *IEEE Trans. Circuits Syst. I Regul. Pap.* **2019**, *66*, 2921–2934. [CrossRef]
10. Kazimierczuk, M.K. *RF Power Amplifiers*; John Wiley & Sons: Hoboken, NJ, USA, 2014.
11. Sheikhi, A.; Hemesi, H. Analysis and design of the novel class-F/E power amplifier with series output filter. *IEEE Trans. Circuits Syst. II Express Briefs* **2021**, *69*, 779–783. [CrossRef]
12. Li, J.; Yuan, Y.; Yuan, B.; Tan, C.; Yu, Z. Analysis and design of broadband 2-W power amplifier based on cascode transistors. *Microw. Opt. Technol. Lett.* **2024**, *66*, e34205. [CrossRef]
13. Babakrpur, E.; Medi, A.; Namgoong, W. Wideband GaAs MMIC driver power amplifiers for X and Ku bands. In Proceedings of the 2017 Texas Symposium on Wireless and Microwave Circuits and Systems (WMCS), Waco, TX, USA, 30–31 March 2017; IEEE: New York, NY, USA, 2017; pp. 1–4.
14. Hua, Y.; Wu, H.; Liao, X.; Liao, C.; Hu, L.; Lv, J. A High-Efficiency 3-Watt GaAs pHEMT X-Band MMIC Power Amplifier. In Proceedings of the 2018 International Conference on Microwave and Millimeter Wave Technology, Chengdu, China, 7–11 May 2018; IEEE: New York, NY, USA, 2018; pp. 1–2.
15. Xie, C.; Wu, P.; Tan, C.; Yuan, Y.; Zeng, J.; Yu, Z. An X/Ku dual-band switchless frequency reconfigurable GaAs power amplifier. *IEEE Microw. Wirel. Compon. Lett.* **2022**, *32*, 539–542. [CrossRef]
16. Yao, G.; Jia, H.; Zhao, Z.; Lu, Y.; Yi, C.; Liu, X.; Feng, T.; Yang, L.A.; Ma, X. A Ku-Band Broadband High-Efficiency GaAs MMIC Power Amplifier. In Proceedings of the 2024 International Conference on Microwave and Millimeter Wave Technology (ICMMT), Beijing, China, 16–19 May 2024; IEEE: New York, NY, USA, 2024; Volume 1, pp. 1–3.
17. Meghdadi, M.; Medi, A. Design of 6–18-GHz high-power amplifier in GaAs pHEMT technology. *IEEE Trans. Microw. Theory Tech.* **2017**, *65*, 2353–2360. [CrossRef]

micromachines

MDPI

Article

Dual-Side Hybrid Embedding Network for Gain Enhancement of Terahertz Amplifiers at Frequencies Near f_{max}

Xiaorui Liu, Jianguo Yu *, Yun Wang, Yibo Huang, Feixiang Zhang, Zhanjiang Wang and Yaqi Cheng

School of Electronic Engineering, Beijing University of Posts and Telecommunications, Beijing 100876, China; lxr1100@bupt.edu.cn (X.L.)
* Correspondence: yujg@bupt.edu.cn

Abstract

This paper proposes a dual-side hybrid embedding network (DHEN) to mitigate gain degradation in terahertz amplifiers at frequencies near f_{max}. The proposed approach employs a pre-embedding network for parasitic absorption, followed by Y-embedding for gain enhancement. A theoretical analysis is conducted to derive the embedding conditions for process-constrained circuit synthesis. In this architecture, a capacitive base-side pre-embedding provides intrinsic DC isolation, while a defected-ground-structure (DGS) inductor realizes the Y-embedding inductive element with reduced layout area. Based on the DHEN, a four-stage amplifier is designed in a 130 nm SiGe BiCMOS process. Electromagnetic co-simulation results demonstrate a power gain of 19.3 dB at 280 GHz, corresponding to an 11.5 dB improvement over a conventional unboosted amplifier. The proposed approach provides a unified synthesis methodology that simultaneously addresses parasitic absorption, DC isolation, and gain enhancement for near-f_{max} THz amplifier design.

Keywords: embedding network; gain enhancement; terahertz amplifier; defected ground structure (DGS)

1. Introduction

Terahertz (THz) technology has demonstrated significant potential in applications such as radio astronomy, high-resolution imaging, and high-speed wireless communications, owing to its high data capacity, enhanced security, and high atmospheric transparency in specific spectral windows [1–7]. However, THz systems suffer from severe free-space path loss and atmospheric attenuation, which necessitate multiple gain stages to achieve long-range wireless transmission [8–13]. These gain stages must employ active devices with sufficiently high maximum oscillation frequency (f_{max}) to operate effectively at THz frequencies [14,15].

Nevertheless, as the operating frequency approaches the transistor f_{max}, the intrinsic gain of the active device decreases rapidly, particularly beyond half f_{max} [16–30]. Consequently, silicon-based THz amplifiers without gain enhancement techniques exhibit insufficient power gain to meet system requirements. One straightforward approach is to cascade multiple stages, which, however, leads to increased chip area and DC power consumption. Alternatively, positive feedback (PF) techniques can be employed to enhance the gain of each amplifier stage. Singhakowinta [21] demonstrated that embedding an active two-port network (A2P) within a linear lossless reciprocal (LLR) network can increase the maximum available gain (G_{ma}) under conjugate matching while maintaining unconditional stability ($K \geq 1$). This process continues until G_{ma} reaches the theoretical upper bound

$G_{max} = (2U - 1) + 2\sqrt{U(U-1)}$, where U denotes the unilateral gain [27]. Based on this principle, various gain enhancement strategies have been reported, including parallel (Y-) embedding, series (Z-) embedding, and their corresponding pre-embedding configurations. These methods enable device gain to approach G_{max} at frequencies near half $f_{\max}$ [22–29].

Despite their effectiveness, conventional LLR-based gain enhancement structures often lack intrinsic DC-blocking capability, resulting in constrained biasing flexibility. In particular, enforcing equal base and collector voltages may degrade key performance metrics [30]. Beyond the DC-blocking issue, the key practical challenge is to jointly handle parasitic absorption, gain enhancement, and process constraints in a unified synthesis flow for near-$f_{\max}$ amplifier design.

To address these limitations, this paper proposes a dual-side hybrid embedding network (DHEN) for gain enhancement in near-$f_{\max}$ THz amplifiers. The DHEN employs a base-side capacitive pre-embedding network and a collector-side Y-embedding network to absorb parasitics, enhance gain, and maintain inherent DC isolation for independent bias control. Unlike prior LLR-based and lossy embedding approaches, the proposed DHEN unifies parasitic absorption and DC isolation within a single synthesis framework, representing the key novelty of this work. A compact DGS inductor is adopted to implement the inductive embedding with reduced layout area. This work develops the analytical framework and synthesis methodology of the DHEN under process constraints, with all passive components EM-modeled and co-simulated with device models. The remainder of this paper is organized as follows. Section 2 introduces the DHEN structure and theoretical analysis; Section 3 presents the amplifier design and simulation results; Section 4 concludes this paper.

2. DHEN Gain Enhancement Technique

Different from conventional dual-side embedding approaches that rely solely on inductive elements, the proposed DHEN introduces a base-side pre-embedding that simultaneously realizes the required reactive embedding and intrinsic DC isolation. This integration enables gain-plane synthesis under explicit bias constraints, rather than treating biasing as an external design consideration.

2.1. Y-Matrix Analysis of the DHEN

To analyze the proposed DHEN structure, the Y-matrix representation of each subnetwork is first examined. Figure 1 illustrates the analytical representation of the DHEN, where reactive elements jX_1 and jX_2 form the pre-embedding network and a shunt susceptance B_p realizes the Y-embedding network.

The transistor Q1 is modeled as a two-port network characterized by its Y-matrix $[Y] = [Y_{11}, Y_{12}; Y_{21}, Y_{22}]$. This Y-matrix can be equivalently transformed into the corresponding ABCD-matrix $[ABCD]$. By inserting these reactive elements at both ports of the A2P, a pre-embedding network is formed. The resulting ABCD-matrix of the pre-embedding network can be written as

$$\begin{aligned}
[ABCD]_{pre} &= \begin{bmatrix} A & B \\ C & D \end{bmatrix}_{Z_1} \cdot \begin{bmatrix} A & B \\ C & D \end{bmatrix} \cdot \begin{bmatrix} A & B \\ C & D \end{bmatrix}_{Z_2} \\
&= \begin{bmatrix} 1 & jX_1 \\ 0 & 1 \end{bmatrix} \cdot \begin{bmatrix} -\frac{Y_{22}}{Y_{21}} & -\frac{1}{Y_{21}} \\ -\frac{\Delta Y}{Y_{21}} & -\frac{Y_{11}}{Y_{21}} \end{bmatrix} \cdot \begin{bmatrix} 1 & jX_2 \\ 0 & 1 \end{bmatrix} \\
&= -\frac{1}{Y_{21}} \begin{bmatrix} jX_1 \cdot \Delta Y + Y_{22} & M_{pre} \\ \Delta Y & jX_2 \cdot \Delta Y + Y_{11} \end{bmatrix}
\end{aligned} \tag{1}$$

where $\Delta Y = Y_{11}Y_{22} - Y_{12}Y_{21}$ and $M_{pre} = 1 + jX_1Y_{11} + jX_2Y_{22} - X_1X_2\Delta Y$.

https://doi.org/10.3390/mi17040432

From (1), the corresponding Y-matrix of the pre-embedding network $[Y]_{pre}$ can be derived as

$$[Y]_{pre} = \frac{1}{M_{pre}} \begin{bmatrix} jX_2 \cdot \Delta Y + Y_{11} & Y_{12} \\ Y_{21} & jX_1 \cdot \Delta Y + Y_{22} \end{bmatrix}. \tag{2}$$

As illustrated in Figure 1, the complete DHEN is obtained by shunt-connecting a Y-embedding network at both the input and output ports of the pre-embedding network. The resulting overall Y-matrix is given in (3)

$$\begin{aligned} [Y]_{tot} &= [Y]_{pre} + \begin{bmatrix} jB_p & -jB_p \\ -jB_p & jB_p \end{bmatrix} \\ &= \begin{bmatrix} \frac{jX_2 \cdot \Delta Y + Y_{11}}{M_{pre}} + jB_p & \frac{Y_{12}}{M_{pre}} - jB_p \\ \frac{Y_{21}}{M_{pre}} - jB_p & \frac{jX_1 \cdot \Delta Y + Y_{22}}{M_{pre}} + jB_p \end{bmatrix}. \end{aligned} \tag{3}$$

Based on the total Y-matrix $[Y]_{tot}$, the influence of the DHEN on the original A2P can be conveniently visualized and analyzed on the gain plane.

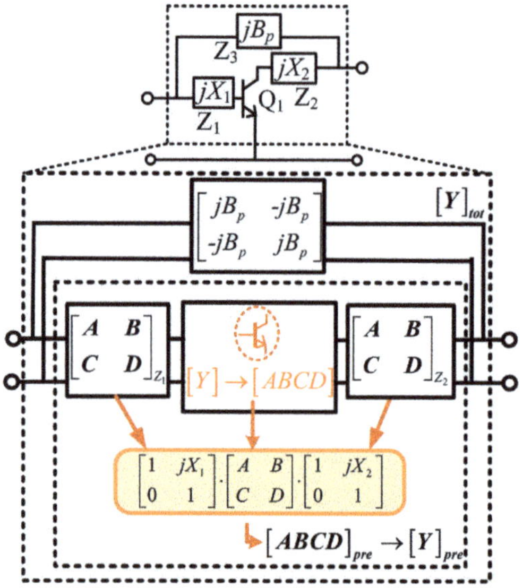

Figure 1. Analytical representation of the proposed DHEN using ABCD- and Y-parameter formulations.

2.2. *Gain-Plane-Based Analytical Solution of the DHEN*

The gain plane is a graphical tool used to analyze the influence of embedding networks on G_{ma} and U, as shown in Figure 2 [21]. On the gain plane, the transistor characteristics, including G_{ma} and the stability factor K, can be mapped onto a two-dimensional coordinate system defined by the ratio between U and the gain parameter A, where $A = Y_{21}/Y_{12}$. In this representation, constant gain values correspond to specific gain circles, while the stability boundary separates the unconditionally stable region ($K \geq 1$) from the unstable region. Since U remains invariant for all LLR networks added to the A2P, variations in G_{ma} and K depend solely on A. Consequently, the movement of the operating point on the gain plane directly reflects the impact of external network tuning on the overall performance.

https://doi.org/10.3390/mi17040432

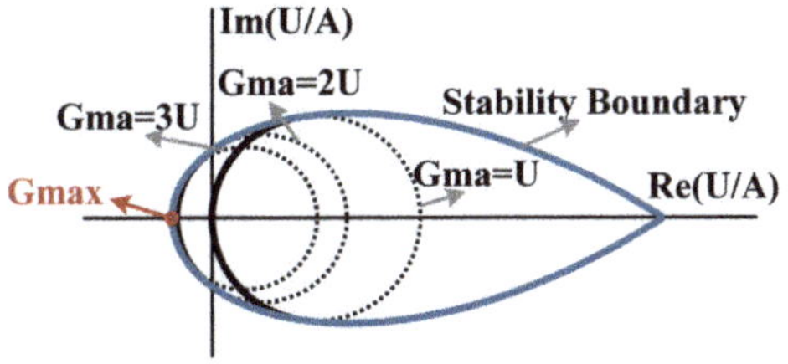

Figure 2. Gain plane and gain circles.

According to [22], the A of the overall network is given by

$$A_{\text{tot}} = \frac{Y_{21\text{tot}}}{Y_{12\text{tot}}} = \frac{Y_{21} - jB_p \cdot M_{pre}}{Y_{12} - jB_p \cdot M_{pre}}. \tag{4}$$

Since all embedding networks are LLRs, the value of U remains constant; therefore, the gain-state point of the overall network becomes

$$\frac{U}{A_{tot}} = \frac{U(Y_{12} - jB_p \cdot M_{pre})}{Y_{21} - jB_p \cdot M_{pre}}. \tag{5}$$

As can be observed from (5), the pre-embedding network itself does not directly shift the gain-state point. Instead, the displacement of the gain-state point is induced by the Y-embedding network. The introduction of the pre-embedding network increases the degrees of freedom governing the trajectory of the gain-state point on the gain plane from a single variable B_p to a combined parameter set (X_1, X_2, B_p), thereby enabling more flexible structural optimization.

Based on the gain-plane method, the following conditions must be satisfied to achieve G_{max} [27]:

$$\begin{cases} \text{Re}\left[\dfrac{U}{A_{\text{tot}}}\right] = \dfrac{-U}{2U - 1 + 2\sqrt{U(U-1)}} \\ \text{Im}\left[\dfrac{U}{A_{\text{tot}}}\right] = 0 \end{cases} \tag{6}$$

By substituting (5) into (6), an infinite set of feasible embedding-element combinations can be obtained, since the number of design variables exceeds that of independent equations. By fixing X_2, the remaining parameters X_1 and B_p can be expressed as follows:

$$\begin{cases} B_p = \dfrac{m_2 m_5 - m_4 m_6}{m_1 m_4 + m_2 m_3} \\ X_1 = \dfrac{m_5 - m_3 B_p}{m_4 B_p} \end{cases}, \tag{7}$$

where

$$\begin{cases} m_1 = [1 - X_2 \cdot \text{Im}(Y_{22})] \cdot (a-1) \\ m_2 = [\text{Im}(Y_{11}) + X_2 \cdot \text{Re}(\Delta Y)] \cdot (a-1) \\ m_3 = X_2 \cdot \text{Re}(Y_{22}) \cdot (a-1) \\ m_4 = [\text{Re}(Y_{11}) - X_2 \cdot \text{Im}(\Delta Y)] \cdot (a-1) \\ m_5 = \text{Re}(Y_{12}) - a \cdot \text{Re}(Y_{21}) \\ m_6 = \text{Im}(Y_{12}) - a \cdot \text{Im}(Y_{21}) \\ a = \dfrac{-1}{2U - 1 + 2\sqrt{U(U-1)}} \end{cases} \tag{8}$$

https://doi.org/10.3390/mi17040432

For a given value of X_2, the corresponding values of X_1 and B_p can be determined using (7).

3. Design Examples

3.1. Amp-Cell

In this section, a representative amp-cell design example is presented to validate the analytical evaluation of the proposed DHEN-based gain enhancement technique under process constraints. The circuit is designed in a 130 nm SiGe BiCMOS process (with $f_{\max} = 450$ GHz), and the target operating frequency is set to 280 GHz, which lies beyond half of $f_{\max}$. The same design kit, as well as a comparable frequency range, has been extensively characterized and experimentally validated in prior works [17–20]. In this process, 280 GHz corresponds to approximately $0.62 f_{\max}$, a regime in which meaningful transistor gain remains available; the practical feasibility of 280 GHz operation in 130 nm SiGe BiCMOS is strongly supported by measured demonstrations across the 206–310 GHz range from these cited works.

In the design of the proposed DHEN, the base-side pre-embedding is formed by a capacitor. This choice inherently provides DC isolation while participating in the reactive embedding required by the analytical synthesis, thereby enabling a direct mapping between the theoretical formulation and circuit-level realization.

The employed HBT features an emitter area of $A_E = 4 \times 0.07 \times 0.9$ µm^2 and is biased at $V_b = 0.94$ V and $V_{cc} = 1.5$ V. At THz frequencies, parasitic effects associated with passive interconnects become non-negligible and must be carefully incorporated into the design procedure. In particular, as illustrated in Figure 3, the parasitic inductance between the HBT base port (M1) and the extrinsic top-metal connection (TM2) is extracted to be approximately 4.9 pH at 280 GHz. The extraction is performed from the EM-simulated interconnect admittance using $L_{\text{ext}} = \frac{\text{Im}(y_{11}^{-1})}{\omega}$ [31], where y_{11} is the interconnect admittance obtained from EM simulation. These parasitics are explicitly included in the pre-embedding network to ensure consistency between the analytical evaluation and the circuit-level design.

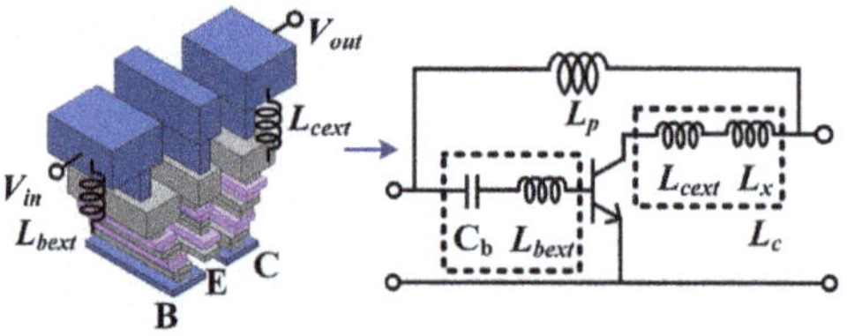

Figure 3. Metal interconnect and associated parasitic inductances, where L_{bext} and L_{cext} denote the extracted base-side and collector-side parasitic inductances, respectively.

By sweeping X_2 from 1 Ω to 27 Ω, the corresponding values of X_1 and B_p were solved from (7) and are plotted in Figure 4a. To mitigate high-frequency loss, the Y-embedding inductance L_p should be minimized. As shown in Figure 4a, decreasing the collector-side reactance X_2 reduces the Y-embedding susceptance B_p, thereby yielding a smaller L_p. Since the collector is biased through the top-metal layer, a purely inductive element is preferred at this node. Consequently, the minimum practical value of the collector embedding element is limited by its constrained parasitic inductance L_{cext}, corresponding to $X_2 \approx 8.62$ Ω at 280 GHz. Under this condition, the resulting parameters are $X_1 = -10.27$ Ω and $B_p = -0.0109$ S. After absorbing the base parasitic inductance L_{bext}, X_1 translates into an additional 30 fF series capacitor, while B_p corresponds to a 52 pH inductor at 280 GHz.

https://doi.org/10.3390/mi17040432

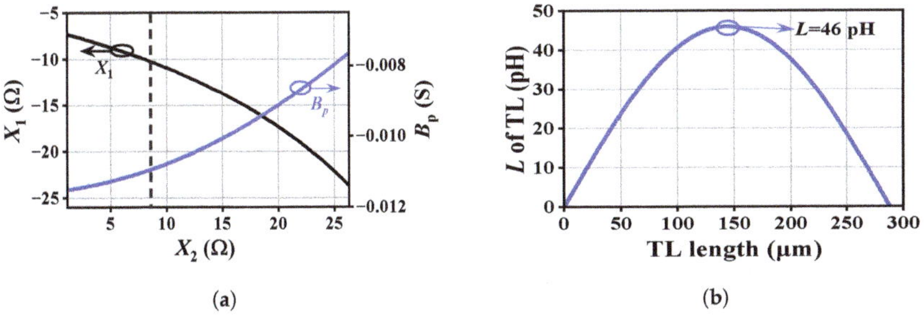

Figure 4. Embedding network solutions and transmission-line realization: (**a**) X_1 (black), B_p (blue), and the selected process-constrained solution (dashed line). (**b**) Equivalent inductance versus TL length at 280 GHz with $Z_0 = 81\ \Omega$; arrow marks the maximum realizable inductance (≈46 pH).

After determining the initial reference values of the embedding network, further optimization is required to account for process constraints. Under these constraints, parameters such as the minimum line width directly limit the maximum achievable characteristic impedance and the realizable inductance range of transmission lines (TL). When the TL length is shorter than half the wavelength ($\lambda/2$), parasitic shunt components can be neglected, and the line can be approximated as a pure inductor, whose inductance satisfies $\omega L_M = Z_0 \sin(\beta l)$.

Figure 4b illustrates the relationship between the equivalent inductance and the physical length of the TL at 280 GHz. As shown in Figure 4b, the maximum inductance achievable with TL under the given process constraints is approximately 46 pH, which is lower than the required Y-embedding inductance of 52 pH obtained from (7). This discrepancy indicates that further optimization is necessary; consequently, the Y-embedding inductance derived from Figure 4a cannot be directly synthesized using conventional TL under the given constraints.

To overcome this process limitation, a re-optimization is performed by adjusting the base-side pre-embedding capacitance C_b, which is initially determined from the analytical solution. By increasing C_b, the required value of the parallel embedding inductance L_p can be effectively relaxed while preserving the targeted gain performance. Figure 5 illustrates the variation of G_{ma} in the Amp-cell as a function of C_b and L_p for a fixed collector inductance of $L_c = 4.9$ pH. As shown in Figure 5, for a given gain level, the required Y-embedding inductance L_p decreases with increasing C_b, enabling a practical re-selection of the embedding parameters within the feasible design space. Accordingly, the final values are chosen as $L_p = 44.5$ pH and $C_b = 43$ fF.

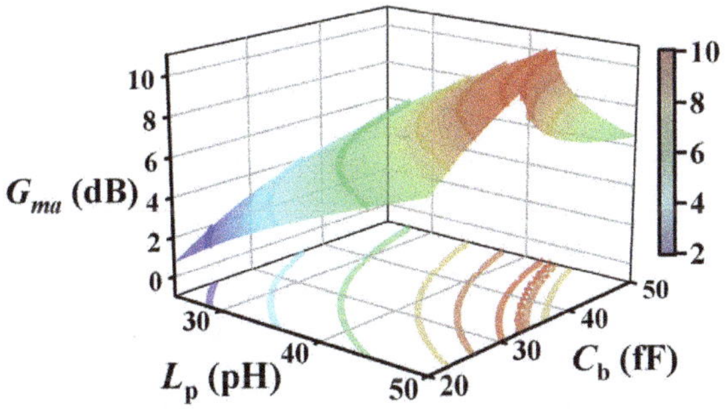

Figure 5. G_{ma} as a function of C_b and L_p in the Amp-cell ($L_c = 4.9$ pH).

https://doi.org/10.3390/mi17040432

The trajectory of the gain-state point of the Amp-cell at 280 GHz under different embedding conditions is shown in Figure 6. Point A corresponds to the original A2P without any embedding network, yielding a G_{ma} of 3.5 dB, which is far below G_{max}. Point B represents the gain-state point with only the Y-embedding network applied, where G_{ma} increases to 6.2 dB and the trajectory moves toward G_{max} yet does not reach it. This behavior highlights the necessity of introducing a pre-embedding network to further steer the gain-state trajectory. Point C corresponds to the theoretical DHEN solution with pre-embedding included, for which the gain-state point reaches G_{max} (10.4 dB), resulting in an overall gain improvement of approximately 6.9 dB compared to the original A2P. Point D denotes the DHEN solution under process constraints, achieving a G_{ma} of 9.5 dB, which is 0.9 dB lower than G_{max} due to process constraints.

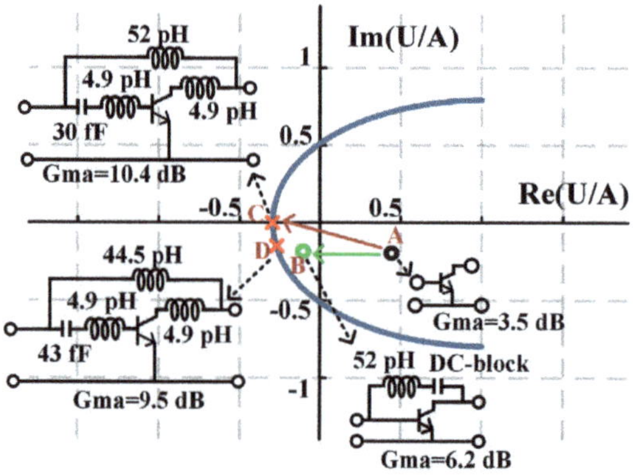

Figure 6. The trajectory of the gain-state point of the Amp-cell at 280 GHz under different embedding conditions: A, original A2P without embedding; B, Y-embedding only; C, ideal DHEN solution reaching G_{max}; D, process-constrained DHEN solution.

Although the process-constrained DHEN solution (Point D in Figure 6) already provides a substantial gain improvement, its design still requires a compact embedding inductor with sufficiently high inductance density. To address this requirement, a defected-ground-structure (DGS) inductor is introduced to enhance the achievable inductance within a limited footprint. Figure 7a shows the three-dimensional view of the DGS inductor, in which a microstrip line (MSL) on the TM2 metal layer is routed above a patterned ground plane on the M3 metal layer. By introducing intentional defects in the ground plane beneath the signal line, the electromagnetic field distribution is perturbed, leading to an increased equivalent inductance compared with a solid-ground inductor. Figure 7b compares the simulated inductance and parasitic resistance of the proposed DGS inductor (solid lines) with those of a conventional solid-ground inductor (dashed lines). As shown, the DGS inductor exhibits a higher equivalent inductance over the frequency range of interest, achieving approximately 44 pH at the target frequency. In contrast, the solid-ground inductor provides about 38 pH under the same physical dimensions. Meanwhile, the DGS structure results in a higher parasitic resistance, reflecting the increased loss associated with the ground-plane defects.

After EM simulation, an equivalent inductance of approximately 44 pH is obtained at 280 GHz using the DGS inductor shown in Figure 7a, with a total effective length of about 118 μm. A 43 fF cross-finger capacitor with a Q of 15 is realized with two thick metal layers. Figure 8 compares the maximum available gain G_{ma} of the conventional common-emitter (CE) amplifier with that of the DHEN-based design. In Figure 8a, where parasitic effects are not considered, the theoretical upper bound of the gain, G_{max}, is 10.4 dB at 280 GHz.

https://doi.org/10.3390/mi17040432

Under the same condition, the conventional CE amplifier exhibits a maximum available gain G_{ma} of only 3.5 dB, which is well below this upper bound. With the introduction of the DHEN, G_{ma} is increased from 3.5 dB to 10.4 dB, reaching the theoretical upper bound and corresponding to a gain enhancement of approximately 6.9 dB per stage. Here, the "ideal" case in Figure 8a does not include transistor interconnect parasitics. In Figure 8b, where parasitic effects are included through EM simulation, the theoretical upper bound G_{max} is reduced to 7.3 dB. Nevertheless, the DHEN remains effective, improving G_{ma} from 2.1 dB to 7.1 dB and yielding a gain enhancement of about 5 dB. The gain reduction from the ideal case (about 3.3 dB) is mainly attributed to parasitic effects of the transistor peripheral interconnects and the passive/layout loss of the realized DHEN network; the CE reference is treated under the same comparison setup.

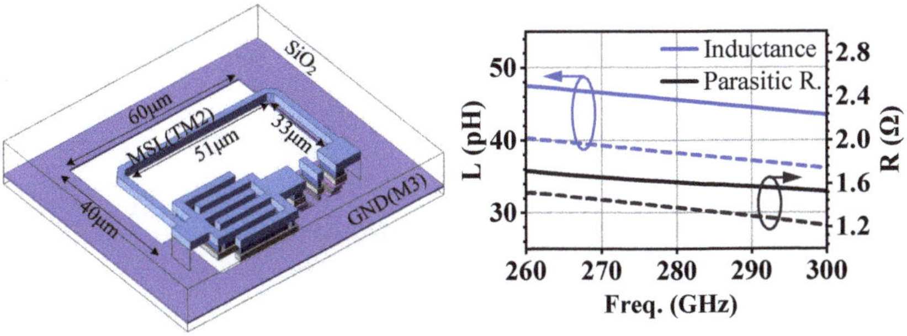

(a) (b)

Figure 7. DGS inductor: (**a**) three-dimensional view; (**b**) simulated inductance and parasitic resistance of the DGS (solid lines) and solid-ground (dashed lines) inductors.

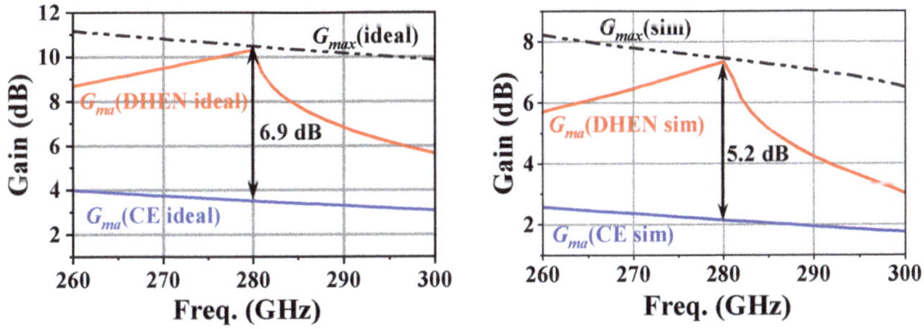

(a) (b)

Figure 8. Comparison of G_{ma} between traditional CE and DHEN architectures: (**a**) ideal lossless embedding network; (**b**) EM-simulation-based embedding network.

3.2. Multi-Stage Amplifier Design

A four-stage amplifier is designed based on the Amp-cell introduced in Section 3.1, as illustrated in Figure 9. All transmission lines used for impedance matching in Figure 9 are realized using 56 Ω MSLs, with metal layer M3 serving as the reference ground plane and TM2 as the signal layer. A 140 μm MSL is employed for DC decoupling to provide RF isolation between the bias network and the signal path. To mitigate gain roll-off at higher frequencies, inter-stage impedance matching is optimized at 285 GHz, thereby maintaining a relatively flat gain response around the target operating frequency.

https://doi.org/10.3390/mi17040432

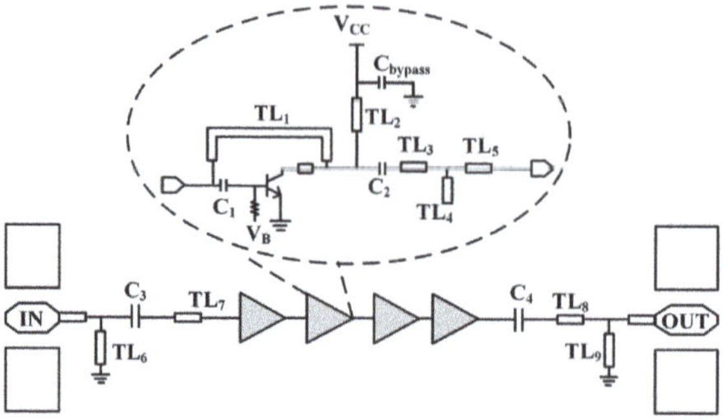

TL1	TL2	TL3	TL4	TL5	TL6	TL7	TL8	TL9	C1	C2	C3	C4	Cbypass
118u	140u	42u	2u	20u	10u	30u	32u	5u	43fF	18fF	20fF	186fF	200fF

Figure 9. The schematic of the four-stage amplifier is based on Amp-cell.

The layout of the four-stage amplifier is shown in Figure 10. The core occupies an area of 390 µm × 275 µm.

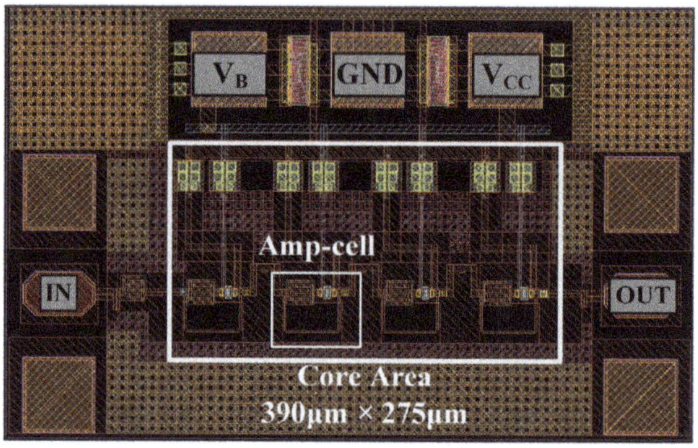

Figure 10. Layout view of the designed four-stage amplifier.

The circuit shown in Figure 9 was simulated, where all passive components, including MSLs and capacitors, were modeled using an EM simulator to account for parasitic effects. As shown in Figure 11, the amplifier achieves an S_{21} of 19.3 dB at 280 GHz, making it suitable for use as a driver amplifier in communication links. To verify the effectiveness of the proposed architecture, a conventional four-stage CE amplifier without any gain enhancement technique was also designed as a reference. This reference amplifier exhibits an S_{21} of only 7.8 dB at 280 GHz. The overall gain improvement in the four-stage amplifier is lower than the ideal per-stage enhancement because cumulative inter-stage matching and passive-network losses are included in the EM-co-simulated multi-stage design. Both amplifiers are well matched, with input/output return losses better than 15 dB. These results confirm that the proposed DHEN technique effectively enhances the power gain of amplifiers operating near f_{max}.

To provide a more complete performance characterization of the proposed four-stage amplifier, additional simulated results are summarized in Figure 12, including stability around the design band, noise figure, passive-network insertion loss, and large-signal output-power behavior.

https://doi.org/10.3390/mi17040432

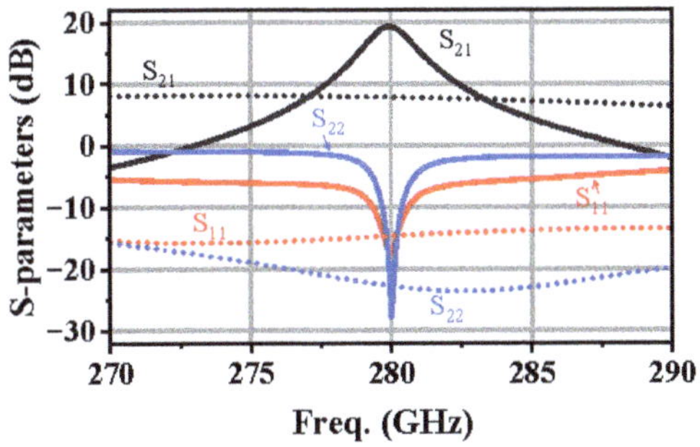

Figure 11. Simulated S-parameters of the proposed DHEN-based four-stage amplifier (solid) and the conventional CE amplifier (dashed), with all passive components EM-modeled.

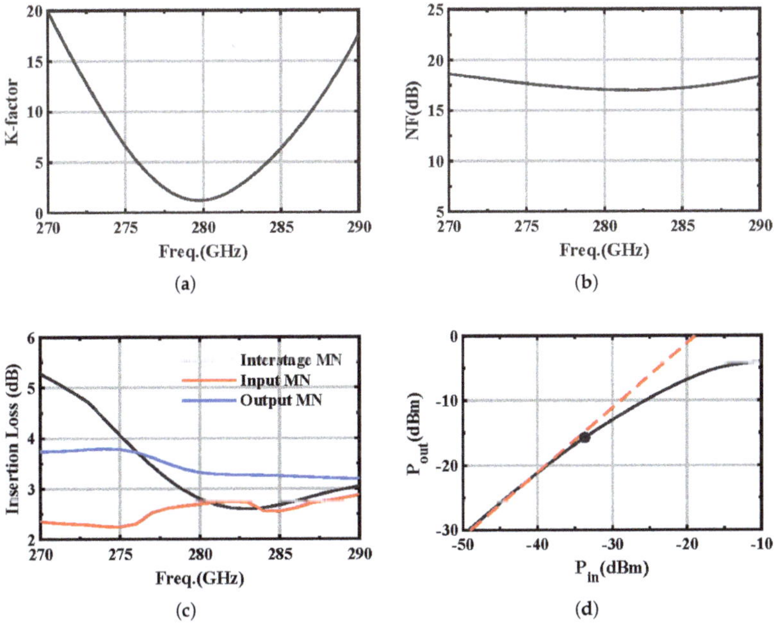

Figure 12. Simulated performance summary of the proposed DHEN-based four-stage amplifier: (**a**) stability factor K; (**b**) noise figure; (**c**) insertion loss of key passive matching networks; (**d**) large-signal P_{out}–P_{in} characteristic, where the solid curve denotes the simulated large-signal response, the dashed line denotes the linear extrapolation, and the dot marks the output 1 dB compression point.

Figure 12a indicates that K remains above 1 (minimum $\approx$ 1.1 near 280 GHz), confirming operation in the unconditionally stable region ($K \geq 1$) defined in Section 2.2. The minimum simulated noise figure is about 16.9 dB near 282 GHz, as shown in Figure 12b. Figure 12c summarizes the insertion loss of key passive networks: the inter-stage matching network shows the largest and most frequency-dependent loss, decreasing from about 5.3 dB at 270 GHz to about 2.6 dB near 290 GHz, while the input/output matching losses vary more mildly. Figure 12d presents the large-signal P_{out}–P_{in} characteristic, from which the output 1 dB compression point and saturation power are approximately -16 dBm and -4 dBm, respectively.

https://doi.org/10.3390/mi17040432

To further benchmark the proposed design, Table 1 compares this work with representative silicon-based sub-THz amplifiers reported near the target frequency range. As shown in Table 1, this work achieves the highest gain per stage among the listed designs (4.83 dB/stage at 280 GHz, simulated), indicating strong gain-enhancement capability under near-$f_{\max}$ operation. The 11.5 dB total gain improvement over the unboosted CE reference—obtained within the same process and at the same frequency—is therefore significant: within the compared set of silicon-based sub-THz demonstrations (Table 1, noting differences in process node and topology), it achieves the highest per-stage gain and directly validates the gain-enhancement efficacy of the proposed DHEN methodology.

Table 1. Comparison of representative silicon-based sub-THz amplifiers near the target frequency range.

Ref.	Technology	Frequency (GHz)	$f_{\max}$ (GHz)	$f/f_{\max}$	Gain (dB)	Gain/Stage (dB)
[22]	65 nm CMOS	260	352	0.74	9.2	2.3
[25]	65 nm CMOS	280	395	0.71	12	4
[30]	65 nm CMOS	298	317	0.94	21	1.31
[32]	130 nm SiGe	283	450	0.63	10.9	0.78
This work	130 nm SiGe	280 *	450	0.62	19.3 *	4.83

* Simulated.

4. Conclusions

This paper presents a gain enhancement approach for amplifiers operating near the transistor $f_{\max}$ based on a dual-side hybrid embedding network. Closed-form expressions derived from gain-plane theory provide analytical guidance for embedding synthesis. The proposed DHEN absorbs interconnect parasitics while a capacitive base-side pre-embedding enables independent bias control without additional DC-blocking capacitors. A four-stage amplifier designed in a 130 nm SiGe BiCMOS process achieves a power gain of 19.3 dB at 280 GHz, corresponding to an 11.5 dB improvement over a conventional unboosted CE amplifier, while maintaining stable operation. These results—fully consistent with the abstract claims—confirm that the proposed DHEN technique effectively enhances the gain of amplifiers operating near f_{max}. It should be noted that this work primarily focuses on validating the upper-limit gain enhancement at the designated design frequency under near-$f_{\max}$ operation. Accordingly, the optimization prioritizes peak gain at the target frequency, while gain–bandwidth co-optimization is beyond the main scope of this paper. Future work will build on the current high-gain core to pursue bandwidth extension, thereby improving usable bandwidth and practical applicability.

Author Contributions: Conceptualization, X.L. and J.Y.; methodology, X.L.; theoretical analysis, X.L.; circuit design and simulation, X.L.; validation, X.L., J.Y., Y.W. and Y.H.; formal analysis, X.L.; investigation, X.L.; resources, J.Y.; data curation, X.L. and Y.C.; writing—original draft preparation, X.L.; writing—review and editing, J.Y., Y.W., Y.H., F.Z., Z.W. and Y.C.; visualization, X.L. and F.Z.; supervision, J.Y.; project administration, J.Y.; funding acquisition, J.Y. All authors have read and agreed to the published version of the manuscript.

Funding: This work was supported by the National Natural Science Foundation of China (62127802) and the National Key Research and Development Program of China (2023YFB2905602).

Institutional Review Board Statement: Not applicable.

Informed Consent Statement: Not applicable.

Data Availability Statement: Data underlying the results presented in this paper are not publicly available at this time but may be obtained from the authors upon reasonable request.

Conflicts of Interest: The authors declare no conflicts of interest.

References

1. Elayan, H.; Amin, O.; Shihada, B.; Shubair, R.M.; Alouini, M.S. Terahertz Band: The Last Piece of RF Spectrum Puzzle for Communication Systems. *IEEE Open J. Commun. Soc.* **2019**, *1*, 1–32. [CrossRef]
2. Wang, C.X.; You, X.; Gao, X.; Zhu, X.; Li, Z.; Zhang, C.; Wang, H.; Huang, Y.; Chen, Y.; Haas, H.; et al. On the Road to 6G: Visions, Requirements, Key Technologies, and Testbeds. *IEEE Commun. Surv. Tutor.* **2023**, *25*, 905–974. [CrossRef]
3. Midili, V.; Nodjiadjim, V.; Johansen, T.K.; Squartecchia, M.; Riet, M.; Dupuy, J.Y.; Konczykowska, A. InP DHBT Technology for Power Amplifiers at mm-Wave Frequencies. *Microelectron. J.* **2017**, *67*, 111–119. [CrossRef]
4. Petricli, I.; Lotfi, H.; Mazzanti, A. Analysis and Design of D-Band Cascode SiGe BiCMOS Amplifiers with Gain-Bandwidth Product Enhanced by Load Reflection. *IEEE Trans. Microw. Theory Tech.* **2021**, *69*, 4059–4068. [CrossRef]
5. Singh, S.P.; Rahkonen, T.; Leinonen, M.E.; Pärssinen, A. A 290 GHz Low Noise Amplifier Operating Above $f_{max}/2$ in 130 nm SiGe Technology for Sub-THz/THz Receivers. In *2021 IEEE Radio Frequency Integrated Circuits Symposium (RFIC)*; IEEE: Piscataway, NJ, USA, 2021; pp. 223–226.
6. Gadallah, A.; Eissa, M.H.; Mausolf, T.; Kissinger, D.; Malignaggi, A. A 300-GHz Low-Noise Amplifier in 130-nm SiGe SG13G3 Technology. *IEEE Microw. Wirel. Components Lett.* **2021**, *32*, 331–334. [CrossRef]
7. Gonzalez-Jimenez, J.; Siligaris, A.; Hamani, A.; Manzillo, F.F.; Courouve, P.; Cassiau, N.; Dehos, C.; Clemente, A. An Energy-Efficient 56-Gb/s D-Band TX-to-RX Link Using CMOS ICs and Transmitarray Antennas. *IEEE Microw. Wirel. Technol. Lett.* **2024**, *34*, 801–804. [CrossRef]
8. Heydari, P. Terahertz Integrated Circuits and Systems for High-Speed Wireless Communications: Challenges and Design Perspectives. *IEEE Open J. Solid-State Circuits Soc.* **2021**, *1*, 18–36. [CrossRef]
9. Song, H.J.; Nagatsuma, T. Present and Future of Terahertz Communications. *IEEE Trans. Terahertz Sci. Technol.* **2011**, *1*, 256–263. [CrossRef]
10. Ghodsi, H.; Kaatuzian, H. Analysis and Design of a SiGe-HBT Based Terahertz Detector for Imaging Arrays Applications. *Microelectron. J.* **2019**, *90*, 156–162. [CrossRef]
11. Wang, L.; Dai, J.Y.; Ding, K.S.; Zeng, H.X.; Cheng, Q.; Yang, Z.Q.; Zhang, Y.X.; Cui, T.J. High-Order Direct Modulation Terahertz Communications with a Wideband Time-Coding Metachip Modulator. *Sci. Adv.* **2024**, *10*, eadq8693. [CrossRef]
12. Liu, X.; Maktoomi, M.H.; Alesheikh, M.; Heydari, P.; Aghasi, H. A 49–63 GHz Phase-Locked FMCW Radar Transceiver for High Resolution Applications. In *ESSCIRC 2023-IEEE 49th European Solid State Circuits Conference (ESSCIRC)*; IEEE: Piscataway, NJ, USA, 2023; pp. 509–512.
13. Hassanzadehyamchi, S.; Alizadeh, A.; Momeni, O.; Niknejad, A.M. Wideband High-Gain Amplifiers in 45-nm CMOS SOI Operating at $0.56f_{\text{max}}$: An Analytical Approach on Reversed Feedback Amplifiers. *IEEE J. Solid-State Circuits* **2025**, *60*, 3816–3829. [CrossRef]
14. Pan, J.; Luo, J.; He, J.; Feng, G.; Apriyana, A.; Zhang, Y.P. A D-Band CMOS Power Amplifier for Short-Range Data Center Communication. *IEICE Electron. Express* **2020**, *17*, 20200159. [CrossRef]
15. Cohen, E.; Degani, O.; Ritter, D. A Wideband Gain-Boosting 8mW LNA with 23dB Gain and 4dB NF in 65nm CMOS Process for 60 GHz Applications. In *2012 IEEE Radio Frequency Integrated Circuits Symposium*; IEEE: Piscataway, NJ, USA, 2012; pp. 207–210.
16. Alizadeh, A.; Hassanzadehyamchi, S.; Moradinia, A.; Nazhad, A.S.; Frounchi, M.; Momeni, O.; Niknejad, A.M.; Cressler, J.D.; Kiaei, S. D-Band Active Transmission Line with 33-GHz Bandwidth and 13-dB Gain at $f_{\text{max}}/2$. *IEEE Trans. Microw. Theory Tech.* **2023**, *72*, 2452–2465. [CrossRef]
17. Li, Z.; Chen, J.; Wang, Z.; Hong, W. A 206–264 GHz Power Amplifier Based on a High-Linearity Cascode Core with 15 dB Gain and 12.2-dBm P_{sat} in 130-nm SiGe BiCMOS. *IEEE Trans. Terahertz Sci. Technol.* **2024**, *14*, 293–297. [CrossRef]
18. Li, H.; Chen, J.; Hou, D.; Zhou, P.; Yu, J.; Yan, P.; Peng, L.; Lu, H.; Hong, W. A 250-GHz Differential SiGe Amplifier with 21.5-dB Gain for Sub-THz Transmitters. *IEEE Trans. Terahertz Sci. Technol.* **2020**, *10*, 624–633. [CrossRef]
19. Ghosh, S.; Zhang, F.; Guo, H.; Kenneth, K.O. 305-GHz Cascode Power Amplifier Using Capacitive Feedback Fabricated Using SiGe HBT's with f_{max} of 450 GHz. In *2023 IEEE Radio Frequency Integrated Circuits Symposium (RFIC)*; IEEE: Piscataway, NJ, USA, 2023; pp. 313–316.
20. Li, X.; Chen, W.; Zhou, P.; Wang, Y.; Huang, F.; Li, S.; Chen, J.; Feng, Z. A 250–310 GHz Power Amplifier with 15-dB Peak Gain in 130-nm SiGe BiCMOS Process for Terahertz Wireless System. *IEEE Trans. Terahertz Sci. Technol.* **2021**, *12*, 1–12. [CrossRef]
21. Singhakowinta, A.; Boothroyd, A. On Linear Two-Port Amplifiers. *IEEE Trans. Circuit Theory* **1964**, *11*, 169. [CrossRef]
22. Bameri, H.; Momeni, O. A High-Gain mm-Wave Amplifier Design: An Analytical Approach to Power Gain Boosting. *IEEE J. Solid-State Circuits* **2017**, *52*, 357–370. [CrossRef]
23. Bameri, H.; Momeni, O. An Embedded 200 GHz Power Amplifier with 9.4 dBm Saturated Power and 19.5 dB Gain in 65 nm CMOS. In *2020 IEEE Radio Frequency Integrated Circuits Symposium (RFIC)*; IEEE: Piscataway, NJ, USA, 2020; pp. 191–194.
24. Yun, B.; Park, D.W.; Lee, S.G. H-Band Power Amplifiers in 65-nm CMOS by Adopting Output Power Maximized G_{max}-Core and Transmission Line-Based Zero-Degree Power Combining Networks. *IEEE J. Solid-State Circuits* **2023**, *58*, 3089–3102. [CrossRef]

https://doi.org/10.3390/mi17040432

25. Park, D.W.; Utomo, D.R.; Lam, B.H.; Hong, J.P.; Lee, S.G. A 280-/300-GHz Three-Stage Amplifiers in 65-nm CMOS with 12-/9-dB Gain and 1.6/1.4% PAE While Dissipating 17.9 mW. *IEEE Microw. Wirel. Components Lett.* **2017**, *28*, 79–81. [CrossRef]
26. Park, D.W.; Utomo, D.R.; Yun, B.; Mahmood, H.U.; Hong, J.P.; Lee, S.G. Design of High-Gain Sub-THz Regenerative Amplifiers Based on Double-G_{max} Gain Boosting Technique. *IEEE J. Solid-State Circuits* **2021**, *56*, 3388–3398. [CrossRef]
27. He, F.; Xie, Q.; Wang, Z. A Novel Gain-Boosting Structure with Z-Embedding and Parallel Pre-Embedding Network for Amplifiers at Near-f_{max} Frequencies. *Int. J. Circuit Theory Appl.* **2023**, *51*, 4465–4476. [CrossRef]
28. Wang, Z.; Heydari, P. A Study of Operating Condition and Design Methods to Achieve the Upper Limit of Power Gain in Amplifiers at Near-f_{max} Frequencies. *IEEE Trans. Circuits Syst. I Regul. Pap.* **2017**, *64*, 261–271. [CrossRef]
29. Xu, Z.; Xie, Q.; Wang, Z. A Study of Collaborative Gain/Noise Optimization for LNAs at Near-Frequencies Based on a Novel Gain-Noise Plane Approach. *IEEE Trans. Circuits Syst. II Express Briefs* **2022**, *70*, 51–55. [CrossRef]
30. Tokgoz, K.K.; Abdo, I.; Fujimura, T.; Pang, J.; Kawano, Y.; Iwai, T.; Kasamatsu, A.; Watanabe, I.; Okada, K. A 273–301-GHz Amplifier with 21-dB Peak Gain in 65-nm Standard Bulk CMOS. *IEEE Microw. Wirel. Components Lett.* **2019**, *29*, 342–344. [CrossRef]
31. Singh, S.P.; Rahkonen, T.; Leinonen, M.E.; Pärssinen, A. Design Aspects of Single-Ended and Differential SiGe Low-Noise Amplifiers Operating Above fmax/2 in Sub-THz/THz Frequencies. *IEEE J. Solid-State Circuits* **2023**, *58*, 2478–2488. [CrossRef]
32. Trinh, V.S.; Song, J.M.; Park, J.D. A 280 GHz 30 GHz bandwidth cascaded amplifier using flexible interstage matching strategy in 130 nm SiGe technology. *Electronics* **2022**, *11*, 3045. [CrossRef]

https://doi.org/10.3390/mi17040432

micromachines

MDPI

Article

Design and Simulation of an Electron Optical System for Terahertz Vacuum Devices

Muhammad Haris Jamil [1,*], Zhiwei Lin [1], Hamid Sharif [1], Nazish Saleem Abbas [2] and Wenlong He [1,*]

[1] College of Electronics and Information Engineering, Shenzhen University, Shenzhen 518060, China
[2] College of Physics and Optoelectronics, Shenzhen University, Shenzhen 518060, China
* Correspondence: harisjamil07@gmail.com (M.H.J.); wenlong.he@szu.edu.cn (W.H.)

Abstract

An electron optic system (EOS) consisting of a sheet electron beam gun (SEB) and a pole offset periodic cusped magnet (PO-PCM) is reported for 340-GHz frequency. A sheet electron beam with a voltage of 29 kV, beam compression ratio of 16, and a beam waist of size 0.17 mm × 0.044 mm was designed and optimized using computer simulation technology (CST). The EOS was capable of transmitting the beam with a current of 6.9 mA through a beam tunnel of size 0.516 mm × 0.091 mm, having a length of 60 mm with the help of a pole offset periodic cusped magnet. The axial magnetic field generated by the PCM was 0.32 T. The EOS was efficient enough to transmit the beam stably through the beam tunnel with a transmission rate of 100%.

Keywords: sheet electron beam; periodic cusped magnet; electron optic system

1. Introduction

Vacuum electronic devices (VEDs) are widely used in commercial and military applications requiring high power at high frequency [1–4]. A core element of a VED is the electron optic system (EOS), which consists of an electron gun and a magnetic focusing structure. Leading to the limitation on the output power, conventional circular beam gun devices do not fulfill the demand of producing high-power coherent radiation at terahertz frequencies. Therefore, the sheet electron beam (SEB) gun devices come to the rescue.

The critical issue in realizing the sheet beam vacuum devices is the generation and transportation of the SEBs, especially at high operating frequencies. This is because the high operating frequency inevitably employs a tiny beam tunnel, which would raise the risk of electron interception in the beam tunnel [5]. Moreover, the SEB with a high current density is necessary for a device working at the millimeter–wave and terahertz–wave, resulting in a huge space charge force, which will bring great difficulty in focusing [6]. To deal with the dilemma, one way is to design an electron gun with a high compression ratio [7,8]. This will increase the current density of the beam and improve the output power of the device directly. Another way is to increase the length of the beam tunnel under the condition of a high electron transmission rate [9,10] so that the beam can have more opportunity to interact with the electromagnetic wave, thus improving the gain, output power and the electron efficiency of the device.

An external magnetic field is used to confine the sheet beam within the tunnel without being intercepted. To avoid the instabilities [6,11], a periodic rather than uniform guiding magnetic field was proposed. Due to miniaturization configuration, PCM has been successfully applied to vacuum electronic devices employing SEB. The PCM focusing is preferable to the terahertz TWTs, where the transverse dimension of the interaction circuit is very limited, and even a small horizontal swing of the beam cannot be tolerated.

Various studies have been conducted that deal with the challenges of designing EOS at terahertz (THz) frequency, out of which a few are cited in the references. For example, in [12], a traveling wave tube with sheet electron beam operating at 263 GHz was discussed, which used PCM–tunable quadrupolar magnet (PCM–TQM) to transport a 19 kV, 0.15 A sheet beam through a tunnel of length 45 mm. The transmission rate was measured to be 97.6% with the RF signal. Yiyang Su et al. in [13] reported an electron optic system working at 0.22 THz frequency. The beam, having a beam voltage of 16.5 kV and a current of 0.5 A, is transported successfully through 18 mm long tunnel using an approximately 0.6 T uniform magnetic field.

In [14,15], TWT amplifiers were discussed at a frequency of 340 GHz. The EOSs used were comprised of sheet electron beam and periodic magnetic structures, having magnetic field values of 0.32 T and 0.46 T, respectively. The SEB in [14] has a voltage of 19.5 kV and a current of 60 mA, while in [15], the SEB used comprised a 15 kV voltage and 25 mA current. Both designs were able to transport the electron beam with 100% transmission rate. The overall output power of the tubes was 25 W and 3.17 W, respectively.

This article presents an electron optic system with a good beam compression ratio and long beam channel length designed for 340 GHz frequency. The device operates at 29 kV and generates a 6.9 mA sheet electron beam, which could be transmitted stably over a length of 60mm with the help of a periodic axial magnetic field of 0.32 T. The current value as low as reported in this article reduces the space charge effect and improves the bunching (interaction of e-beam and RF signal) in the high frequency structure, hence giving more uniform gain for the wider frequency band. Low current also reduces the thermal stress and boosts the cathode life. The voltage of 29 kV improves coupling with the high-frequency structure, therefore providing maximum energy transfer to the RF signal. The organization of this paper is as follows. In Section 2, we will discuss the design and simulation results of the SEB gun. Section 3 discusses the design and fabrication results of the PCM structure. Co-simulation and transmission properties of EOS are discussed in Section 4, and the conclusion is presented in Section 5.

2. Sheet Electron Beam—Design and Simulation

Based on the design of the Pierce-type gun [16], an electron gun was designed in this study. The gun consisted of an elliptical cathode, a beam-focusing electrode and the anode with the beam tunnel, as shown in Figure 1. The electron gun parameters, including 29 kV operating voltage, a 6.9 mA beam current, and a compact 0.516 × 0.091 mm beam channel, were specifically optimized for integration with our 340 GHz high-frequency system, which constitutes the next stage of this research project. The output power of the beam gun was expected to be around 200 W.

An SEB gun was designed based on the above-mentioned parameters using computer simulation technology (CST). The dimensional values of the gun are given in Table 1, where w_c is the half width and h_c is the half height of the cathode, respectively, θ_x and θ_y are the beam-focusing angles in the xoz-plane and yoz-plane, respectively. w_t is the width of the tunnel, and h_t is the height of the beam channel. The cathode of size 0.31 mm × 0.39 mm and focusing electrodes were biased at −29 kV, and the anode was grounded (0 V). The electrostatic trajectory of the beam was studied using a particle tracking solver, and the simulation was conducted under the limitation of space charge emission. The mesh size of the model was more than 20 million, and the number of emission points on the cathode were 2646.

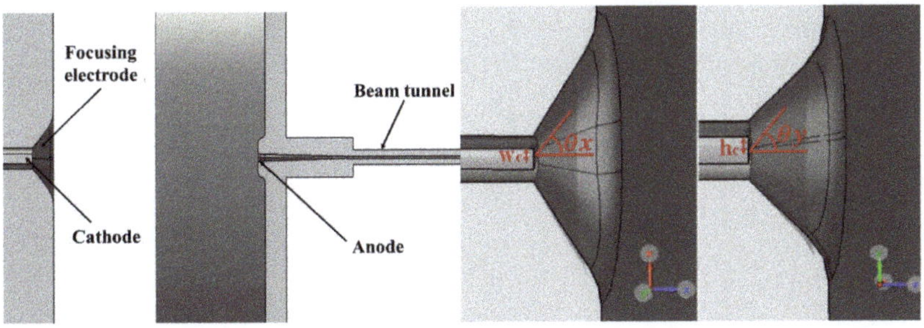

(**a**) (**b**)

Figure 1. (**a**) Electron gun; (**b**) left: x–z (wide side), right: y-z (narrow side) plan views of the emitter and focusing electrodes.

The cathode emits electrons that move in the form of a beam towards the anode. The current of the emitted beam was 6.9 mA as depicted by the simulation results, making the cathode current density 7.4 A/cm^2. The focusing electrode was used to converge the electron beam into an approximately elliptical shape. The anode has a rectangular opening slit that allows the beam to pass through. Thus, the beam converged near the anode opening and then passed through the beam channel towards the collector. The simulated beam trajectory and the electric field distribution of the e–gun are shown in Figure 2. The electric field value at the cathode and the focusing electrode was approximately equal to 62 kV/cm.

(**a**) Y-Z (narrow side) plan view of sheet electron beam trajectory

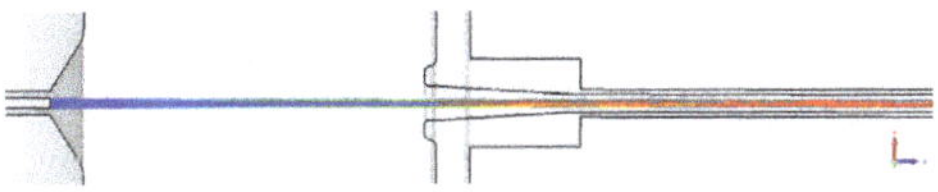

(**b**) X-Z (wide side) plan view of sheet electron beam trajectory

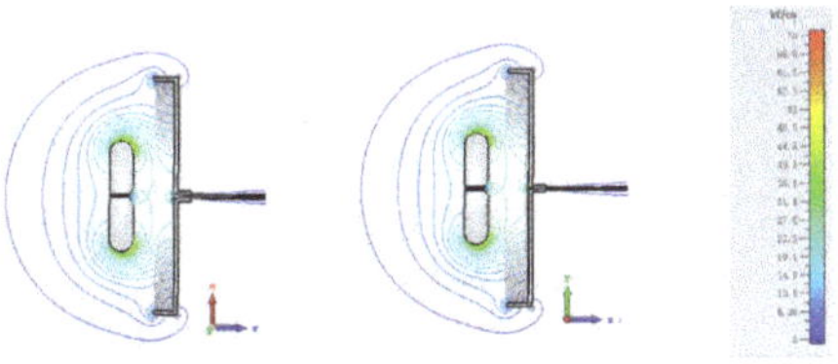

(c) Electric field distribution

Figure 2. Beam trajectory and electric field distribution: (**a**) Y-Z plane, (**b**) X-Z plane, and (**c**) electric field in both planes.

Table 1. Dimensional values of the sheet beam gun.

w_c	h_c	θ_x	θ_y	w_t	h_t
0.155 mm	0.195 mm	35.5°	30.5°	0.516 mm	0.091 mm

Simulation results showed that the size of the beam at the beam waist was 0.17 mm × 0.044 mm. This made the overall compression ratio M of the beam approximately equal to 16. The elliptical shape of the beam at the beam waist is shown in Figure 3, making it admissible for the beam wave interaction. The current density of the beam was calculated to be 117 A/cm^2. A low-current electron beam, as shown in Figure 4, offers significant advantages, including enhanced beam laminarity, better RF interaction, reduced space charge effects, and lower thermal load, increasing device longevity and energy efficiency.

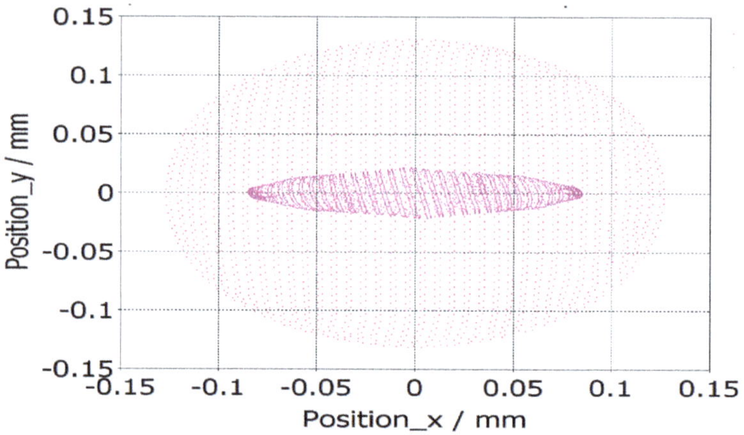

Figure 3. Beam cross-section (red) at cathode and beam waist (pink).

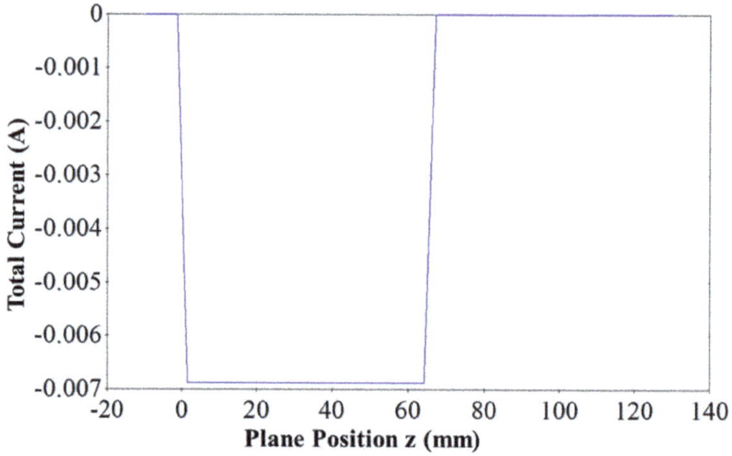

Figure 4. Total beam current.

3. Design and Simulation of Pole Offset Periodic Cusped Magnet

The electron beam, after passing the beam waist, starts to diverge due to the space charge effect. To confine the beam inside the beam tunnel, an external magnetic field was required. The pole offset–periodic cusped magnetic (PO–PCM) structure was used for this purpose. The transmission of a sheet beam under the PO–PCM satisfies the following expressions.

$$m_o \frac{d\dot{x}}{dt} = qE_{\mathrm{sc},x} + q\left(\dot{y}B_z - \dot{z}B_y\right) \tag{1a}$$

$$m_o \frac{d\dot{y}}{dt} = qE_{\mathrm{sc},y} + q\left(\dot{z}B_x - \dot{x}B_z\right) \tag{1b}$$

where m_o is the rest mass of the electron and q is the charge of the electron, $E_{sc,x}$, $E_{sc,y}$ are the components of the space charge field and B_x, B_y and B_z are the components of the focusing magnetic field. B_z is used to balance the space charge force in the y-direction (narrow), and B_y is used to balance the force in the x-direction (wider).

The period of the magnetic structure is L_m, and B_o is the peak value of axial magnetic field B_z. L_m should be well smaller than the plasma oscillation wavelength $\geq_p$ typically ratio $\frac{L_m}{\lambda_p} < 0.33$. To confine the SEB in the Y-direction, the magnetic amplitude B_o of the PCM field should range from $1.5B_{bri}$ to $2B_{bri}$ of the Brillouin value, as given in [17].

$$B_{bri} = \sqrt{\frac{1}{\gamma} \cdot \frac{\sqrt{2}I_o}{w_b h_b \epsilon_o \eta^{3/2} U^{1/2}}} \tag{2}$$

where γ is the relativistic constant, w_b is the half width of the beam, h_b is the half height of the beam, η is the charge-to-mass ratio of an electron, and U is the voltage of the beam. Another restriction between B_o and periodic length L_m is [18].

$$\left(\frac{\eta B_o^2 L_m^2}{U}\right) \leq \frac{417}{4} \tag{3}$$

To produce the Lorentz force in order to balance the space charge force in the x-direction, B_y should satisfy the following conditions

$$-eE_{sc,x} \leq \left|-ev_z \times B_y\right| \tag{4}$$

where v_z denotes the electron velocity in the z-direction. Based on the above analysis, a pole offset periodic cusped magnet was designed and optimized using the CST magnetostatic solver, as shown in Figure 4. The mesh size used for simulation was more than 16 million. The pole pieces were set as pure iron with a nonlinear B-H relationship. The magnetic blocks used for the simulation have remanence of 1.1 T and coercive force of 833 kA/m. Using the half-height and half-width of the beam, which were 0.022 mm and 0.085 mm, respectively, along with the other parameters discussed in Section 2, the Brillouin magnetic field was calculated to be 0.2 T. The period of the magnetic system L_m was 3.6 mm. The PCM structure consisted of a magnetic block and an iron pole piece with dimensions $w_m \times h_m \times d_m$ and $w_p \times h_p \times d_p$ having identical values of 15 mm × 7 mm × 1.8 mm was simulated. The value of the axial magnetic field B_z that best matched the beam was found to be 0.32 T, which is 1.6 times the Brillouin magnetic field. The staggered distance of pole–piece in the x-direction D_X was 3 mm. The distance between the upper and lower halves of the PCM was 3mm, and the length of the structure was kept around 100 mm just for experimental purpose.

To verify the simulation design, the above-discussed PCM structure with the mentioned parameters was physically fabricated. The magnetic material used was Samarium Cobalt, and the DT8A iron was used as the pole piece. Figure 5a,b shows the fabricated assembly of the periodic permanent magnet and testing platform used for the measurements. The results showed that the PCM focusing structure could provide a periodic B_z of ~0.32 T. A graphical comparison between the simulated and experimentally measured data was made using MATLAB R2023a. The B_z˜z graph in Figure 6 shows a good agreement between the compared values of B_z.

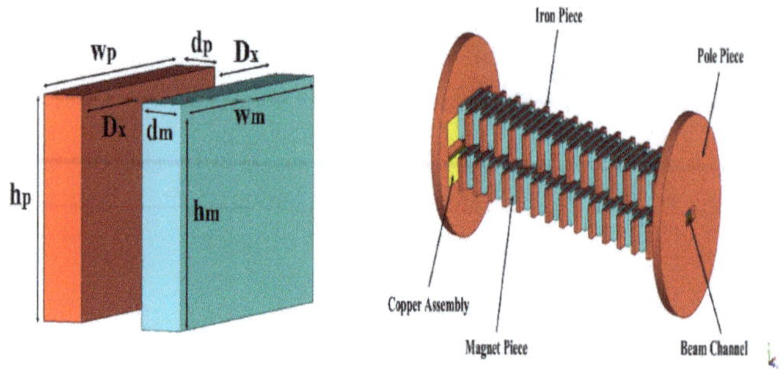

(**a**)

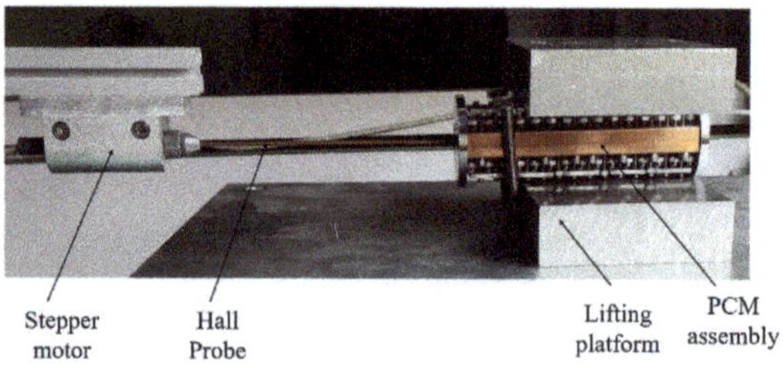

(**b**)

Figure 5. (**a**) PCM assembly CST model. (**b**) PCM assembly testing platform.

B_y, which was necessary to confine the beam into the tunnel in a wider direction, was also measured and compared in the MATLAB graph, as shown in Figure 7. Since B_y is sensitive to the vertical position of the Hall probe, a little deviation in the y-direction changes B_y significantly. Figure 7 shows the measured B_y and simulated B_y at (0.2 mm, 0.18 mm), which are consistent. The numerical value of the simulated B_y was 0.05 T.

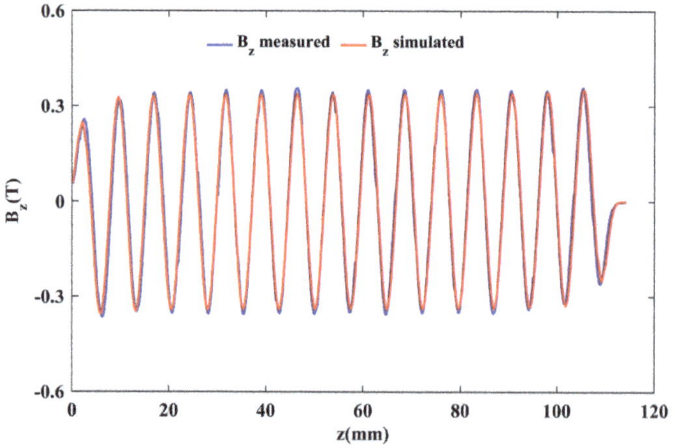

Figure 6. Graph of B_z verses z axis.

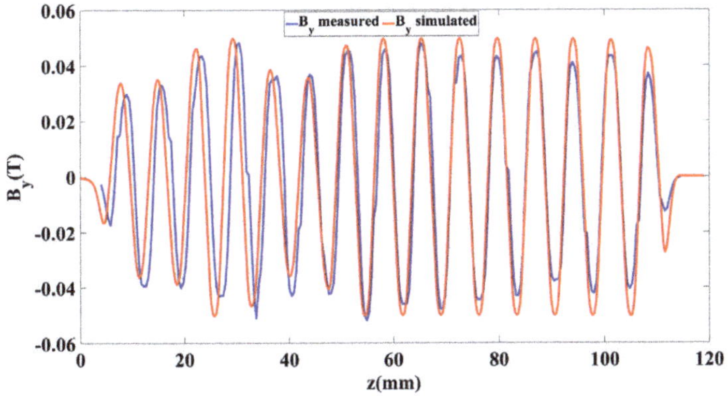

Figure 7. Graph of B_y verses z axis.

Thermal expansion due to RF heating or beam interception in high frequency structure adversely affects the alignment of pole pieces and weakens the effectiveness of the transverse magnetic field. Such a degradation can be mitigated by active cooling and thermal shielding.

4. Electron Optic System and Transmission Property of SEB

An electron optic system is discussed in this section. EOS was composed of a sheet beam e-gun and a periodic magnetic focusing structure designed above. The calculated magnetostatic field of the simulated magnetic structure was imported to the sheet beam simulation model. One of the crucial problems during integration is the matching between the e-gun and the periodic focusing structure. The beam should enter the magnetic field at the beam waist, which was 12.8 mm away from the cathode. After importing the external magnetic field, the sheet beam e-gun model was simulated using the particle tracking solver. The beam envelope observed from the narrow side and wide side is shown in Figure 8. The simulation results show that the beam is confined well inside the dimensions of the beam tunnel, hence achieving a 100% transmission rate for a 60 mm long beam tunnel, as shown in Figure 4.

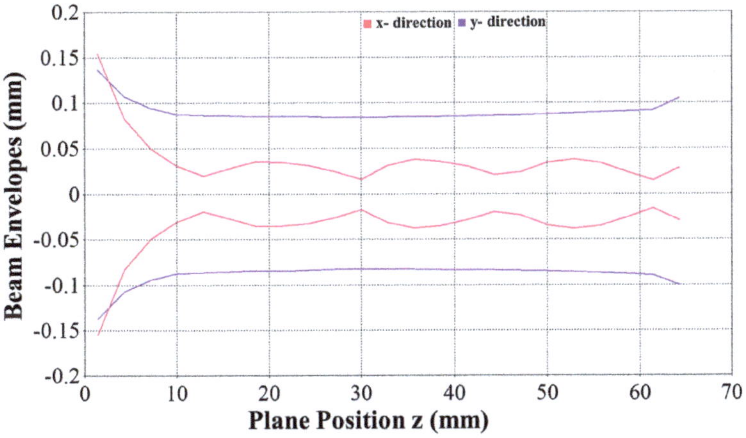

Figure 8. Beam envelope in narrow (x) and wide (y) directions.

5. Conclusions

This paper presents the design and optimization of a periodic cusped magnet-focused sheet beam optical system for application in a 340 GHz traveling wave tube amplifier, representing the next phase of our research project. The simulated design achieved a beam current density of 117 A/cm^2 with a 100% transmission rate for a sheet beam having a voltage of 29 kV, a current of 6.9 mA and a beam waist of 0.17 mm × 0.044 mm. A PCM with an axial magnetic field of 0.32 T was used to transmit the beam through a relatively long beam tunnel of length up to 60 mm. The simulated PCM structure was fabricated, and the B_z and B_y magnetic fields were measured. Although slight variations were observed in the measured B_y, the axial magnetic field B_z closely matched the simulated results, confirming the accuracy of the model. The device dimensions are compatible with standard microfabrication techniques, including CNC machining and LIGA processes. Furthermore, the optimized electric field distribution across the emitter, combined with an achievable current density of 117 A/cm^2, which is well within the practical operating range for scandate cathodes, enhances the design's real-world feasibility.

Author Contributions: M.H.J. led the study's conceptualization, designed the experiments, and directed the overall planning and execution of the project. He also drafted the initial version of the manuscript and played a major role in the revisions. Z.L. helped in software simulations. H.S. contributed to data visualization, preparing all figures and tables. N.S.A. helped in conceptulizing the idea and assisted in writing, reviewing and revising the draft. W.H. provided supervision and offered critical insights and guidance throughout the research. All authors reviewed the manuscript, provided constructive feedback, and approved the final version for submission. All authors have read and agreed to the published version of the manuscript.

Funding: This research did not receive any specific grant from funding agencies in the public, commercial, or not-for-profit sectors.

Data Availability Statement: The data that support the findings of this study are available from "the corresponding author" upon reasonable request.

Acknowledgments: I would like to express my sincere gratitude to Wenlong He for their invaluable guidance and support throughout this research.

Conflicts of Interest: The authors declares no conflict of interest.

References

1. Navrotsky, I.A.; Burtsev, A.A.; Emelyanov, V.V.; Titov, V.N.; Ryskin, N.M. Electron-Optic System with a Converged Sheet Electron Beam for a 0.2-THz Traveling-Wave Tube. *IEEE Trans. Electron Devices* **2021**, *68*, 798–803. [CrossRef]
2. Sharif, H.; Jamil, M.H.; He, W. Optimized Quasi-Optical Mode Converter for TE33,12 in 210 GHz Gyrotron. *Micromachines* **2025**, *16*, 308. [CrossRef] [PubMed]
3. Wang, D.; Wang, G.; Wang, J.; Li, S.; Zeng, P.; Teng, Y. A high-order mode extended interaction klystron at 0.34 THz. *Phys. Plasmas* **2017**, *24*, 023106. [CrossRef]
4. Sharif, H.; Jamil, M.H.; He, W.; Abbas, N.S. Design of Quasi-Optical TE17,6 Mode Converter for High-Power 170-GHz Gyrotrons. *IEEE Trans. Plasma Sci.* **2025**, *53*, 834–841. [CrossRef]
5. Yin, P.C.; Xu, J.; Yang, R.C.; Luo, J.J.; Zhang, J.; Jia, D.D. Sheet Electron Optical System for a 1.03-THz Traveling-Wave Tube. *IEEE Electron Device Lett.* **2022**, *43*, 1343–1346. [CrossRef]
6. Booske, J.H.; Basten, M.A.; Kumbasar, A.H.; Antonsen, T.M.; Bidwell, J.S.W.; Carmel, Y.; Destler, W.W.; Granatstein, V.L.; Radack, D.J. Periodic magnetic focusing of sheet electron beams. *Phys. Plasmas* **1994**, *1*, 1714–1720. [CrossRef]
7. Pasour, J.; Wright, E.; Nguyen, K.; Balkcum, A.; Wood, F.; Atkinson, J. 3.4: Sheet beam stick for low-voltage W-band extended interaction klystron (EIK). In Proceedings of the 2010 IEEE International Vacuum Electronics Conference (IVEC), Monterey, CA, USA, 18–20 May 2010; pp. 43–44. [CrossRef]
8. Li, Y.; Wen, Z.; Wang, X.; Fan, Y.; Luo, J.; Zhu, M. A Sheet Beam Electron Gun with High Compression Ratio and Long Transmission Distance Performance for W-Band TWT Applications. *IEEE Trans. Plasma Sci.* **2014**, *49*, 734–741. [CrossRef]

9. Zhao, D.; Lu, X.; Liang, Y.; Yang, X.; Ruan, C.; Ding, Y. Researches on an X-Band Sheet Beam Klystron. *IEEE Trans. Electron Devices* **2014**, *61*, 151–158. [CrossRef]
10. Shi, X.; Wang, Z.; Tang, T.; Gong, H.; Wei, Y.; Duan, Z. Theoretical and Experimental Research on a Novel Small Tunable PCM System in Staggered Double Vane TWT. *IEEE Trans. Electron Devices* **2015**, *62*, 4258–4264. [CrossRef]
11. Booske, J.H.; McVey, B.D.; Antonsen, T.M., Jr. Stability and confinement of nonrelativistic sheet electron beams with periodic cusped magnetic focusing. *Phys. Plasmas* **1993**, *73*, 4140–4155. [CrossRef]
12. Zheng, Y.; Gamzina, D.; Popovic, B.; Luhmann, N.C. Electron Beam Transport System for 263-GHz Sheet Beam TWT. *IEEE Trans. Electron Devices* **2016**, *63*, 4466–4472. [CrossRef]
13. Su, Y.; Wang, P.; Wang, W.; Ruan, C.; He, W. Theoretical Analysis of Sheet Beam Electron Gun for Terahertz Vacuum Electron Devices. *IEEE Trans. Electron Devices* **2022**, *69*, 5865–5870. [CrossRef]
14. Guo, J.; Dong, Y.; Shi, N.; Wang, S.; Wang, Z.; Liu, X. A 0.34-THz Standing Wave Enhanced Sheet Electron Beam Traveling-Wave Tube. *IEEE Trans. Electron Devices* **2023**, *70*, 307–312. [CrossRef]
15. Hu, P.; Lei, W.; Jiang, Y.; Huang, Y.; Song, R.; Chen, H. Demonstration of a Watt-Level Traveling Wave Tube Amplifier Operating Above 0.3 THz. *IEEE Electron Device Lett.* **2019**, *40*, 973–976. [CrossRef]
16. Vaughan, J.R.M. Synthesis of the Pierce gun. *IEEE Trans. Electron Devices* **1981**, *28*, 37–41. [CrossRef]
17. Zhao, D. Validity of closed periodic magnetic focusing for sheet electron beams. *Phys. Plasmas* **2009**, *16*, 113102. [CrossRef]
18. Zhang, C.; Feng, J.; Cai, J.; Pan, P.; Su, S.; Gong, Y. Focusing of the Sheet Electron Beam with Two-Plane Periodic Cusped Magnetic System for Terahertz TWTs. *IEEE Trans. Electron Devices* **2021**, *68*, 3056–3062. [CrossRef]

micromachines

MDPI

Article

A Wideband D-Band Frequency Sextupler Chain with High Harmonic Rejection in 100 nm GaAs pHEMT Technology

Pinqing Wang, Zhe Chen *, Yubin Guo, Yue Qi and Peng Yang

State Key Laboratory of Millimeter Waves, School of Information Science and Engineering, Southeast University, Nanjing 210096, China; wangpinqing@seu.edu.cn (P.W.); guoyubin@seu.edu.cn (Y.G.); qiyue@seu.edu.cn (Y.Q.); 220220997@seu.edu.cn (P.Y.)

* Correspondence: zhechen@seu.edu.cn

Abstract

This paper presents a wideband D-band frequency sextupler chain implemented in a 100 nm GaAs pHEMT process. The proposed circuit comprises an input-stage frequency tripler, an inter-stage harmonic-rejection power amplifier, and an output-stage frequency doubler. The tripler adopts a balanced topology, which effectively suppresses the fundamental frequency component. The inter-stage power amplifier not only delivers sufficient drive power to the doubler but also enhances suppression of undesired harmonics. The output doubler employs a single-balanced configuration to suppress odd-order harmonics while extracting the second harmonic. The measured peak output power of the sextupler chain is 2.33 dBm, corresponding to an input power of 2 dBm, resulting in a conversion gain of 0.33 dB. The 3 dB output bandwidth spans from 126.3 to 152.7 GHz, corresponding to a relative bandwidth of 18.9%. Owing to the balanced multiplier topology and harmonic-rejection PA, the 5th and 7th harmonics are suppressed by more than 20 dBc. The combination of high output power, wide operating bandwidth, and excellent harmonic suppression makes the design well suited for wideband D-band signal generation in diverse applications.

Keywords: frequency sextupler; D-band; GaAs; balanced topology

1. Introduction

With the rapid development of mobile communication technologies, particularly the widespread deployment of 5G and the upcoming 6G systems, spectrum resources are becoming increasingly scarce. Consequently, research into millimeter-wave (mmWave) frequency bands has become increasingly important [1,2]. Within this domain, the D-band has emerged as a key research focus in mmWave communications due to its substantial available bandwidth [3,4].

D-band signal generation is typically realized by cascading a low-frequency source with frequency multipliers [5]. However, undesired harmonic signals generated within the multiplier can degrade the spectral purity of the output. These spurious harmonics are usually suppressed using filters [6]. Nevertheless, in broadband frequency multiplier designs, some undesired harmonics may fall within the target frequency band, rendering conventional filtering less effective.

In this paper, a wideband D-band frequency sextupler chain in a 100 nm GaAs pHEMT process is designed. Both the cascaded frequency tripler and doubler adopt balanced topologies, which inherently suppress unwanted harmonics. A harmonic rejection PA is adopted as well. This configuration eliminates the reliance on additional filters and enhances suitability for broadband applications. The frequency sextupler chain exhibits a peak output power of 2.33 dBm with a 3 dB bandwidth from 126.3 to 152.7 GHz. The circuit topology is given in Section 2 and the measurement results are presented in Section 3.

2. Circuit Design

As shown in Figure 1, the architecture of the frequency sextupler chain is presented. The tripler is employed as the input stage, while the doubler serves as the output stage. A power amplifier is inserted to provide sufficient drive power for the doubler and to suppress the unwanted harmonics from the tripler.

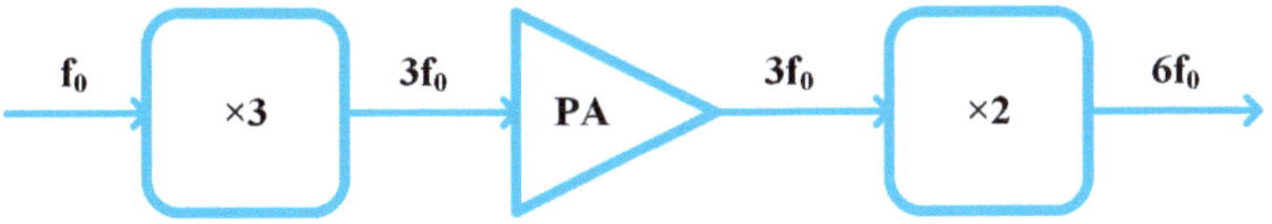

Figure 1. Architecture diagram of the frequency sextupler chain.

2.1. Design of the Balanced Tripler

The balanced tripler depicted in Figure 2 consists of a fundamental-wave coupler, tripling units, and a third-harmonic coupler [7]. This configuration effectively suppresses the fundamental component and facilitates the extraction of the third harmonic.

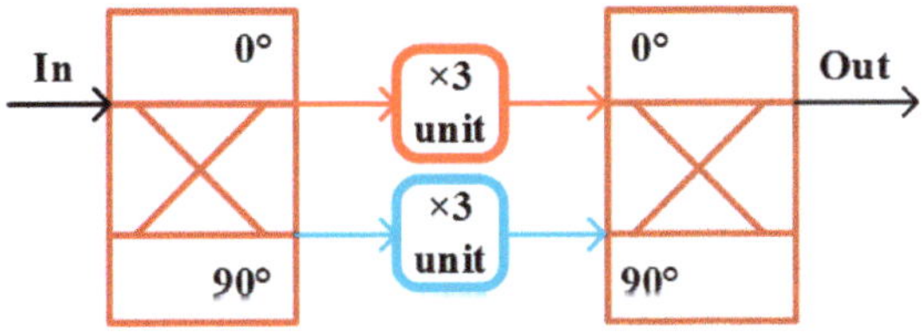

Figure 2. Architecture diagram of the balanced tripler.

In the design, the coupler splits the single-end input signal into two branches with a 90° phase difference. Each branch independently performs the frequency tripling, and their outputs are subsequently combined through the third-harmonic coupler to achieve efficient power summation.

As shown in Figure 3, the Lange coupler is composed of several $\lambda/4$ transmission line sections. The simulated amplitude and phase imbalances of the two Lange couplers are shown in Figure 4. The amplitude and phase imbalances of the fundamental frequency coupler and the third-harmonic coupler are less than 0.13 dB and 1.5°; 0.4 dB and 4°, respectively.

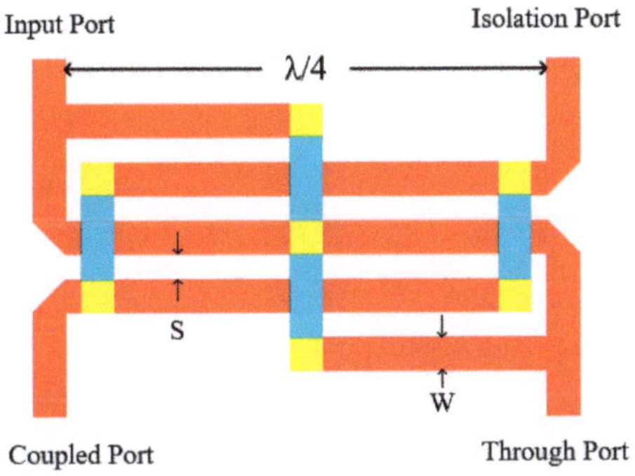

Figure 3. Layout of Lange coupler.

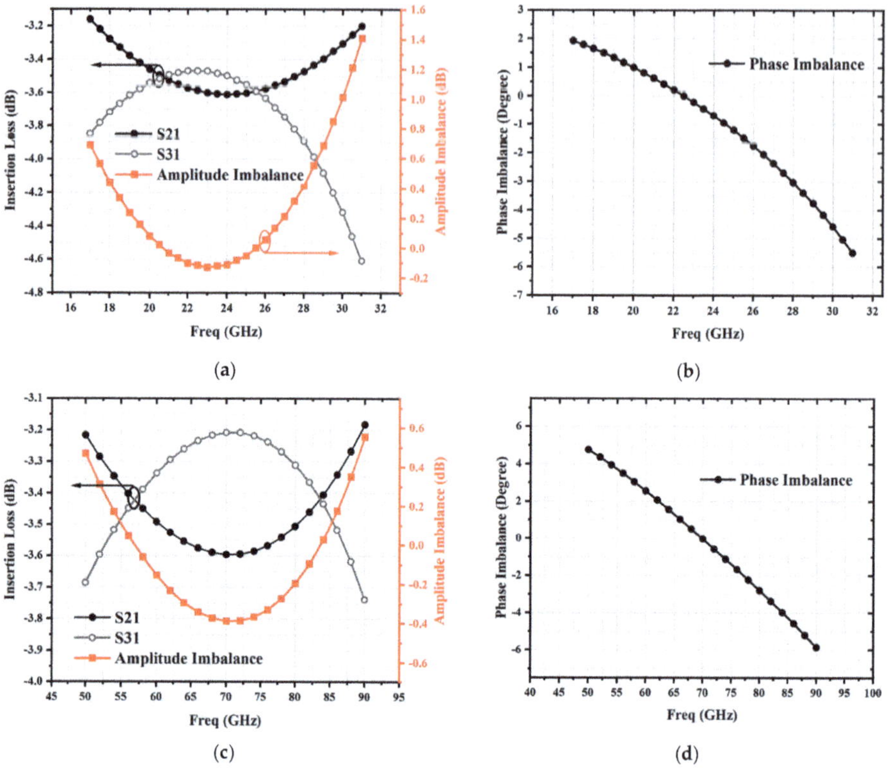

Figure 4. Simulated amplitude and phase imbalance of two Lange couplers: (**a**) amplitude imbalance of fundamental-wave coupler; (**b**) phase imbalance of fundamental-wave coupler; (**c**) amplitude imbalance of third-harmonic coupler; (**d**) phase imbalance of third-harmonic coupler.

The schematic of the tripler unit is shown in Figure 5, where Q1 serves as a driver amplifier to provide sufficient input power, enabling the tripler transistor Q2 to achieve a high third-harmonic conversion gain. Q2 generates the third harmonic, which is biased in Class A mode to suppress the second harmonics. R1 and R4 are connected in series with the transistor gates to reduce the conversion gain and improve low-frequency stability.

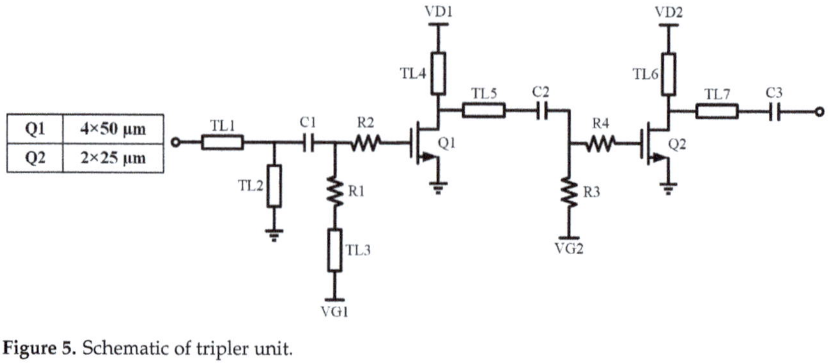

Figure 5. Schematic of tripler unit.

The output power versus the frequency of the proposed balanced tripler is simulated with an input power level of 0 dBm, and the results are shown in Figure 6. The fundamental power is suppressed by more than 20 dBc, and the second harmonic can be further attenuated through post-stage circuit.

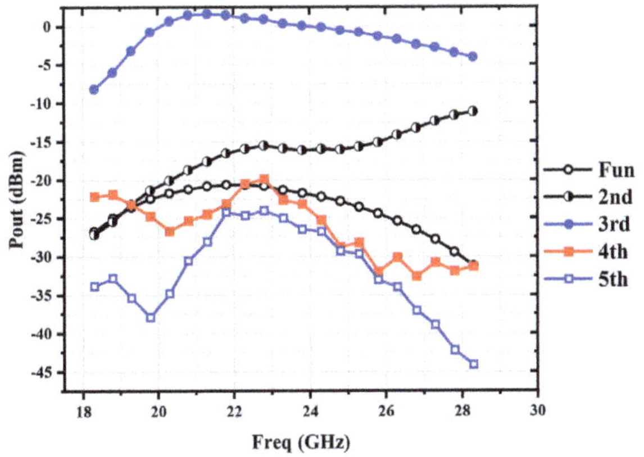

Figure 6. Simulated output power of the balanced tripler with an input level of 0 dBm.

2.2. *Design of the Inter-Stage E-Band PA*

An inter-stage power amplifier is designed to amplify the third-harmonic output of the preceding tripler stage and to provide sufficient drive power for the subsequent passive frequency doubler, as well as adding extra rejection to other unwanted harmonic outputs. This ensures enhanced second-harmonic conversion gain at the output of the doubler. Similar to the tripler, R2 is serially connected to the gate to improve the stability of the PA. The schematic diagram of the designed power amplifier is presented in Figure 7.

Q1	4×25 μm
Q2	4×50 μm
Q3	4×75 μm

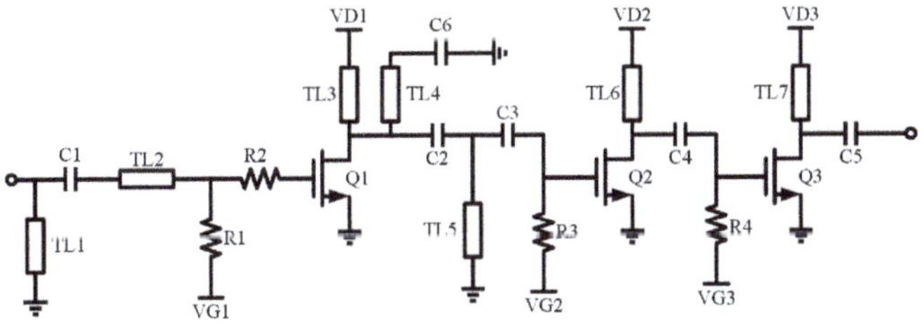

Figure 7. Schematic of PA.

As shown in Figure 8, the small-signal gain S21 exceeds 19 dB, with a 3 dB bandwidth ranging from 57.7 to 80.3 GHz. Under a 0 dBm input power, the output power exceeds 18 dBm, corresponding to a large-signal gain greater than 18 dB, with a bandwidth extending from 57.2 to 81.1 GHz.

To further suppress the harmonics generated by the preceding tripler stage, the PA matching circuit is designed as a band-pass frequency-selective network. This network is composed of capacitors and microstrip lines, which simultaneously serves the dual purpose of impedance matching and harmonic rejection. As shown in Figure 9, a significant improvement in the suppression of the fundamental, second-order harmonic, and fourth-order harmonic is observed after the PA, from −21.2~−32.5 dBc to −29.6~−50.7 dBc for input frequency of 21.05~25.45 GHz.

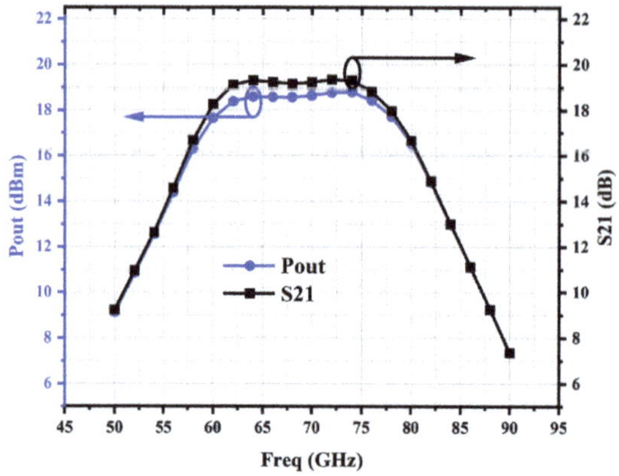

Figure 8. Simulated small-signal gain and large-signal gain of PA.

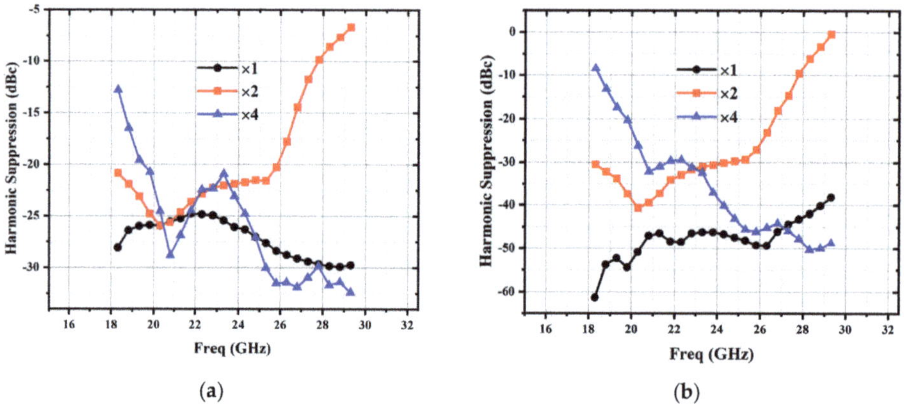

Figure 9. Harmonic suppression of PA: (**a**) harmonic suppression without PA; (**b**) harmonic suppression with PA.

2.3. Design of the Single-Balanced D-Band Doubler

The designed single-balanced doubler consisting of a pair of anti-parallel diodes and a Marchand balun is shown in Figure 10. This architecture effectively utilizes the nonlinear characteristics of diodes to achieve frequency doubling, while suppressing the odd-order harmonics and extracting even-order harmonics [5].

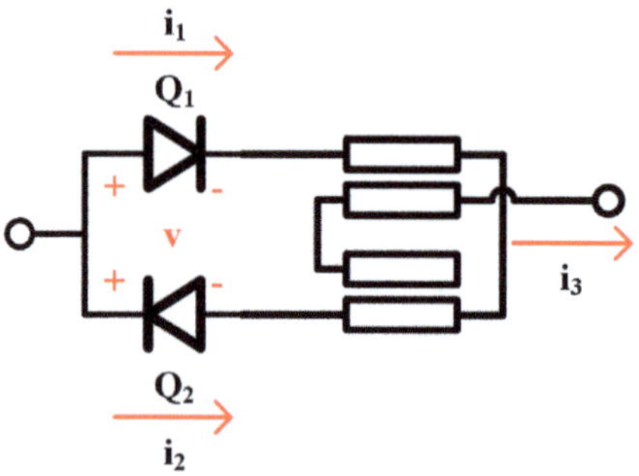

Figure 10. Schematic of single-balanced doubler.

In the single-balanced doubler, the anti-parallel diodes, which are realized using two discrete diode elements, require input signals of equal amplitude and in-phase relationship to effectively cancel the odd-order harmonics. However, because one diode is driven at the anode and the other at the cathode, this inevitably introduces asymmetry in both the physical structure and the output impedance. The asymmetrical Marchand balun is employed to compensate for this imbalance. The layout is shown in Figure 11; two unequal-length coupled-line sections are utilized for this compensation. All the design parameters are optimized with Advanced Design System.

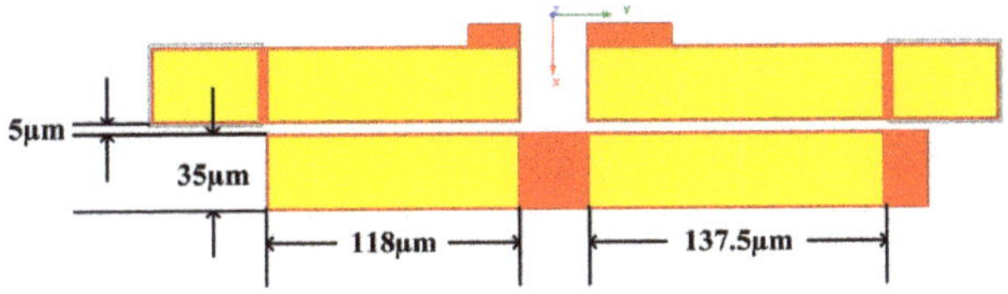

Figure 11. Layout of the asymmetrical Marchand balun.

As shown in Figure 12, with this asymmetrical Marchand balun, the two anti-parallel diodes receive equal input power, approximately 14 dBm. In total, the single-balanced doubler obtains an input power of 17 dBm from the PA. All the large-signal simulations were conducted using the harmonic balance with Advanced Design System.

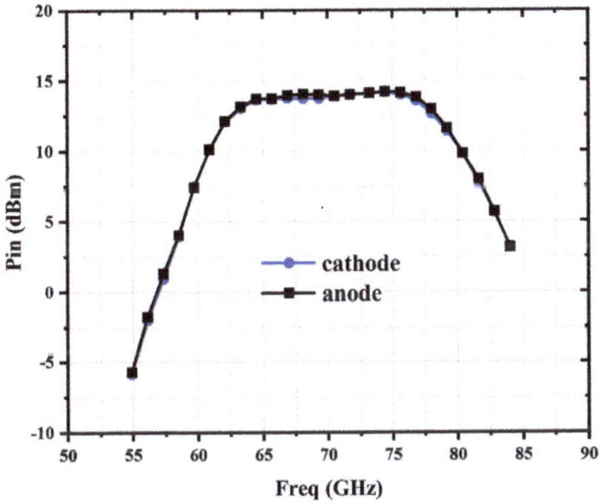

Figure 12. Simulated input power of the single-balanced doubler.

3. Measured Results

Figure 13 shows a microphotograph of the proposed six-times frequency multiplication chain. The chip occupies a total area of 1600 μm × 2000 μm, including all pads. The circuit is measured using on-wafer probing at room temperature (~26 D.C.). S11 and S22 are measured with GGB probe Model 50A-GSG and Cascade probe I170-T-GSG, using a vector network analyzer (Keysight N5245A) and with VNA extenders (OML V06VNA2-T/R-A_RLA 110-170G), respectively, using SOLT calibration sets. In large-signal measurements, the input signal is provided by a signal generator (R&S SMW200A), while the output power is recorded by an Erickson PM4 power meter and spectrum analyzer (Keysight N9030A) with D-band mixer SFB-06-N1.

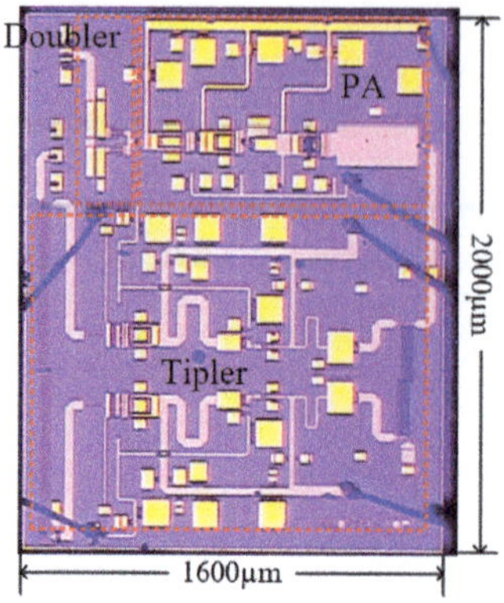

Figure 13. The microphotograph of six-times frequency multiplication chain.

As shown in Figure 14, the measured S11 coincides well with the simulation, remaining below −10 dB across the 10 to 37 GHz frequency range. However, the measured S22 deviates from the simulation, shifting by 13.5 GHz towards higher frequencies. This deviation may be attributed to the fact that the device models in the PDK used in this work are not based on measured data above 90 GHz, but rather derived from curve fitting, which may not be entirely accurate. Meanwhile, the utilized ADS EM simulation tool may also introduce simulation inaccuracies in D-band. And the transistors' metal figures are added in the EM simulation, which may overestimate the parasitic effects for the passive circuits, leading to a frequency shift in measurement.

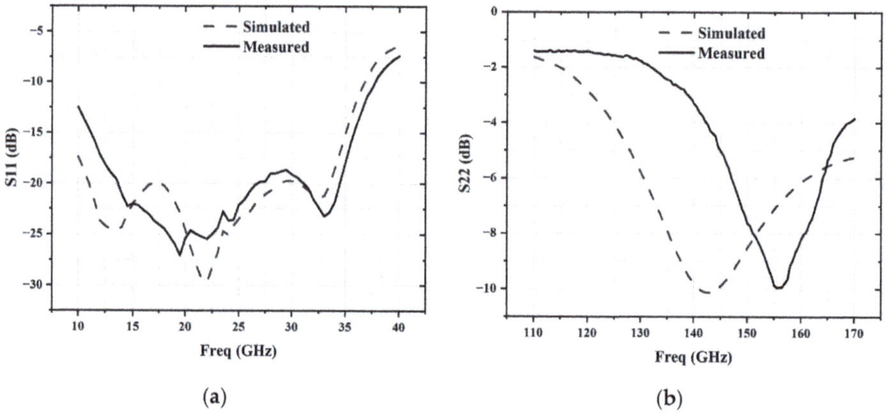

Figure 14. Simulated and measured S-parameters: (**a**) S11; (**b**) S22.

Figure 15 shows a comparison between the measured and simulated output power of the frequency sextupler chain under input power levels of 0 dBm, 1 dBm, 2 dBm, and 3 dBm. The measured saturated output power is 3.04 dBm, corresponding to an input power of 3 dBm. At an input power of 1 dBm, the peak output power reaches 2.14 dBm, resulting in a conversion gain of 1.14 dB. The 3 dB output bandwidth spans from 126.9 to 151.8 GHz, giving a relative bandwidth of 17.9%. At an input power of 2 dBm, the peak output power is 2.33 dBm, with a conversion gain of 0.33 dB and an output bandwidth of 3 dB ranging from 126.3 to 152.7 GHz, corresponding to a relative bandwidth of 18.9%.

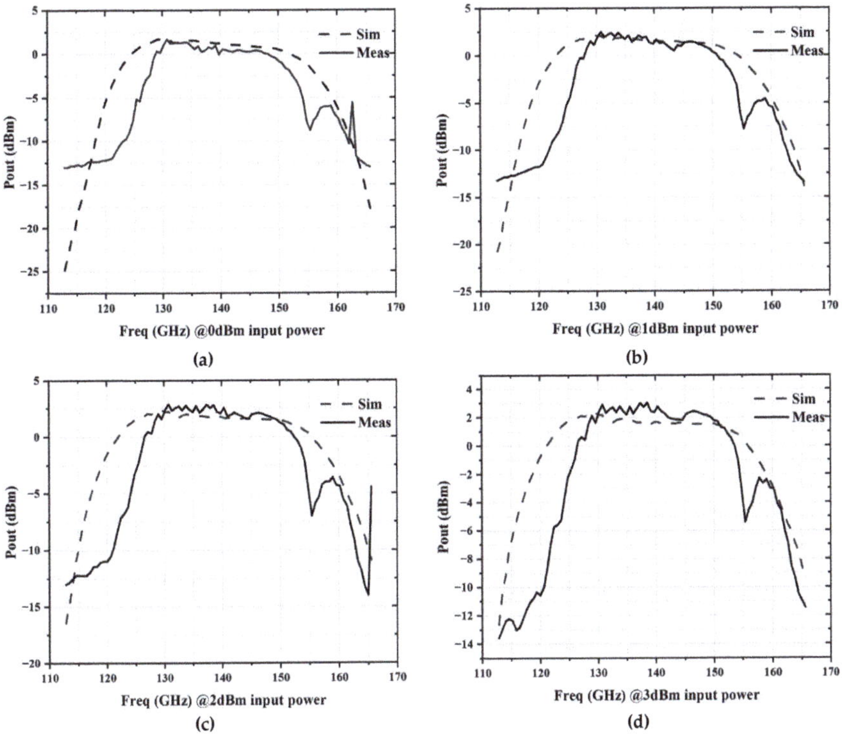

Figure 15. Simulated and measured large-signal versus input power: (**a**) output power @0dBm input power; (**b**) output power @1dBm input power; (**c**) output power @2dBm input power; (**d**) output power @3dBm input power.

Figure 16 shows the simulated and measured 5th and 7th harmonic suppression of the frequency sextupler chain at an input power of 2 dBm. The measured results show that within the 3 dB output bandwidth of 126.3–152.7 GHz, both the 5th and 7th harmonics are suppressed by more than 20 dBc. Table 1 compares this work with other reported multiplier designs in GaAs process, where this work shows the widest 3 dB output frequency bandwidth in D-band with high harmonic rejections.

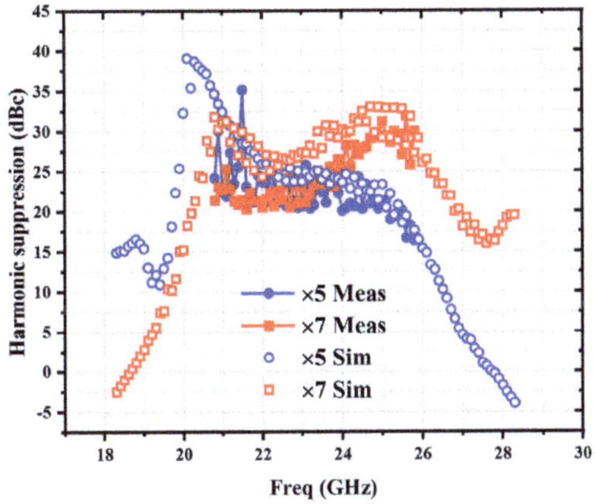

Figure 16. Simulated and measured harmonic suppression.

Table 1. Comparison with other reported GaAs multiplier designs.

Reference	[7]	[8]	[9]	[10]	[11]	[12]	[13]	This Work
Process	100 nm AlGaAs mHEMT	250 nm GaAs pHEMT	150 nm GaAs pHEMT	100 nm GaAs pHEMT	150 nm GaAs pHEMT	150 nm GaAs pHEMT	100 nm GaAs pHEMT	100 nm GaAs pHEMT
Freq (GHz)	79–100	90–99	88–99.5	85–110	34.44–42.56	37–43	71–90	126.3–152.7
Multi Factor	8	3	2	2	2	2	2	6
Gain (dB)	8.9	−3	−4.3	3.2	−4	0.9	−3.3	0.33
Pout (dBm)	6.9	8	7.1	8.2	8	0.9	10.5	2.33
Size (mm^2)	6	2.21	2	1.35	NA	0.72	1.9	3.2

4. Conclusions

In this paper, the design of a D-band frequency sextupler chain is presented in a 100 nm GaAs pHEMT process. The presented circuit consists of balanced tripler, harmonic-rejection PA and output doubler. At an input power of 2 dBm, the peak output power of the sextupler reaches 2.33 dBm, resulting in a conversion gain of 0.33 dB. The 3 dB output bandwidth spans from 126.3 to 152.7 GHz, corresponding to a relative bandwidth of 18.9%. Moreover, the 5th and 7th harmonics suppression are better than 20 dBc across the operating frequency range, making it suitable for the wideband signal generation in D-band applications.

Author Contributions: Conceptualization, P.W.; Writing original draft preparation, P.W.; Conceptualization, Z.C.; Validation, Z.C.; Supervision, Z.C.; Methodology, Y.G.; Formal analysis, Y.G.; Methodology, Y.Q.; Software, Y.Q.; Data curation, P.Y.; Resources, P.Y.; Writing—review and editing, Z.C. and P.W. All authors have read and agreed to the published version of the manuscript.

Funding: This work was funded in part by the National Natural Science Foundation of China under Grant 62171128 and Grant 62188102, in part by the Top-Ten Technological Projects for Hunan Province in 2025 under Grant 2025QK1009, and in part by the Fundamental Research Funds for the Central Universities under Grant 2242022k60008.

Data Availability Statement: The original contributions presented in the study are included in the article, further inquiries can be directed to the corresponding author.

Conflicts of Interest: The authors declare no conflicts of interest.

References

1. Zhou, P.; Chen, J.; Yan, P.; Yu, J.; Hu, J.; Dong, H.; Wang, L.; Hou, D.; Gao, H.; Hong, W. A 28.5-dB EVM 64-QAM 45-GHz Transceiver for IEEE 802.11aj. *IEEE J. Solid State Circuits* **2021**, *56*, 3077–3093. [CrossRef]
2. Rappaport, T.S.; Xing, Y.; Kanhere, O.; Ju, S.; Madanayake, A.; Mandal, S.; Alkhateeb, A.; Trichopoulos, G.C. Wireless Communications and Applications Above 100 GHz: Opportunities and Challenges for 6G and Beyond. *IEEE Access* **2019**, *7*, 78729–78757. [CrossRef]
3. Li, C.; Liu, P.; Zhu, X.W.; Zhu, A. A SIW-based Wideband Circularly Polarized 4 × 4 Patch Array Antenna in 45 GHz Band. In Proceedings of the 2019 12th UK-Europe-China Workshop on Millimeter Waves and Terahertz Technologies (UCMMT), London, UK, 20–22 August 2019; pp. 1–4.
4. Wang, L.; Chen, J.; Hou, D.; Hong, W. A 45 GHz Low-Voltage Cascode Power Amplifier Based on Capacitor Coupling Technology. In Proceedings of the 2021 IEEE MTT-S International Wireless Symposium (IWS), Nanjing, China, 23–26 May 2021; pp. 1–3.
5. Zhu, F.; Luo, G. A W-Band Frequency Quadrupler MMIC. In Proceedings of the 2020 International Conference on Microwave and Millimeter Wave Technology (ICMMT), Shanghai, China, 20–23 September 2020; pp. 1–3.
6. Chandra-Prabhu, A.; Grzyb, J.; Hillger, P.; Heinemann, B.; Rücker, H.; Pfeiffer, U. A Wideband W-Band Frequency Tripler with a Novel Mode-Selective Filter for High Harmonic Rejection. In Proceedings of the 2023 18th European Microwave Integrated Circuits Conference (EuMIC), Berlin, Germany, 18–19 September 2023; pp. 197–200.

7. Kallfass, I.; Tessmann, A.; Massler, H.; Lewark, U.; Kuri, M.; Riessle, M.; Zink, M.; Leuther, A. Balanced active frequency multipliers for W-band signal sources. In Proceedings of the 2011 6th European Microwave Integrated Circuit Conference, Manchester, UK, 10–12 October 2011; pp. 101–104.
8. Hu, Y.; Xie, J.; Bi, X. A W-band Frequency Tripler with 8-dBm Output Power Using 0.25-m GaAs pHEMT Process. In Proceedings of the 2022 IEEE MTT-S International Microwave Workshop Series on Advanced Materials and Processes for RF and THz Applications (IMWS-AMP), Guangzhou, China, 12–14 December 2022; pp. 1–3.
9. Kim, M.; Choi, K.; Kim, J. W -Band Backward Distributed Frequency Doubler Using GaAs 0.15-µm pHEMT Process. *IEEE Microw. Wirel. Compon. Lett.* **2019**, *29*, 400–402. [CrossRef]
10. Ye, Y.; Zhang, J.; Sun, X.W. A W-Band Traveling-Wave Frequency Doubler With Output Power of 9 dBm and Power Efficiency of 11.2%. *IEEE Microw. Wirel. Compon. Lett.* **2014**, *24*, 704–706. [CrossRef]
11. Campos-Roca, Y.; Verweyen, L.; Fernandez-Barciela, M.; Bischof, W.; Curras-Francos, M.; Sanchez, E.; Hulsmann, A.; Schlechtweg, M. 38/76 GHz PHEMT MMIC balanced frequency doublers in coplanar technology. *IEEE Microw. Guid. Wave Lett.* **2000**, *10*, 484–486. [CrossRef]
12. Lu, K.H.; Huang, J.J.; Chen, W.C.; Chang, H.Y.; Wang, Y.C. A K-Band Frequency Doubler in 0.15-µm GaAs pHEMT with an Autonomous Circuit for Stability Analysis. In Proceedings of the 2018 IEEE International Symposium on Radio-Frequency Integration Technology (RFIT), Melbourne, Australia, 15–17 August 2018; pp. 1–3.
13. Shi, Q.H.; Liu, W.L.; Lu, Y.; Liu, X.; Feng, T.; Zhao, Z.Y.; Yi, C.P.; Ma, X.H.; Hao, Y. An E/W-band Frequency Doubler with 10.5-dBm Output Power Using 0.1-m GaAs pHEMT Process. In Proceedings of the 2024 IEEE MTT-S International Microwave Workshop Series on Advanced Materials and Processes for RF and THz Applications (IMWS-AMP), Nanjing, China, 9–11 November 2024; pp. 1–3.

micromachines

MDPI

Article

A Dual-Band Bandpass Filter with Wide Upper Stopband Using Stepped-Impedance Resonators and an Integrated Low-Pass Filter

Liqin Liu [1,2,*], Yuanmo Lin [1,2], Qun Chen [1,2], Li Zhang [1,2] and Minhang Weng [1,2]

1 College of Artificial Intelligence, Putian University, Putian 351100, China
2 Electronic Information Industry Technology Research Institute of Putian, Putian 351100, China
* Correspondence: llqsmile006@ptu.edu.cn

Abstract

In this paper, a dual-band bandpass filter with a wide upper stopband is proposed and designed by integrating stepped-impedance resonators (SIRs) and a low-pass filter. The operating center frequencies of the designed dual-band filter are targeted at 2.5 GHz and 5.35 GHz, respectively, to meet the frequency requirements of typical wireless communication scenarios. Notably, the filter achieves a wide upper stopband ranging from 6.1 GHz to 25 GHz, which can effectively suppress unwanted high-frequency interference signals within this frequency range and avoid mutual interference with other high-frequency communication systems. And it exhibits insertion losses of 0.12 dB (2.5 GHz) and 0.6 dB (5.35 GHz) in its two passbands to ensure minimal useful signal attenuation. The simulation results agree well with the measured results.

Keywords: wide upper stopband; low loss; stepped impedance resonator (SIR); quality value

1. Introduction

The rapid advancement of modern communication systems has spurred significant developments in planar filters. Consequently, the design criteria for wide stopband filters, demanding compact size, high selectivity, and superior stopband performance, have become increasingly stringent.

A key requirement is the effective elimination of spurious harmonics and noise, which necessitates wide stopband filters with excellent out-of-band rejection. Conventionally, wide stopbands are realized by cascading multiple resonators [1–4]. However, this approach often leads to excessive physical dimensions and significant insertion loss. This has driven many researchers to investigate novel techniques for creating wide stopbands. One popular approach is the Defected Ground Structure (DGS), which achieves a wide stopband by modifying the ground plane. A single DGS unit can typically generate one or two transmission zeros, and advantageous filtering characteristics can be achieved through careful structural optimization [5–7]. For instance, Guoan Wu et al. [8] introduced a bandpass filter based on a half-wavelength SIR, employing open stubs and a DGS to shape the stopband and achieve good performance. Nevertheless, DGSs can compromise the ground plane's integrity and increase the overall size, hindering integration. Deshan Tang et al. [9] developed a dual-passband filter using two Substrate Integrated Defected Ground Structure (SIDGS) resonant units, achieving a wide upper stopband extending to 5.5 f_1 with low radiation loss. Fuchang Chen [10] designed wide stopband filters and diplexers using Uniform Impedance Resonators (UIRs), generating multiple transmission zeros by tuning the electrical lengths of parallel-coupled lines and stubs to suppress harmonics. Pin Wen et al. [11] designed a dual-band wide stopband filter by analyzing an open-stub-loaded, short-circuited stepped-impedance resonator with hybrid electromagnetic coupling. Minhang Weng et al. [12] created a dual-band filter with stub-loaded stepped-impedance ring resonators, adding rectangular stub-loaded resonators (RSLRs) at the I/O ports for upper stopband suppression. Xueyu Huang et al. [13] fabricated a dual-band filter using a dual-mode dual-composite right/left-handed (D-CRLH) resonator, achieving a stopband up to 4.45 f_1. Zhang, Y. et al. [14] utilized an inverse design model based on a conditional deep convolutional generative adversarial network (CDC-GAN) to optimize a dual-band filter, achieving a stopband extension to 4.67 f_1. Mingming Ma et al. [15] proposed a wide stopband dual-band filter based on a self-coupled resonant cavity with asymmetric parallel microstrip lines, extending its stopband to 4.82 f_1. Weisheng Tang [16] adopted symmetric coupled quarter-wavelength stepped-impedance resonators (SIRs) to design a dual-band ultra-wide stopband filter. By adjusting the feeding position and electrical length ratio of the SIRs, the desired suppression level is achieved without affecting the characteristics of the fundamental passband, exhibiting excellent performance. Yijun Li et al. [17] utilized a slot-loaded substrate-integrated waveguide (SIW) cavity design with microstrip line feeding. Two quarter-wavelength resonators were introduced into the upper and lower transverse slots, which were, respectively, coupled with two shorted patches, achieving a −20 dB stopband range of up to 2.86 f_1. In Bowei Liu et al.'s study [18], the top layer of the filter employed hook-shaped microstrip feeding lines (HSMFLs) composed of stepped-impedance variation microstrip lines (SIVMLs) and folded microstrip lines with an enhanced coupling effect at the ends. A T-junction resonator was arranged on the second layer, thus realizing a bandpass filter based on low-temperature co-fired ceramic (LTCC) technology with an upper stopband width of 3.35 f_1, which exhibited excellent performance. Despite these advances, the stopbands in many existing designs are insufficient to suppress all spurious harmonics, and achieving harmonic suppression across an ultra-wide frequency range remains a formidable challenge.

In this work, a dual-band filter featuring a wide upper stopband is proposed. The filter is composed of a stepped-impedance resonator (SIR) section combined with a low pass filter (LPF). The dual-band response is achieved through interdigital coupling between the SIRs. Subsequently, the integrated LPF suppresses the upper passband of the dual-band filter, resulting in a wide upper stopband capable of effectively rejecting high-frequency interference.

2. Design Process

Figure 1 presents the geometric configuration of a dual-band filter with upper wide stopband. This BPF is capable of meeting WLAN standards while concurrently delivering a wide upper stopband, high band selectivity, and low insertion loss.

The substrate employed here is Rogers RT/duroid 5880, characterized by a thickness of 0.787 mm, a relative dielectric constant (ε_r) of 2.2, and a loss tangent of 0.0009. The dual-band BPF prototype primarily consists of two key components: a set of cross-coupled stepped-impedance resonators (SIRs) and a low-passband filter. Input and output (I/O) lines are coupled to the cross-coupled SIRs via taps on both sides. As a result, two signal transmission paths are formed between the input and output ports—a design feature is the intentional introduction of the cross-coupling effect, which is critical to the performance of this dual-band bandpass filter. The simulation is conducted using Ansys HFSS 2023 R1, a high-frequency structural simulation tool suitable for microwave passive components.

https://doi.org/10.3390/mi17010075

Regarding boundary conditions, the upper surface of the substrate (microstrip structure) adopts a Radiation Boundary to avoid boundary reflection; the lower surface (ground copper foil) uses a Perfect E Boundary to simulate a complete ground plane; and the substrate sides employ a Perfect H Boundary to mimic infinite extension. For port configuration, the input and output ports are Wave Ports, with parameters matching the microstrip lines, a characteristic impedance of 50 Ω, and excitation in TEM mode.

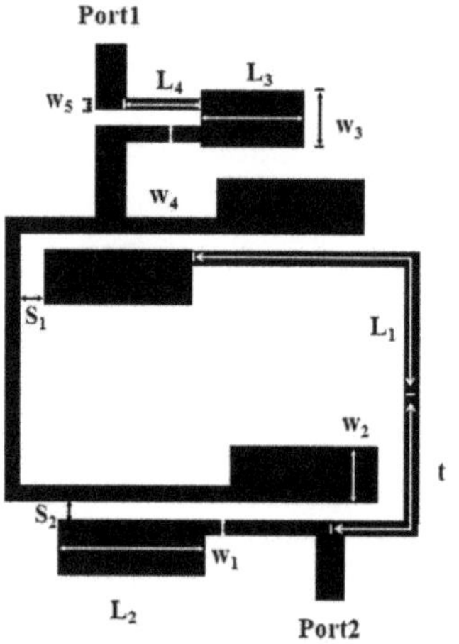

Figure 1. Geometrical diagram.

2.1. Dual-Band Filter Design

The dual-band filter proposed in this work is targeted for operation at 2.5 GHz and 5.35 GHz. For its two passbands, the 3 dB fractional bandwidths (FBWs) are configured as follows: the first passband (centered at 2.5 GHz) has an FBW1 of 34.8%, while the second passband (centered at 5.35 GHz) features an FBW2 of 12.1%. Notably, both passbands share an identical passband ripple of 0.01 dB.

Prior to analyzing the filter's overall performance, it is necessary to first discuss the resonant behavior of the stepped-impedance resonator (SIR) illustrated in Figure 2. This SIR is structured by cascading two high-impedance segments (denoted as Z_2) with a single low-impedance segment (denoted as Z_1). The impedance ratio (R)—a key parameter of the SIR—is defined as $R = Z_2/Z_1$. The input admittance (Y_{in}) of the SIR is expressed by the equation below:

$$Y_{in} = \frac{2(R\tan\theta_1 + \tan\theta_2)\cdot(R - \tan\theta_1\cdot\tan\theta_2)}{R(1-\tan^2\theta_1)\cdot(1-\tan^2\theta_2) - 2(1+R^2)\cdot\tan\theta_1\cdot\tan\theta_2} \tag{1}$$

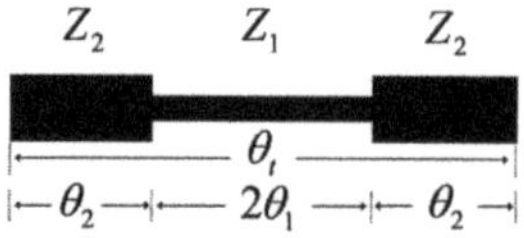

Figure 2. Schematic of the stepped-impedance resonator.

The resonant conditions of the SIR can be derived when its Y_{in} equals zero. Additionally, by adjusting R and α of the SIR, the positions of its higher-order resonant modes can

https://doi.org/10.3390/mi170100075

be tuned—either bringing these modes closer to each other or spacing them farther apart. The definition of the length ratio (α) is given as follows:

$$\alpha = \frac{\theta_2}{\theta_1 + \theta_2} = \frac{2\theta_2}{\theta_t} \tag{2}$$

The introduction of α enriches the resonant mode characteristics of the SIR, thereby providing greater design flexibility for the filter. Here, θ_t represents the total electrical length of the SIR. When Equation (2) is substituted into Equation (1), multiple resonant modes—whose properties are dependent on both the R and α—can be derived, and these modes are visualized in Figure 3.

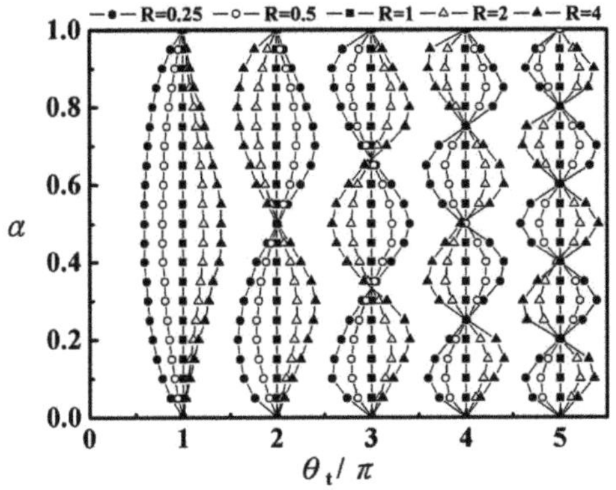

Figure 3. Fundamental and higher-order resonant modes as function of the length ratio α and impedance ratio R = 0.25, 0. 5, 1, 2, and 4.

Notably, in the non-stepped-impedance scenario (i.e., when R = 1, which is equivalent to α = 0 or α = 1), the relationship $\theta_t = n\pi$ holds. This result indicates that the nth resonant mode of the resonator is excited when its total length corresponds to n times a half-wavelength.

It can be clearly observed that the spacing between certain higher-order resonant modes and the fundamental resonant mode (i.e., whether they are far apart or close to each other) depends on the selection of the R and α. Therefore, the same fundamental SIR structure can be extended to design dual-band, three-band, or even four-band filters, highlighting the scalability and flexibility of this approach for more complex multi-band applications. Thus, by appropriately choosing R and α of the SIR, it is feasible to achieve a dual-band response. Following the above guidelines, we mapped the requirements for the two target center frequencies onto Figure 3. By adjusting the impedance ratio R and length ratio α, the positions of high-order resonant modes can be flexibly tuned (Figure 3). For example, when R = 0.5 and α = 0.26, the resonant modes are exactly at 2.5 GHz (fundamental mode) and 5.35 GHz (second-order mode). In contrast, the resonant modes of uniform impedance resonators (UIRs) are fixed at $n \times$ fundamental frequency (e.g., 2.5/5/7.5 GHz), which cannot match the 5.35 GHz requirement. This specific value allows the filter to attain two key performance traits at the same time: a dual-band operating response suitable for WLAN uses, and an extensive stopband. The electrical sizes of the SIR at the operating resonance frequencies that are calculated using the free-space wavelength are 120 mm at 2.5 GHz and 56 mm at 5.35 GHz. A single SIR can provide two or more resonant modes eliminating the need for multiple independent resonators. The total length of the SIR in this work is 39.5 mm, which is shorter than the length of two independent UIRs. Also, there are some alternative routes to compact multiband resonators [19], for example, using

https://doi.org/10.3390/mi170100075

dielectric materials with very high relative permittivity as the core resonator and relying on subwavelength resonators (e.g., split-rings, split-loops) to confine electromagnetic energy at the subwavelength scale. Constructed into the designed filter are high-impedance line sections (Z_1 = 100 Ω) featuring a 0.7 mm strip width, as well as low-impedance line sections (Z_1 − 50 Ω) with a 2.4 mm strip width. In the initial design stage, the target performance specifications were first defined. Based on these specifications, the lumped-element values of the low-pass prototype filter were calculated, yielding g_1 = 0.90467 and g_2 = 1.25868 (where g_n denotes the n-th lumped-element value for n = 1,2) [20]. Once these lumped-element values were obtained, the required coupling matrices M and external quality factor (Q_e) of the bandpass filter (BPF) were derived. This derivation was conducted on the basis of coupling theory, while also adhering to the standard design protocol presented in reference [20].

$$M_{ij}^{I} = \begin{bmatrix} 0 & 0.326 \\ 0.326 & 0 \end{bmatrix} \text{ at 2.5 GHz} \tag{3}$$

and

$$M_{ij}^{II} = \begin{bmatrix} 0 & 0.113 \\ 0.113 & 0 \end{bmatrix} \text{ at 5.35 GHz} \tag{4}$$

$$Q_e^{I} = 3.27 \text{ and } Q_e^{II} = 9.41$$

where M_{ij} denotes the coupling coefficient, with $M_{12} = M_{21} = FBW/\sqrt{g_1 g_2}$ and $Q_e = g_1 g_2/FBW$.

To realize the pre-set band performance, the coupling levels between adjacent SIRs should be suitably selected to correspond with the intended coupling matrix M. Particularly, the coupling level between Resonator 1 and Resonator 2 is primarily shaped by the spacing S_2; hence, S_2 is adjusted to regulate the coupling coefficient M_{12}. In addition, the external quality factor Q_e of the dual-band BPF is affected by the distance t—that is to say, varying the distance t makes it possible to adjust Q_e. It should be emphasized that the computed coupling between resonators can be distinguished by two main resonant frequencies. These frequencies arise from the splitting of the original resonance condition, which is induced by electromagnetic coupling. Based on this principle, the simulated coupling coefficient M is computed as follows [20]:

$$M = \frac{f_H^2 - f_L^2}{f_H^2 + f_L^2} \tag{5}$$

The calculated quality factor values between the resonators can be characterized by the two threshold resonant frequencies. Accordingly, the simulated external quality factor Q_e is computed as follows [20]:

$$Q_e = \frac{f_0}{f_{H\,3dB} - f_{L\,3dB}} \tag{6}$$

where f_{H3dB} denotes the higher frequency of the 3 dB fractional bandwidth (FBW), and f_{L3dB} denotes the lower one. The coupling coefficient M and external quality factor Q_e of the dual-band BPF, as illustrated in Figure 4, were calculated using a full-wave electromagnetic (EM) simulator.

Figure 5 is the structure diagram of the dual-frequency filter. Optimal simulation of the filter's performance was carried out via a full-wave electromagnetic (EM) simulator, with subtle structural tweaks applied over the course of the process. The simulation results are presented in Figure 6, using the optimized structural parameters: L_1 = 14.65 mm, w_1 = 0.7 mm, L_2 = 5.1 mm, w_2 = 2.4 mm, t = 7.5 mm; S_1 = 0.3 mm, and S_2 = 0.25 mm. The dual-band filter delivers dual-band filtering performance with center frequencies at

https://doi.org/10.3390/mi170100075

2.5 GHz and 3.53 GHz, boasting 3 dB bandwidths of 34.8% and 12.1% for its respective passbands. Employing a 0-degree feed arrangement for both input and output, the filter creates transmission zeros at 1.83 GHz, 3.56 GHz, and 6 GHz, thus achieving high selectivity. Nevertheless, it has been noted that a significant number of harmonics from higher-order modes manifest within the 6.1 GHz to 25 GHz range.

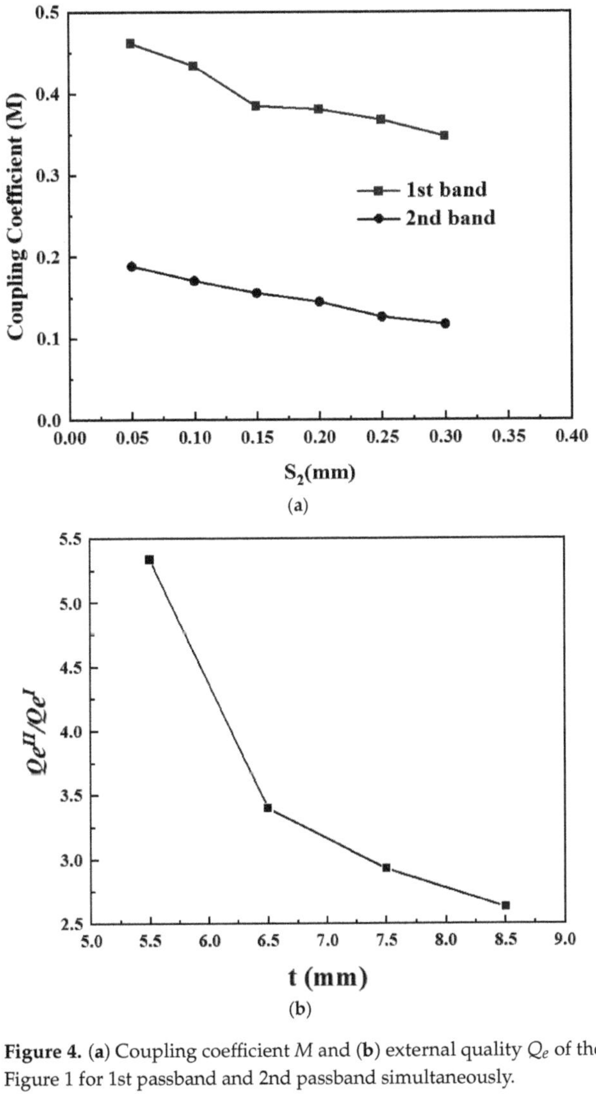

Figure 4. (**a**) Coupling coefficient M and (**b**) external quality Q_e of the dual-band BPF as shown in Figure 1 for 1st passband and 2nd passband simultaneously.

2.2. Wide Upper Stopband Design

A wide upper stopband is essential in multi-service wireless systems (e.g., WLAN, 5G, satellite communications) to prevent interference from higher-frequency signals and harmonics generated by the filter itself or other active components in the RF front-end. To implement the wide upper stopband of the proposed dual-band filter, it is necessary to suppress harmonics at 6.1 GHz and 25 GHz. Additionally, the rejection of unwanted interference in the higher frequency range (6.1 GHz to 25 GHz), as indicated in Figure 6, needs to be improved. This is achieved by incorporating a low-pass filter (LPF), as depicted

https://doi.org/10.3390/mi17010075

in Figure 7. The LPF is designed to provide an adjustable stopband for effective high-frequency signal suppression. The structure of the LPF consists of a pair of coupled lines and a shunt open-circuited resonator, realized as a high-impedance–low-impedance–high-impedance transmission line configuration. As demonstrated in [21], a wide stopband with effective out-of-band rejection can be obtained when the lengths of the coupled lines and the shunt resonator are approximately a quarter-wavelength ($\lambda g/4$) at a specific center frequency. Figure 8 presents a comparative analysis of the LPF's S-parameters for different electrical length ratios (θ_2/θ_1). The analysis reveals that the sharpest attenuation slope is achieved when θ_3/θ_4 = 1.37, which offers superior suppression of adjacent channels and spurious signals.

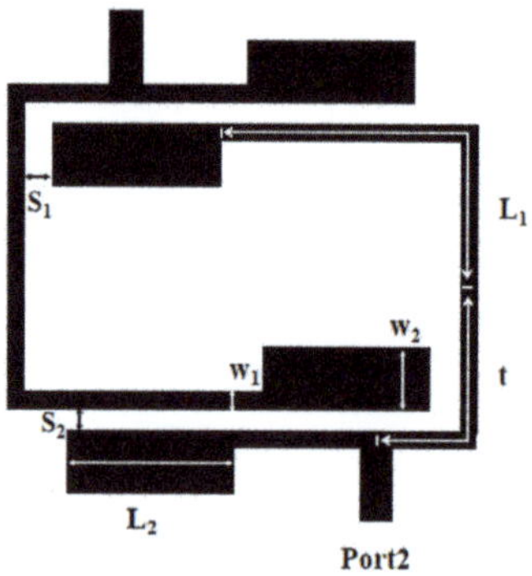

Figure 5. Structure diagram of the dual-frequency filter.

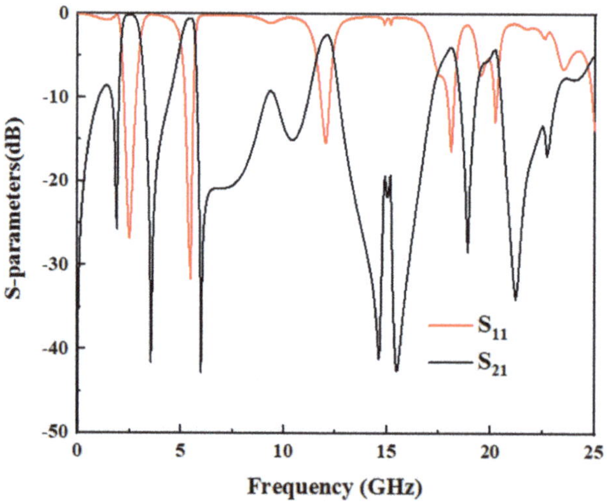

Figure 6. Simulation S-parameter diagram of the dual-frequency filter.

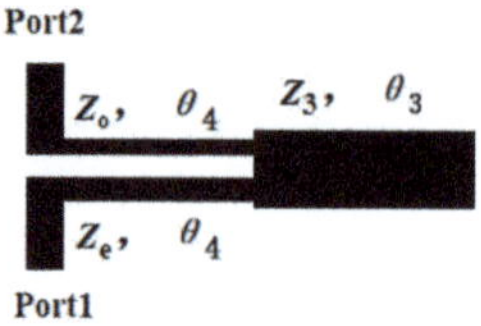

Figure 7. The low-pass filter used in this design.

https://doi.org/10.3390/mi170100075

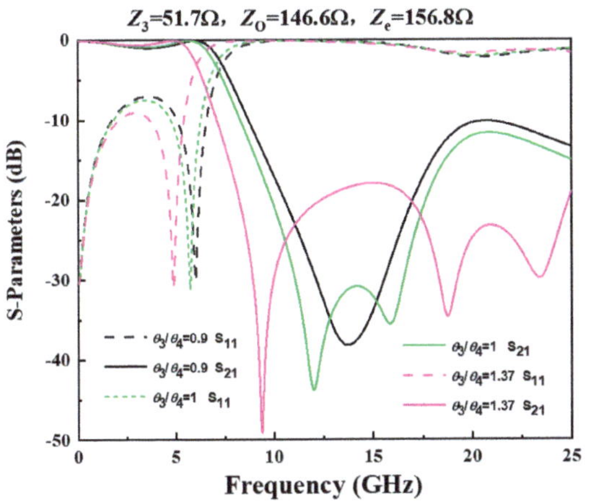

Figure 8. Comparison chart of S-parameters of low-pass filters under different electron length ratios.

2.3. Design of the Dual-Band Bandpass Filter with a Wide Upper Stopband

To achieve the desired wide stopband, the low-pass filter (LPF) is integrated with the input transmission line of the dual-band filter, as depicted in the final layout in Figure 1. The final physical dimensions have been optimized as follows: L_1 = 14.65 mm, w_1 = 0.7 mm, L_2 = 5.1 mm, w_2 = 2.4 mm, L_3 = 3.36 mm, w_3 = 2.3 mm, L_4 = 2.45 mm, w_4 = 0.2 mm, w_5 = 0.25 mm, t = 7.5 mm, S_1 = 0.3 mm, and S_2 = 0.25 mm. As shown in Figure 9, the integration of the LPF significantly enhances the filter's performance by providing a lower insertion loss and a substantially wider upper stopband. The filter exhibits an insertion loss of 0.12 dB in the first passband and 0.6 dB in the second passband. Notably, the upper stopband is extended to cover the frequency range from 6.1 GHz to 25 GHz.

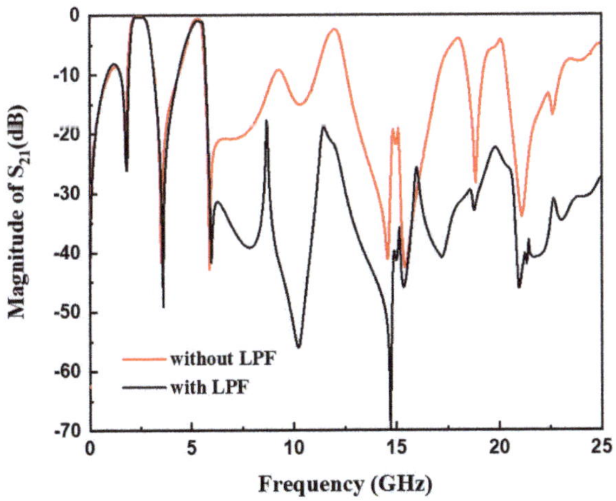

Figure 9. Comparison chart of S_{21} dual-frequency filter with low-pass filter and dual-frequency filter without low-pass filter.

3. Experimental Test Results

The designed filter is fabricated and then measured by a Network Analyzer. A photograph of the fabricated dual-band BPF with a wide upper stopband is shown in Figure 10. The dimensional parameters of the filter are as follows: L_1 = 14.65 mm, w_1 = 0.7 mm,

https://doi.org/10.3390/mi17010075

L_2 = 5.1 mm, w_2 = 2.4 mm, L_3 = 3.36 mm, w_3 = 2.3 mm, L_4 = 2.45 mm, w_4 = 0.2 mm, w_5 = 0.25 mm, t = 7.5 mm, S_1 = 0.3 mm, and S_2 = 0.25 mm. The input and output ports are connected via a 50 Ω broadband transmission line with a width of 0.8 mm. Additionally, the whole size is only 15.46 mm × 16.46 mm, i.e., approximately] 0.16 λg × 0.17 λg, where λg is the guided wavelength at the first frequency. The simulated results and the measured results are shown in Figure 11. The measured results have a low insertion loss ($|S_{21}|$) less than 0.12 dB, a bandwidth of 34.8%, and a Q_e value of 3.27 for 2.5 GHz, and a low insertion loss ($|S_{21}|$) less than 0.6 dB, a bandwidth of 12.1%, and a Q_e value of 9.41 for 5.35 GHz. And the range of the upper stopband is from 6.1 GHz to 25 GHz. The transmission zeros can be obviously introduced on the two-side skirt of each passband, which is due to the coupling effect between two SIRs, causing the multi-path propagation mode and 0-degree feed. The transmission zeros significantly improve the selectivity of the fabricated dual-band BPF with a wide upper stopband. The mean absolute error between simulated and measured curves in Figure 11 can be calculated as 0.57 dB. The minor discrepancy primarily stems from substrate etching tolerance (±0.05 mm) and SMA connector insertion loss (~0.1 dB), confirming good consistency between simulation and measurement.

Figure 10. Photograph of the fabricated dual-band BPF with wide upper stopband.

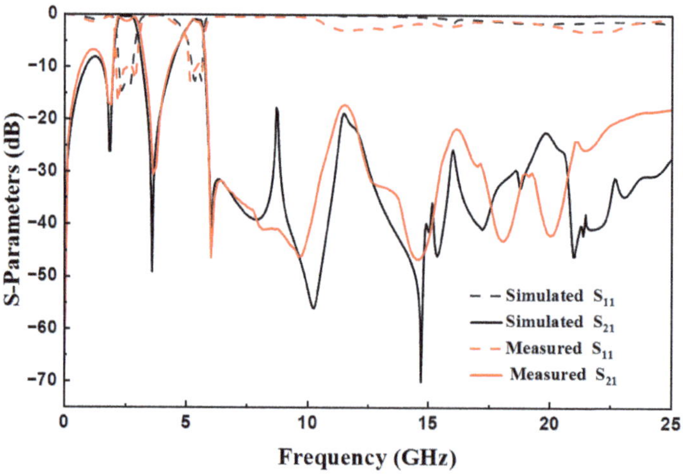

Figure 11. Comparison between simulation and actual measurement.

Table 1 shows the comparison of major performance characteristics of the proposed filter against data presented in other research reports. The proposed filter achieves lower insertion loss (0.12 dB at 2.5 GHz, 0.6 dB at 5.35 GHz), outperforming prior studies [11,15],

https://doi.org/10.3390/mi170100075

and a wider upper stopband (6.1–25 GHz, 10 f_1), surpassing [13,15] to suppress harmonics below $10f_1$. Structurally, it directly integrates "cross-coupled SIRs" with a "$\lambda g/4$-loaded LPF" (not cascaded), reducing size to 63% of [12]. Applicable to WLAN dual bands, it suppresses 5G millimeter-wave (24/28 GHz) and satellite communication (12–18 GHz) interference, enabling direct use in RF front-ends to simplify systems.

Table 1. Comparison of major performance characteristics of the proposed filter against data presented in other research reports.

	1st/2nd Passbands (GHz)	$\lvert S_{21} \rvert$ (dB)	FBW (%)	Upper Stopband	Circuit Size (mm^2) (λg^2)
11	1.51/3.68	1.55/1.63	4.25/2.6	6.4 f_1	0.0065
12	2.4/4	1.4/1	8/39	8.3 f_1	0.0432
13	3.5/5.8	0.73/1.43	18/8.3	4.45	0.010282
14	3/5	1.19/1.77	17.9/4.1	4.67	0.26
15	2.34/4.72	1.05/1.86	15.8/8.33	4.82	0.0224
Proposed Filter	2.5/5.35	0.12/0.6	34.8/12.1	10 f_1	0.0272

4. Conclusions

This paper designs a dual-band filter with a wide upper stopband and implements it by using stepped-impedance resonators and a low passband filter. Analysis of the odd and even modes reveals the resonant frequency distribution of the stepped-impedance resonators. The proposed filter's design entails two key steps: first, developing the dual-band filter, and then integrating it with a low passband filter to achieve harmonic suppression. The resulting filter features dual passbands centered at 2.5 GHz and 5.35 GHz, with 3 dB bandwidths of 34.8% and 12.1%, respectively. It has three transmission zeros, ensuring excellent selectivity. Notably, it achieves $\lvert S_{21} \rvert$ attenuation exceeding 20 dB across 6.1 GHz to 25 GHz, demonstrating effective harmonic suppression up to ten times the fundamental frequency. And the designed filter is also compact. This proposed dual-band BPF with a wide stopband is well-suited for WLAN systems.

Author Contributions: Conceptualization, M.W.; methodology, L.L.; software, L.L.; validation, L.L. and Y.L.; formal analysis, L.L. and L.Z.; investigation, Q.C. and Y.L.; resources, L.L.; data curation, L.L. and L.Z.; writing—original draft preparation, M.W.; writing—review and editing, L.L. and Y.L.; visualization, L.L. and M.W.; supervision, L.L.; project administration, L.L.; funding acquisition, L.L. and Y.L. All authors have read and agreed to the published version of the manuscript.

Funding: This study was supported by the Natural Science Foundation of Fujian Province (2025J011039, 2023J011012, 2020J01921) and Electronic Information Industry Technology Research Institute of Putian (2023GJGZ003).

Data Availability Statement: All the material conducted in the study is mentioned in article.

Conflicts of Interest: The authors declare no conflicts of interest.

References

1. Kuo, T.N.; Li, W.C.; Wang, C.H.; Chen, C.H. Wide-stopband microstrip bandpass filters using quarter wavelength stepped impedance resonators and bandstop embedded resonators. *IEEE Wirel. Compon. Lett.* **2008**, *18*, 389. [CrossRef]
2. Tang, C.W.; Chen, M.G. Wide stopband parallel-coupled stacked SIRs bandpass filters with open-stub lines. *IEEE Microw. Wirel. Compon. Lett.* **2006**, *16*, 666. [CrossRef]
3. Tu, W.H.; Chang, K. Compact microstrip bandstop filter using open stub and spurline. *IEEE Microw. Wirel. Compon. Lett.* **2005**, *15*, 268. [CrossRef]
4. Tang, C.W. Miniaturised wide stopband rejected microstrip filter with coupled spur-lines. *Electron. Lett.* **2006**, *42*, 286. [CrossRef]

5. Balalem, A.; Ali, A.R.; Machac, J.; Omar, A. Quasi-elliptic microstrip low-pass filters using an interdigital DGS slot. *IEEE Microw. Wirel. Compon. Lett.* **2007**, *17*, 586–588. [CrossRef]
6. Luo, X.; Ma, J.G.; Li, E.P.; Ma, K. Hybrid microstrip T-stub/defected ground structure cell for electromagnetic interference bandpass filter design. *IEEE Trans. Electromagn. Compat.* **2011**, *53*, 717–725. [CrossRef]
7. Rao, Y.; Qian, H.J.; Yang, B.; Gomez-Garcia, R.; Luo, X. Dual-band bandpass filter and filtering power divider with ultra-wide upper stopband using hybrid microstrip/DGS dual-resonance cells. *IEEE Access* **2020**, *8*, 23624–23637. [CrossRef]
8. Wu, G.A.; Zhang, X.; Li, W.G. Novel bandpass filter with high selectivity and very wide stopband using open stub loaded and DGS. *IEICE Electron. Express* **2017**, *16*, 20170316. [CrossRef]
9. Tang, D.; Han, C.X.; Deng, Z.X.; Qian, H.J.; Luo, X. Substrate-Integrated Defected Ground Structure for Single- and Dual-Band Bandpass Filters with Wide Stopband and Low Radiation Loss. *IEEE Trans. Microw. Theory Tech.* **2021**, *69*, 659–670. [CrossRef]
10. Riaz, M.; Virdee, B.S.; Shukla, P.; Onadim, M. Quasi-elliptic ultra-wideband bandpass filter with super-wide stopband. *AEÜ Int. J. Electron. Commun.* **2019**, *105*, 171–176. [CrossRef]
11. Wen, P.; Ma, Z.W.; Liu, H.W.; Zhu, S.S.; Ren, B.P.; Guan, X.H.; Ohira, M. Individually controllable dual-band bandpass filter with multiple transmission zeros and wide stopband. *IEICE Electron. Express* **2019**, *16*, 20190127. [CrossRef]
12. Weng, M.H.; Lan, S.W.; Chang, S.J.; Yang, R.Y. Design of Dual-Band Bandpass Filter with Simultaneous Narrow- and Wide-Bandwidth and a Wide Stopband. *IEEE Access* **2019**, *7*, 147694–147703. [CrossRef]
13. Huang, X.Y.; Feng, W.J.; Tang, K.; Cao, Y.F.; Che, W.Q.; Xue, Q. Miniaturized Dual-Band Filter Using D-CRLH Resonators with Wide Stopband and Controllable Frequencies/Bandwidths. *IEEE Microw. Wirel. Technol. Lett.* **2025**, 1–4. [CrossRef]
14. Zhang, Y.; Xu, J. Inverse design of dual-band microstrip filters based on generative adversarial network. *IEEE Microw. Wirel. Technol. Lett.* **2024**, *34*, 29–32. [CrossRef]
15. Ma, M.M.; You, F.; Qian, T.; Shen, C.; Qin, R.X.; Wu, T.; He, S.B. A wide stopband dual-band bandpass filter based on asymmetrical parallel-coupled transmission line resonator. *IEEE Trans. Microw. Theory Techn.* **2022**, *70*, 3213–3223. [CrossRef]
16. Tang, W.S.; Zheng, S.Y.; Pan, Y.M. A compact dual-band bandpass filter with large frequency ratio based on ultra-wide-stopband bandpass filter. *China Commun.* **2025**, *22*, 156–169. [CrossRef]
17. Li, Y.J.; Ni, J.X.; Li, S.Y.; Han, Q.; Tian, J.; Li, N.; Ma, T.Y. Miniaturized dual-band bandpass filter with closely spaced passbands and wide stopband. *Front. Phys.* **2025**, *13*, 1637651. [CrossRef]
18. Liu, B.W.; Shi, G.; Wang, Z.C.; Hu, S.W. A N77 band bandpass filter for 5G applications using LTCC technology with wide upper stopband. *Frequenz* **2025**, *67*. [CrossRef]
19. Serebryannikov, A.E.; Gokkavas, M.; Gundogdu, T.F.; Volski, V.; Vandenbosch, G.A.E.; Vasylchenko, A.; Ozbay, E. Ultraminiature Antennas Combining Subwavelength Resonators and a Very-High-ε Uniform Substrate: The Case of Lithium Niobate. *IEEE Trans. Antennas Propag.* **2020**, *68*, 5071–5081. [CrossRef]
20. Hong, J.S.; Lancaster, M.J. *Microstrip Filters for RF/Microwave Applications*; John & Wiley: New York, NY, USA, 2001.
21. Hsieh, M.Y.; Wang, S.M. Compact and wideband microstrip band-stop filter. *IEEE Microw. Wirel. Compon. Lett.* **2005**, *15*, 472–474. [CrossRef]

https://doi.org/10.3390/mi170100075

micromachines

MDPI

Article

Metamaterial Incident Photon Reconstruction Theory Based on Resonant Dipole Phase

Boli Xu and Renbin Zhong *

Terahertz Research Center, School of Electronic Science and Engineering, University of Electronic Science and Technology of China, Cooperative Innovation Centre of Terahertz Science, Chengdu 610054, China; 202211022430@std.uestc.edu.cn
* Correspondence: rbzhong@uestc.edu.cn

Abstract

In this study, a Metamaterial Incident Photon Reconstruction Theory (MIPRT) is developed to describe the modulation process of metamaterials on incident photons. The theory includes the Invariant Incident Photon Hypothesis and Resonant Phase Deconstruction and Quantification; it reveals the modulation characteristics of metamaterials on incident photons, not by first absorption and then re-emission but by inducing coherent destructive interference, which brings about redistribution of the spatial probability of photon occurrence. This theory is validated in a single-layer metamaterial, and a unique relationship between the resonant phase and amplitude is derived and confirmed by simulation. The proposed MIPRT brings a comprehensive understanding of the electromagnetic (EM) response characteristics of metamaterials and provides a new idea for metamaterial theory from another perspective.

Keywords: metamaterial; amplitude–phase relationship; resonant phase

1. Introduction

Metamaterials designed to control the amplitude, phase, or wavefront of electromagnetic (EM) waves have been widely studied in recent years [1–3]. Researchers have proposed various theories to reveal the laws underlying the phenomenon of the controlling of EM waves by metamaterials. The principle of multilayer metamaterials modulating phenomena is well described by transmission matrix method theory [4]. The theory of the photon band diagram can interpret the characteristics of EM wave transmission inside metamaterials [5], and these kinds of metamaterials are also called photonic crystals occasionally. Considering the surface phase gradient distribution, the generalized Snell equations describe the formation of the EM wavefront controlled by metamaterials [6]. Ahmed H. et al. used two-port equivalence to derive the relationship of transmission phase limitations by mathematical reasoning [7]. These methods rely heavily on numerical simulations but lack insight into the physical mechanisms. As for the analytical methods, such as impedance analysis [8], transmission line equivalence [9], and general dipole analysis [10,11], they hold significant practical value in the study of metamaterials, especially in design, optimization, and engineering applications.

In this paper, with the Invariant Incident Photon Hypothesis, the MIPRT theory is proposed with the definition of the plane wave phase (PWP) and resonant dipole phase (RDP); MIPRT outlines the modulation process of metamaterials' EM waves via resonant dipoles. The analytical relationship between the resonance phase and the amplitude of the reflected wave is derived for a single-layer metamaterial, and the results are verified through simulation. It provides a deeper understanding of the electromagnetic response characteristics of metamaterials, which could play a key role in the design of artificial materials. Particularly in the context of micro- and nanoscale device design, the modulation processes and underlying regularities revealed in this work provide clear directional guidance for the analysis and design of target devices. In addition, the explicit relationship between amplitude and phase modulation offers an intuitive constraint on admissible parameter spaces, thereby helping designers avoid physically unattainable device configurations.

2. Metamaterial Incident Photon Reconstruction Theory

2.1. The Invariant Incident Photon Hypothesis

The model of the resonant dipole is commonly adopted to qualitatively explain the interaction of metamaterials and electromagnetic waves (incident photons). The interaction between the metamaterial and the incident photons includes two steps: the excitation of resonant dipoles in the metamaterial and the modulation of the incident photons by the metamaterial. Once the interaction reaches a steady state, the latter step of modulation of the incident photons becomes the dominant effect. At this stage, previous theories usually suggested that incident photons are firstly absorbed by resonant dipoles upon striking the metamaterial, after which the dipoles emit new photons. However, the Invariant Incident Photon Hypothesis asserts that incident photons are not absorbed by resonant dipoles, but instead undergo photon interference with the photons emitted by the dipoles, resulting in a new spatial probability distribution of the photons. Therefore, there are invariant incident photons in the whole space, including the incident half space and the transmission half space separated by the metamaterial. The new spatial probability distribution of photons is described as the probability amplitude A_{new} in both half spaces:

$$A_{new} = A_i + A_s \tag{1}$$

Here, A_i represents the probability amplitude of incident photons (IPs) and A_s is the probability amplitude of resonant dipole-emitted photons (RPs). Thus, the RPs correspond to the probability amplitude of reflected photons in the incident space, while in the transmission space, the direction of momentum of RPs is same as that of IPs, so the A_{new} of IPs and RPs in the transmission space represents the probability amplitude of the photons modified by the metamaterial.

From the perspective of energy conservation, the coherent destructive interference between IPs and RPs leads to a reduction in energy in the transmission space. This energy decrease is just equal to the RP energy in the incident space, which is considered as an energy increase. In other words, the incident photons and dipole-emitted photons undergo coherent destructive interference, resulting in a new spatial distribution of photons. Therefore, the spatial distribution probability of the modulated photons decreases in the transmission half space and increases in the reflection half space. During the modulation process, the incident photons remain invariant throughout the entire space rather than being absorbed by the resonant dipoles.

2.2. Resonant Phase Deconstruction and Quantification

The key parameters of a photon are amplitude and phase: amplitude represents energy, and phase differences of the coherent photons will bring about destructive interference and then govern the conservation of energy. The resonant dipole produces near-field photons (localized EM field) and far-field photons (radiated field). Near-field photons are spatially confined around the dipole, while far-field photons propagate outward, carrying energy

https://doi.org/10.3390/mi17010130

and phase information to the far field. Therefore, to discuss the modulation process in detail, we define the phase associated with near-field photons to be the resonant dipole phase (RDP, related to the near-field radiation pattern of resonant dipoles), while the phase of far-field photons is defined as the plane wave phase (PWP, representing the far-field radiation pattern of resonant dipoles). To facilitate calculation, the incident photon phase is taken as the zero reference phase; then the PWP and RDP can be deduced.

We take a single-layer metamaterial as a case; it has strip-shaped units, with a strip length of 50 μm and a period of 80 μm, and its resonance frequency is 2.62 THz. The RDP and PWP of the strip-shaped metamaterial can be determined quantitatively (detailed derivation is proposed in the Supplementary Materials).

Figure 1 shows the variation curves of the RDP and PWP at different incident photon frequencies. They exhibit the same trend of variation but a unit phase difference of 90 degrees.

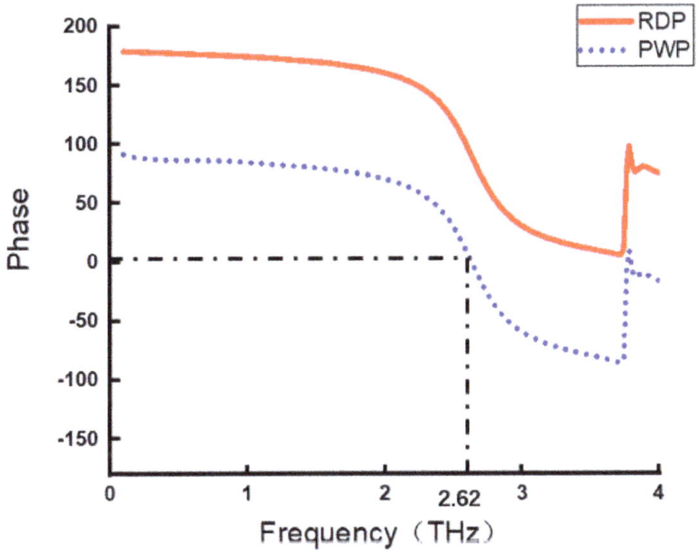

Figure 1. PWP and RDP curves vary with incident frequencies (resonance frequency is 2.62 THz).

When the incident photon frequency is relatively lower than the resonant frequency of the metamaterial, the PWP is positive. This means that the metamaterial will bring about a positive phase shift because the intrinsic frequency of the metamaterial is higher than that of the incident photon. However, when the incident photon frequency is higher than the intrinsic frequency, the PWP becomes negative. This means that the metamaterial will cause a negative phase shift because the response delay of the metamaterial will cause a lagging effect as it drags the photos along. The abnormal phase changes at the high-frequency band of the PWP curves are caused by non-geometric optical scattering, which is also a common principle for most metamaterial absorbers [12,13].

According to the definitions of RDP and PWP, it can be considered that near-field photons (with RDP) are the precursor of the far-field photons (with PWP). The photons with RDP emitted by the resonant dipole form around the array element atoms of the metamaterial and induce spatial coherence and phase alignment, resulting in far-field photons with PWP. The 90-degree phase difference between the RDP and PWP confirms their causal relationship. Energy also transitions from the near-field photons to the far-field photons: when the near-field photons reach their maximum amplitude, the far-field

https://doi.org/10.3390/mi17010130

photon amplitude is zero and vice versa, which is in accordance with the law of energy conservation. In comparison to the RDP, the PWP is easier to measure to reveal the metamaterial's response to incident photons.

3. Verification of the MIPRT on a Single-Layer Metamaterial

As mentioned above, energy conservation within MIPRT is consistent with photon interference in the transmitted half space and reflection in the incident half space. When the number of photons involved is sufficiently large, the photon description can be effectively replaced by a classical electric field representation. In this regime, the energy is strongly dependent on the field amplitude, whereas photon interference is predominantly governed by the phase difference of the electric field. Therefore, it can be anticipated that the field amplitude and phase are intrinsically linked.

Applying MIPRT on a single-layer metamaterial, Figure 2 shows the diagram of the field decomposition, assuming that the incident electric field wavevector is $\vec{k}_i$ throughout the entire space. In the steady state, the excited resonant dipole in the metamaterial will radiate the electric field outward. The radiation electric field wavevector is $\vec{k}_s$, and it is represented as $\vec{k}_{s-}$ in the incident half space and $\vec{k}_{s+}$ in the transmitted half space with symmetric direction. More generally, for the case of multi-lobe radiation of the resonant dipole, the electric field wavevector of each lobe can be expressed as $\vec{k}_{s2}$, $\vec{k}_{s3}, \ldots$, and the amplitude–phase relationship of Equation (2) can be obtained by solving Maxwell's equations under the boundary conditions (detailed derivation is presented in the Supplementary Materials)

$$\frac{\sum_{m=1}^{n}[E_{sm0} \times cos(\alpha_m)]}{E_{i0}} = -cos(\phi) \quad (n = 1,2,3\ldots) \tag{2}$$

where E_{i0} is the electric field amplitude of the incident electric field, E_{sm0} is the electric field amplitude of the mth radiation lobe whose electric field wavevector is $\vec{k}_{sm}$, α_m is the angle between $\vec{k}_{sm}$ and $\vec{k}_i$, and ϕ is the PWP.

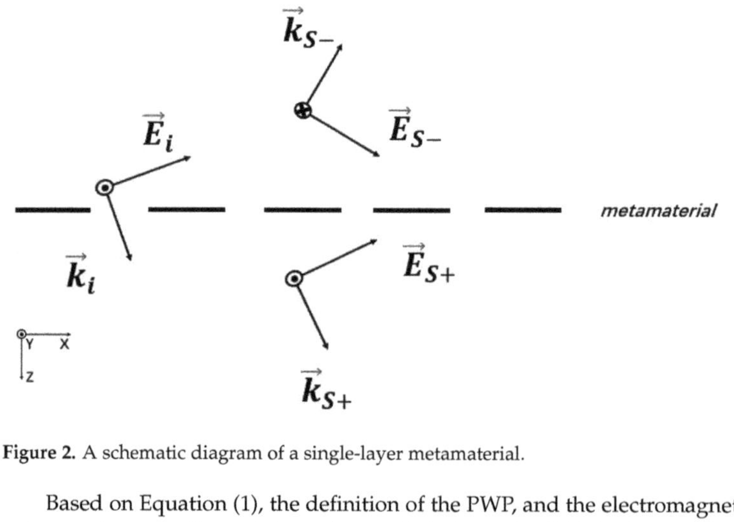

Figure 2. A schematic diagram of a single-layer metamaterial.

Based on Equation (1), the definition of the PWP, and the electromagnetic field distribution shown in Figure 2, it can be concluded that the PWP in this case is equal to the phase of the reflected wave. In addition, further investigation within this model is expected to enable the derivation of the relationship between the transmitted wave phase and the aforementioned parameters.

For the simplest case, there is only one dipole radiation lobe; when thermal loss is negligible (the most common treatment for metal structural metamaterials), Equation (2) can be simplified as Equation (3):

$$R = cos^2(\phi) \tag{3}$$

R is reflectivity of the metamaterial. Equation (3) indicates that the PWP and the reflectivity are strongly coupled and highly correlated, implying that the phase of the reflected wave directly determines its amplitude.

Two typical metamaterial structures are randomly selected to verify the amplitude–phase relationship of Equation (3). One comes from the COMSOL 5.4 case library [14] with open ring slit unit structure, as shown in Figure 3a; Figure 3b is another one, which has long metal strip unit structures.

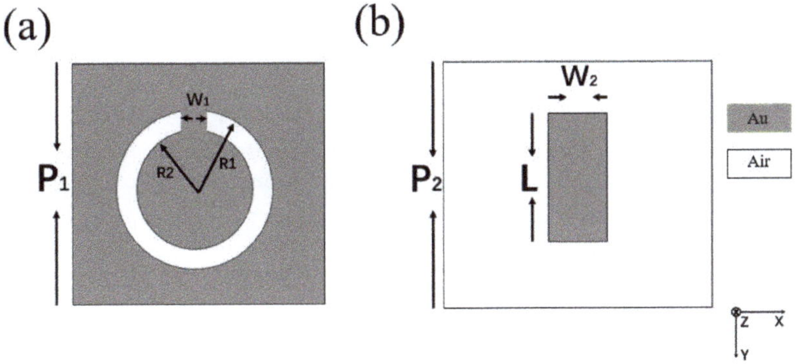

Figure 3. Unit structures of metamaterial. (**a**) Open ring slit unit with P_1 = 15 mm, W_1 = 1 mm, $R1$ = 3.5 mm, and $R2$ = 5 mm. (**b**) Long strip unit with P_2 = 80 μm, W_2 = 8 μm, and L = 50 μm.

The two metamaterial structures are incident by EM waves at the angles of 0° and 30°, respectively, and the polarization of the incident electric field is perpendicular to the opening of the open ring slit or parallel to the long edge of the metal strip (+Y axis direction); the simulated curves of reflection and absorption of them are shown in Figure 4.

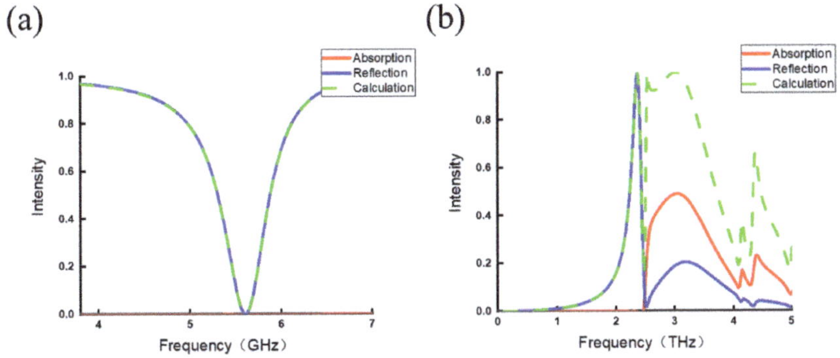

Figure 4. Simulated absorption curves (red line) of metamaterial structures and the comparison of reflection curves between the simulated (solid line) and theoretically calculated (dashed line) results. (**a**) Open ring slit. (**b**) Long strip.

In metamaterials, absorption is commonly defined in the context of absorbers. Since such absorbers are predominantly developed for radar-related applications, the analysis

https://doi.org/10.3390/mi17010130

typically focuses on the transmitted and reflected waves, and the absorption is consequently defined as one minus the transmittance and reflectance. In a broader physical sense, however, absorption usually refers to the conversion of electromagnetic energy into other forms of energy within the device. The absorption calculation in metamaterials therefore mixes two distinct contributions: energy dissipated through scattering and energy lost through thermal dissipation. Only the latter corresponds to absorption in the conventional sense. To more accurately describe the physical nature of the calculated absorption, we distinguish these two contributions as scattering absorption and thermal loss absorption, respectively.

In the metamaterial system considered here, the metal is treated as an ideal conductor. As a result, the absorption obtained from the absorption spectrum corresponds purely to scattering absorption.

Figure 4a shows a zero scattering absorption curve (red line) across the entire 0–7 GHz frequency range, representing no scattering absorption for the open ring slit metamaterial, so Equation (3) is applicable in the entire frequency band. The simulated reflectivity curve (blue solid line) fits well with the theoretically calculated results (green dashed line).

However, for the metal strip metamaterial, as depicted in Figure 4b, the scattering absorption curve shows a more complex variation: the no scattering absorption frequencies are below 2.5 THz, rendering Equation (3) applicable to this structure only within a partial frequency range (0–2.5 THz). So the correspondence between the simulated and theoretically calculated results are limited to the no absorption frequency range of 0–2.5 THz, while a notable deviation appears when the frequency range surpasses 2.5 THz. Hence, Equation (3) is verified to be the effect for the no scattering absorption case. However, since Equation (2) contains electric field vectors of all radiation lobes, it is expected to solve the scattering absorption problem.

In the preceding sections, metamaterials with split-ring slit geometries and simple bar-shaped structures were discussed. In what follows, Equation (3) is further verified using a more complex metamaterial configuration. The corresponding meta-atom geometry and its transmission characteristics are shown in Figure 5:

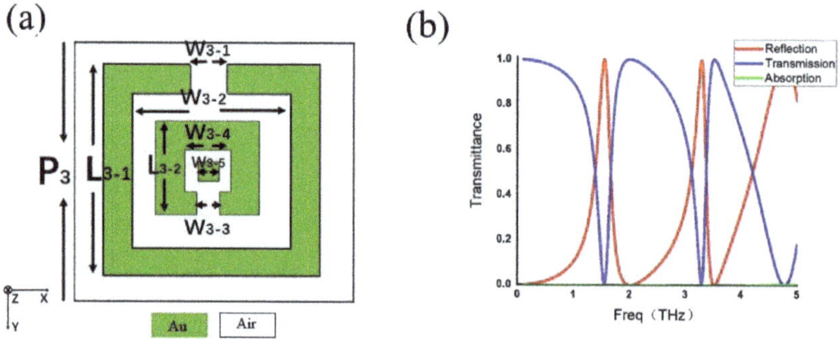

Figure 5. Meta-atom unit-cell geometry and transmission characteristics of the metamaterial. (**a**) $P_3 = 40$ μm, $W_{3-1} = 4.8$ μm, $W_{3-2} = 20.4$ μm, $W_{3-3} = 2.4$ μm, $W_{3-4} = 10.2$ μm, $W_{3-5} = 4$ μm, $L_{3-1} = 30$ μm, and $L_{3-2} = 15$ μm. (**b**) Transmission, reflection, and absorption spectra of the metamaterial.

Figure 5a illustrates the meta-atom structure of this metamaterial, where the green regions denote metal and the white regions denote free space. Owing to the geometric complexity of the structure, multiple resonant features are expected to appear within the frequency spectrum. This expectation is confirmed in Figure 5b, where the red curve represents reflection, the blue curve represents transmission, and the green curve represents

https://doi.org/10.3390/mi17010130

absorption. It can be observed that transmission minima coincide with reflection maxima, while absorption remains negligible over the frequency range of interest (0–5 THz).

According to Equation (3), the reflection can be calculated from the phase of the resonant polariton, and the transmission spectrum can then be obtained accordingly. A comparison between the transmission spectrum calculated from the polariton phase and the directly simulated transmission spectrum is shown in Figure 6.

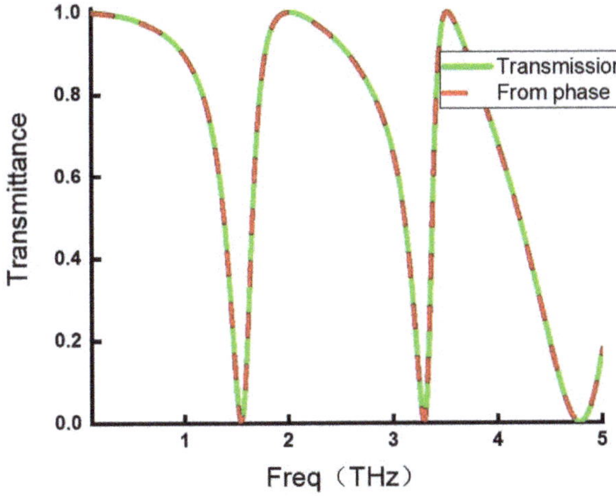

Figure 6. Comparison between the phase-calculated transmission spectrum and the actual transmission spectrum.

In Figure 6, the solid green line represents the directly obtained transmission spectrum, while the dashed red line corresponds to the transmission spectrum calculated from the phase. The two curves exhibit excellent agreement over the entire frequency range, demonstrating the validity of the phase-based transmission calculation. This result directly confirms the correctness of Equation (3) and the MIPRT framework.

To examine whether this agreement is merely coincidental for a specific choice of structural parameters, the overall dimensions of the metamaterial are uniformly scaled by factors of 0.6, 0.8, and 1.2. The corresponding comparisons between the calculated and actual transmission spectra are presented in Figure 7. Figure 7a shows the case with a scaling factor of 0.6, where only a single transmission dip appears within the 0–5 THz range and the calculated transmission curve remains consistent with the actual one. In Figure 7b, corresponding to a scaling factor of 0.8, two transmission dips emerge, and again the calculated and actual transmission spectra coincide well. For the scaling factor of 1.2 shown in Figure 7c, the enlarged meta-atom gives rise to three transmission dips within the considered frequency range, and the calculated transmission curve matches the actual transmission spectrum across the entire band.

In summary, for this geometrically complex metamaterial, the transmission spectra calculated from the resonant polariton phase consistently agree with the actual transmission spectra across multiple structural scaling parameters. This demonstrates that the observed agreement is not accidental but instead reflects the intrinsic validity of Equation (3) and the MIPRT framework. Combined with the results obtained for the two simpler metamaterial configurations discussed earlier, these findings indicate that MIPRT successfully describes electromagnetic wave modulation in simple and complex metamaterials, slit-based structures, and cases with nonzero incidence angles. The amplitude–phase relationships revealed by Equation (3) are consistently satisfied, further indicating that MIPRT provides a

https://doi.org/10.3390/mi17010130

reliable theoretical framework for uncovering the underlying physical principles governing electromagnetic modulation in metamaterials.

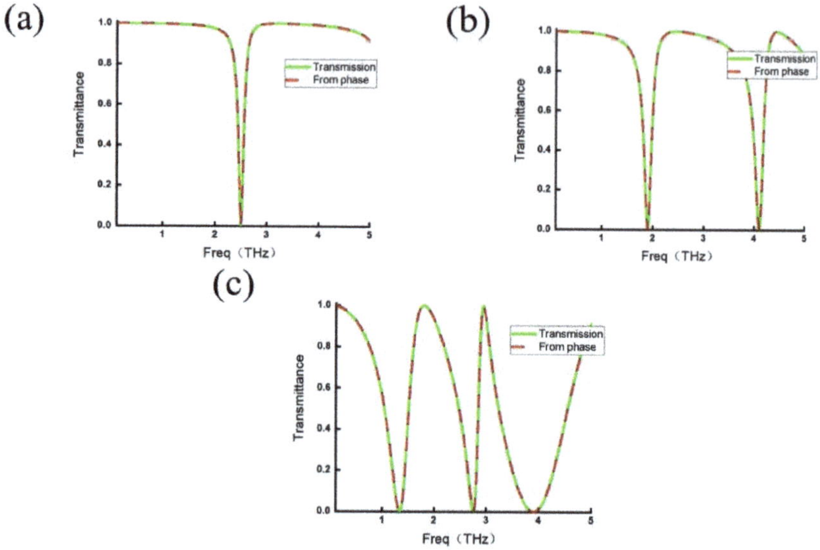

Figure 7. Comparison between the phase-calculated and actual transmission spectra for different scaling factors. (**a**) Scaling factor of 0.6. (**b**) Scaling factor of 0.8. (**c**) Scaling factor of 1.2.

Although the present study is limited to numerical validation and does not include experimental cross-verification, the validation performed is sufficiently comprehensive. It is well established that, when properly configured, numerical simulations show good agreement with experimental results. In this work, the proposed theory is verified across multiple distinct metamaterial configurations, including representative structures drawn from the simulation software's benchmark library, which further confirms the correctness of the simulation settings. These results collectively indicate that experimental validation is not essential for supporting the conclusions of this study.

Nevertheless, future experimental verification, if pursued, could be readily achieved by extracting the phase and amplitude of the reflected wave to directly test the validity of Equation (3). Alternatively, the theory could be examined by retrieving the resonant phase of the additional resonant dipolar field associated with the meta-atoms, although extracting near-field phase and field intensity experimentally remains technically challenging.

In addition, all metamaterial structures considered in this work are assumed to be ideally infinite and periodic. While practical experiments are necessarily limited to finite-sized samples, sufficiently large finite metamaterials can be well approximated as infinite when illuminated by realistically finite sources. Therefore, the proposed model and conclusions remain applicable to finite-sized metamaterial implementations under such conditions.

4. Indirect Verification via the Extinction Theorem

In optical theory, there exists a fundamental result known as the extinction theorem, first established by Ewald [15] and Oseen [16]. This theorem states that, at any point inside a medium, the incident wave is extinguished through destructive interference with the dipole fields induced in the medium and is replaced by another wave that propagates with a different phase velocity. It reveals that the dipole field inside a medium can be regarded

https://doi.org/10.3390/mi17010130

as the superposition of a negative incident field and a secondary wave with a modified propagation speed.

When the medium is replaced by a metamaterial, this concept becomes distorted: the dipoles in the medium are no longer elementary material dipoles but instead transform into the resonant dipoles discussed in the previous sections. Consequently, the essence of the extinction theorem is intrinsically embedded in the MIPRT. In this section, we provide a derivation based on this perspective.

We begin by expressing the fundamental form of the extinction theorem [17] using Green's function $G(r, r')$:

$$\int G(r, r')[-i\omega P(r')]d^3r' = -\vec{E}_i(r) + \vec{E}_{med}(r) \tag{4}$$

where r denotes the observation point, r' the source point, $\vec{E}_i(r)$ the incident field, $\vec{E}_{med}(r)$ the field inside the medium, ω the angular frequency, and $P(r')$ the electric dipole moment density of the medium.

When this formulation is applied to metamaterials, the left-hand side of the equation represents the superposition of radiation emitted by the resonantly excited resonant dipoles:

$$\sum G_{meta}(r, r')(-i\omega)p(r') = -\vec{E}_i(r) + \vec{E}_{med}(r) \tag{5}$$

For a single-layer metamaterial, we assume that the meta-atoms are distributed periodically on the plane $z = 0$. Under this assumption, the spatial and temporal radiation patterns of each meta-atom are identical and differ only by a phase shift determined by their reference position. This leads to

$$(-i\omega)p_0 \sum G_{meta}(r, r')e^{jk_{||}\cdot(r'-r)} = -\vec{E}_i(r) + \vec{E}_{med}(r) \tag{6}$$

$$k_{||}^2 = k_i^2 - k_z^2 \tag{7}$$

where the subscript $||$ denotes the direction parallel to the metamaterial surface and p_0 is a constant representing the resonant electric dipole moment of a meta-atom.

We now examine the modulation of the electromagnetic wave by the metamaterial. Since the observation point is taken to be in the far field, the radiation term on the left-hand side can be decomposed into a plane wave propagating along the principal radiation direction ($A_0 e^{j\vec{k}_{main}\cdot\vec{r}}$) and higher-order diffraction components $\vec{E}_D(r)$, where A_0 is a constant. Thus,

$$A_0 e^{j\vec{k}_{main}\cdot\vec{r}} + \vec{E}_D(r) = -\vec{E}_i(r) + \vec{E}_t(r) \tag{8}$$

According to the definition of resonant dipole radiation introduced earlier, we have

$$\vec{E}_s(r) = A_0 e^{j\vec{k}_i\cdot\vec{r}} + \vec{E}_D(r) \tag{9}$$

Substituting this expression into Equation (8), we obtain

$$\vec{E}_t(r) = \vec{E}_i(r) + \vec{E}_s(r) \tag{10}$$

This equation constitutes the electric field representation of Equation (1). It indicates that, in the transmission region, there exists a virtual incident field component. After coherent destructive interference between this virtual field and the radiation emitted by the meta-atoms, the resulting field manifests as the transmitted wave. In this manner, the phys-

https://doi.org/10.3390/mi17010130

ical meaning of the extinction theorem is incorporated into the metamaterial framework, and the validity of MIPRT is verified through the concept of extinction.

5. Summary and Conclusions

This paper aims to provide a clear and concise insight into the interaction between metamaterials and incident photons. Based on the Invariant Incident Photon Hypothesis and Resonant Phase Deconstruction and Quantification, MIPRT is proposed to clarify the complex modulation process of metamaterials on incident photons. This theory assumes the coexistence of invariant incident photons and resonant dipoles within the metamaterial; then photons emitted from the resonant dipoles will coherently interact with incident photons. The PWP and RDP are defined and quantified to describe the resonant phase modulation process of metamaterials on incident photons. The theoretical relationship between the PWP and reflectivity of the metamaterials is derived by applying MIPRT on a single-layer metamaterial model; the results are verified by simulation.

The MIPRT introduces a new metamaterial research approach and is expected to provide a more detailed and reasonable physical explanation for the interaction between metamaterials and incident photons. In addition, the development of theoretical frameworks can effectively facilitate the expansion of potential applications. In micro- and nanoscale devices, the present theory reveals that the amplitude and phase responses obey trigonometric relationships. This feature suggests that such devices may enable nonlinear functional modulation through appropriate transformations in optical modulation processes. Consequently, this mechanism indicates potential applicability in photonic neural network architectures, where it could serve a role analogous to an activation function.

Supplementary Materials: The following supporting information can be downloaded at: https://www.mdpi.com/article/10.3390/mi17010130/s1.

Author Contributions: B.X. completed the research and writing of the entire article, and R.Z. provided guidance for the research and writing of the article. All authors have read and agreed to the published version of the manuscript.

Funding: This paper was supported by the Sichuan Science and Technology Program (No. 2024NSFSC0465), the National Natural Science Foundation of China (No. 62471119), and the National Key Research and Development Program of China (No. 2018YFF01013001 and 2017YFA0701000).

Data Availability Statement: The data that support the findings of this study are available from the corresponding author upon reasonable request.

Conflicts of Interest: The authors declare no conflicts of interest.

References

1. Luo, X. Sub-Wavelength Electromagnetics. *Front. Optoelectron.* **2016**, *9*, 141–196. [CrossRef]
2. Chen, H.T.; Taylor, A.J.; Yu, N. A Review of Metasurfaces: Physics and Applications. *Rep. Prog. Phys.* **2016**, *79*, 076401. [CrossRef] [PubMed]
3. Yu, N.; Capasso, F. Flat Optics with Designer Metasurfaces. *Nat. Mater.* **2014**, *113*, 139–150. [CrossRef] [PubMed]
4. Grady, N.K.; Heyes, J.E.; Chowdhury, D.R.; Zeng, Y.; Reiten, M.T.; Azad, A.K.; Taylor, A.J.; Dalvit, D.A.; Chen, H.T. Terahertz Metamaterials for Linear Polarization Conversion and Anomalous Refraction. *Science* **2013**, *340*, 1304–1307. [CrossRef] [PubMed]
5. Joannopoulos, J.D.; Villeneuve, P.R.; Fan, S. Photonic Crystals: Putting a New Twist on Light. *Nature* **1997**, *389*, 143–149. [CrossRef]
6. Remnev, M.A.; Klimov, V.V. Metasurfaces:A New Look at Maxwell's Equations and New Ways to Control Light. *Physics-Uspekhi* **2018**, *61*, 157–190. [CrossRef]
7. Abdelrahman, A.H.; Elsherbeni, A.Z.; Yang, F. Transmission Phase Limit of Multilayer Frequency-Selective Surfaces for Transmit array Designs. *IEEE Trans. Antennas Propag.* **2014**, *62*, 690–697. [CrossRef]
8. Costa, F.; Monorchio, A.; Manara, G. Analysis and Design of Ultra Thin EM Absorbers Comprising Resistively Loaded High Impedance Surfaces. *IEEE Trans. Antennas Propag.* **2010**, *58*, 1551–1558. [CrossRef]

9. Baena, J.D.; Bonache, J.; Martín, F.; Sillero, R.M.; Falcone, F.; Lopetegi, T.; Laso, M.A.; Garcia-Garcia, J.; Gil, I.; Portillo, M.F.; et al. Equivalent-Circuit Models for Split-Ring Resonators and Complementary Split-Ring Resonators Coupled to Planar Transmission Lines. *IEEE Trans. Microw. Theory Tech.* **2005**, *53*, 1451–1461. [CrossRef]
10. Kaelberer, T.; Fedotov, V.A.; Papasimakis, N.; Tsai, D.P.; Zheludev, N.I. Toroidal Dipolar Response in a Metamaterial. *Science* **2010**, *330*, 1510–1512. [CrossRef] [PubMed]
11. Yin, X.; Schäferling, M.; Metzger, B.; Giessen, H. Interpreting Chiral Nanophotonic Spectra: The Plasmonic Born-Kuhn Model. *Science* **2013**, *13*, 6238–6243. [CrossRef] [PubMed]
12. Singh, K.; Afzal, M.U.; Esselle, K.P.; Kovaleva, M. Towards Decreasing Side Lobes Produced by Near-Field Phase Gradient Metasurfaces. In Proceedings of the 2019 IEEE International Symposium on Antennas and Propagation and USNC-URSI Radio Science Meeting, Atlanta, GA, USA, 7–12 July 2019.
13. Zaker, R.; Sadeghzadeh, A. Passive Techniques for Target Radar Cross Section Reduction: A Comprehensive Review. *Int. J. RF Microw. Comput. Aided Eng.* **2020**, *30*, e22411. [CrossRef]
14. *COMSOL Multiphysics*, Version 2018. Available online: https://cn.comsol.com/model/frequency-selective-surface-periodic-complementary-split-ring-resonator-15711 (accessed on 1 December 2025)
15. Ewald, P.P. Zur begründung der kristalloptik. *Ann. Phys.* **1916**, *354*, 1–38. [CrossRef]
16. Oseen, C.W. Über die Wechselwirkung zwischen zwei elektrischen Dipolen und üer die Drehung der Polarisationsebene in Kristallen und Flüssigkeiten. *Ann. Phys.* **1915**, *353*, 1–56. [CrossRef]
17. Born, M. *Principles of Optics*, 7th ed.; Cambridge University Press: Cambridge, UK, 1999; Volume 89.

https://doi.org/10.3390/mi17010130

micromachines

MDPI

Article

Intrinsic Relations Between Transmission and Reflection in Metamaterials

Boli Xu and Renbin Zhong *

Terahertz Research Center, School of Electronic Science and Engineering, University of Electronic Science and Technology of China, Chengdu 610054, China; 202211022430@std.uestc.edu.cn
* Correspondence: rbzhong@uestc.edu.cn

Abstract

Metamaterials possess high freedom on structural design, yet their ability to modulate electromagnetic waves is subject to intrinsic constraints that are independent of specific meta-atom geometries. The constraints are revealed by analyzing the statistical amplitudes and phases of transmission and reflection wave in some representative metamaterials. Based on scattering theory, a reconstructed and more general description of the electromagnetic modulation process in metamaterials is established. Two explicit and geometry-independent corollaries concerning the coupling between transmission and reflection waves are further obtained and verified. The results provide a new perspective on the fundamental modulation mechanism of metamaterials on electromagnetic waves.

Keywords: metamaterials; artificial structure; electromagnetic waves; transmission-reflection coupling; amplitude-phase coupling

1. Introduction

In recent years, metamaterials—artificially engineered materials composed of designed meta-atoms—have become widely recognized [1]. Among them, electromagnetic metamaterials, which are employed to manipulate various properties of electromagnetic waves, occupy a prominent position [2]. Owing to their high design freedom and flexible electromagnetic control, electromagnetic metamaterials have been applied across diverse fields. Representative examples include absorbers used as stealth layers in antenna and radar applications, such as the absorber proposed by Yong Zhi Cheng [3] and by Long Yang [4]; optical imaging devices, including the invisibility cloak demonstrated by D. Schurig et al. [5] and the perfect lens with focusing functionality proposed by Wei-Ting Chen et al. [6]; and information-related applications, such as information-carrying metamaterials proposed by Tie Jun Cui [7], intelligent programmable metamaterials developed by Qian Ma et al. [8], and multilayer metasurfaces functioning as physical neural networks introduced by Che Liu et al. [9].

Across these seemingly diverse metamaterial devices, several non-negligible yet peculiar characteristics are implicitly manifested. For instance, in the frequency-selective surface (FSS) designed by Yunus E. et al. that enhances the performance of broadband reconfigurable antenna apertures [10], all reflection peaks are found to share the same phase value of π. In particular, for reflection spectra exhibiting three distinct resonance peaks, the phase of each peak remains π, hinting that reflection maxima in such metamaterials are associated with a fixed phase value.

This behavior is not an isolated case. In the dielectric Huygens' surfaces designed by Manuel Decker et al., the transmission peak also occurs at a specific phase of π [11]. Similarly, for the rod antenna metasurface designed by Nanfang Yu et al., the transmission peak corresponds to a phase of $\pi/2$ [12]. Notably, in all these cases, the peak phase appears near the midpoint of the overall phase distribution. Furthermore, in the birefringent reflect array metasurface proposed by Mohsen Farmahini-Farahani et al., it is observed that when structural parameters are varied, the transmission peak shifts in frequency and the zero-phase point shifts accordingly [13]. These observations collectively suggest that transmission/reflection amplitudes and their corresponding phases are constrained by an underlying relationship.

Consistent with this intrinsic constraint, the transmission phase of single-layer metamaterials is typically confined within the range of $-90°$ to $90°$, and phases beyond this range are observed only in sufficiently thick structures where multi-interface effects become significant [14]. Correspondingly, reported phase-gradient metamaterials are almost exclusively realized using multilayer architectures [15].

All these features indicate that a single-layer metamaterial does not possess unlimited capability in modulating electromagnetic waves. Instead, the achievable amplitudes and phases of transmitted and reflected waves must satisfy certain intrinsic rules that cannot be broken by structural variations alone, unless a single-layer structure is extended to a multilayer configuration with multiple coupled resonances.

Existing descriptions based on S-parameters, transmission-line models, and transfer-matrix methods [16] mainly impose general constraints such as energy conservation and reciprocity or relate transmission and reflection indirectly through intermediate parameters (e.g., effective impedance), and they are typically associated with multilayer coupling effects. However, an explicit and direct constraint on the amplitudes and phases of transmitted and reflected waves at the single-layer level has not been clearly established. These rules have not yet been explicitly identified nor recognized as possessing independent physical significance.

In this study, these intrinsic rules are explicitly identified and formulated as direct coupling relations between the amplitudes and phases of transmitted and reflected waves at the single-layer level.

To address this gap, in this study, we systematically summarize the value distributions of reflection and transmission waves in metamaterials, demonstrating that the amplitudes and phases of incident, transmitted, and reflected waves are not arbitrary. While these distributions appear distinctive, they remain fully consistent with the scattering theory, with periodic structures making these effects more evident for analysis. The interaction between metamaterials and electromagnetic waves is systematically analyzed using Green's function, allowing us to provide a reconstructed and more general description of the metamaterial-based electromagnetic modulation, which can deepen the understanding of metamaterial behavior and enrich the fundamental theory of metamaterials.

Moreover, two explicit corollaries are derived: an amplitude-phase relation for reflected and transmitted waves and a transmission–reflection relation, with their physical significance, observable phenomena, and potential applications analyzed and the corollaries validated using several representative metamaterials.

This study provides a clearer physical understanding of wave modulation in micro-/nano-structured media, which may offer theoretical guidance for the design of micro/nano electromagnetic devices such as metasurfaces, nano-resonators, and integrated photonic components.

https://doi.org/10.3390/mi17040493

2. Relation Analysis of the Amplitudes and Phases of Reflection and Transmission Coefficients

In this section, three representative metamaterial structures are selected for an analysis of the amplitudes and phases of their reflection and transmission coefficients. The meta-atom geometries are shown in Figure 1. The green regions denote the metallic components (Au), while the white regions represent air.

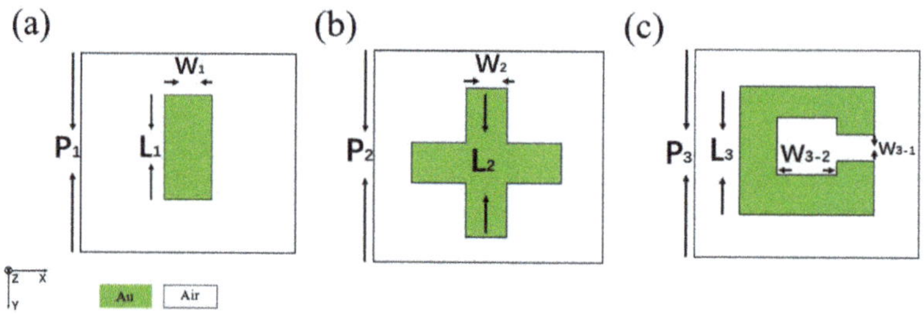

Figure 1. Unit structures of metamaterial. (**a**) Strip-shaped metallic meta-atom with $P_1 = 80$ μm, $L_1 = 50$ μm, and $W_1 = 8$ μm. (**b**) Cross-shaped meta-atom with $P_2 = 80$ μm, $L_2 = 50$ μm, and $W_2 = 8$ μm. (**c**) Split ring meta-atom with $P_3 = 40$ μm, $L_3 = 15$ μm, $W_{3\text{-}1} = 2.4$ μm, and $W_{3\text{-}2} = 10.2$ μm.

For the strip-shaped metallic meta-atom metamaterial shown in Figure 1a, the spectra of the transmission and reflection coefficients, their corresponding phase responses, and the transmission and reflection efficiencies can be obtained by a simulation, as presented in Figure 2. The red curves denote reflection, while the blue curves denote transmission. Figure 2a,c show the transmission/reflection coefficients and transmission/reflection efficiencies; it can be observed that the positions of the transmission minima coincide with those of the reflection maxima, which is an evident consequence of energy conservation. As for the phase responses shown in Figure 2b, the transmission phase is confined within a range from −90° to +90°, whereas the reflection phase is distributed within intervals of −180° to −90° and 90° to 180°. They have a similar variation trend along frequencies, and the absolute phase difference between them remains constant, revealing nontrivial behavior.

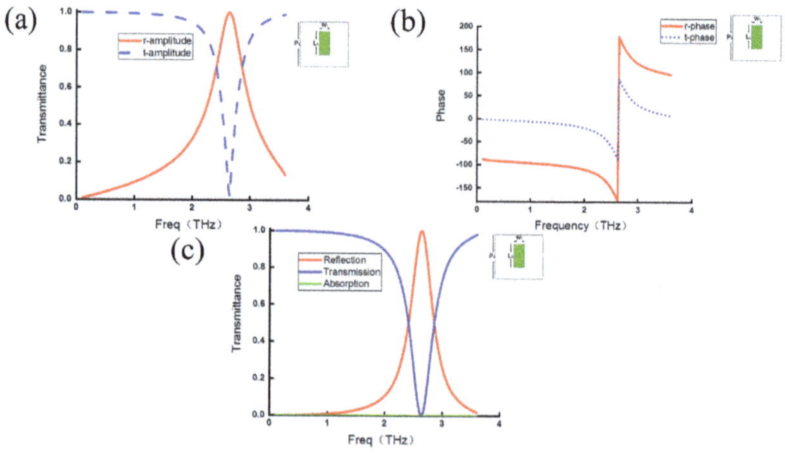

Figure 2. Strip-shaped metallic metamaterials. (**a**) Reflection coefficient amplitude curve (red line) and transmission coefficient amplitude curve (blue line). (**b**) Reflection coefficient phase curve (red line) and transmission phase coefficient curve (blue line). (**c**) Reflection efficiency curve (red line), transmission efficiency curve (blue line), and absorption efficiency curve (green line).

https://doi.org/10.3390/mi17040493

In addition, abrupt phase changes (when the phase is within the interval from −180° to 180°) are observed at the transmission minima/reflection maxima. A specific relationship between phases and the corresponding amplitudes can be observed. Following a cosine relationship, the transmission phase reaches ±90° at zero amplitude, while the reflection phase reaches ±180° at unit amplitude.

The same analysis is performed for the other two metamaterial structures, with the results shown in Figures 3 and 4. The amplitudes and phases of the transmitted and reflected waves remain consistent with the previously identified characteristics, indicating that these features are universal.

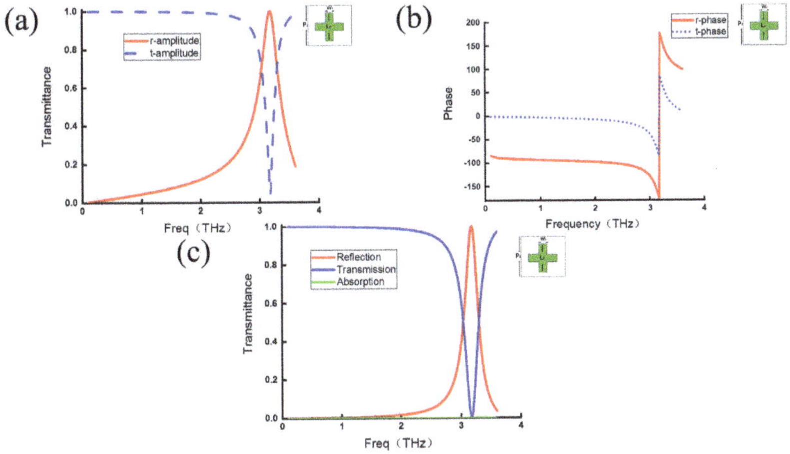

Figure 3. Cross-shaped metamaterials. (**a**) Reflection coefficient amplitude curve (red line) and transmission coefficient amplitude curve (blue line). (**b**) Reflection coefficient phase curve (red line) and transmission phase coefficient curve (blue line). (**c**) Reflection efficiency curve (red line), transmission efficiency curve (blue line), and absorption efficiency curve (green line).

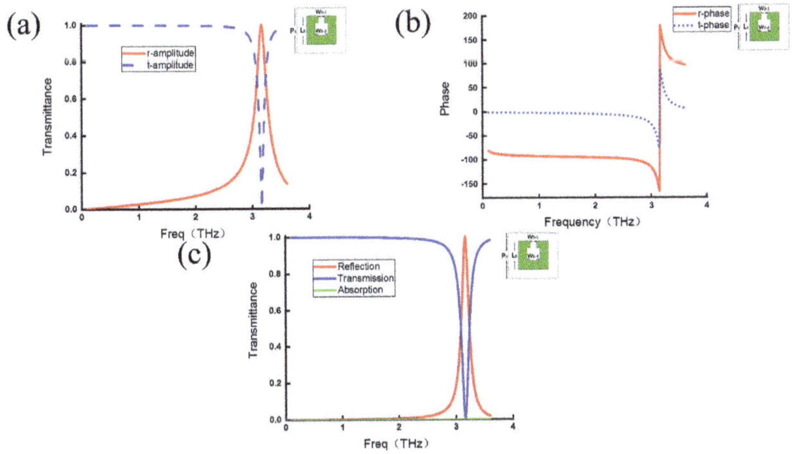

Figure 4. Split ring metamaterials. (**a**) Reflection coefficient amplitude curve (red line) and transmission coefficient amplitude curve (blue line). (**b**) Reflection coefficient phase curve (red line) and transmission phase coefficient curve (blue line). (**c**) Reflection efficiency curve (red line), transmission efficiency curve (blue line), and absorption efficiency curve (green line).

https://doi.org/10.3390/mi17040493

To illustrate the distributions and relationships of the amplitudes and phases of the transmission/reflection coefficients more clearly, the coefficients within the relevant frequency band are described in the complex plane, as shown in Figure 5. The blue dots correspond to reflection coefficients, while green dots correspond to transmission coefficients for a different frequency. All the above-mentioned three metamaterial structures exhibit identical features: the reflection coefficients form a circular distribution on the left side of the imaginary axis, while the transmission coefficients form a circular distribution on the right side, further confirming the cosine-type relationship between the amplitudes and phases of the transmission/reflection coefficients intuitively.

Furthermore, three representative frequency points are marked in Figure 5: 2 THz (yellow arrows), 2.6 THz (red arrows), and 3 THz (black arrows). A notable property can be observed: at a certain frequency, the arrow of the reflection coefficient in the blue circle is always perpendicular to that of the transmission coefficient in the green circle, indicating their phase difference—maintaining a fix value of 90°. Apparently, these phenomena are not accidental but rather an inevitable consequence of electromagnetic wave modulated by metamaterials.

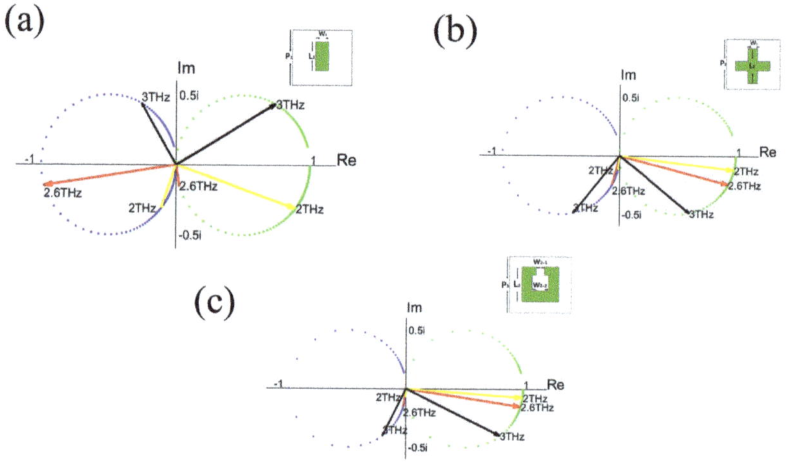

Figure 5. The complex plane distribution diagram of the transmission and reflection coefficients. (**a**) Strip-shaped metallic metamaterials. (**b**) Cross-shaped metamaterial. (**c**) Split ring metamaterials.

From a physical perspective, both transmitted and reflected electromagnetic waves can be regarded as originating from resonant dipoles excited by the incident wave in metamaterials. It can be defined as material scattering phenomena and be described by Green's function formulations in electromagnetism.

The scattering theory already contains nontrivial first-order field correlations beyond energy-conservation constraints. A representative example is the extinction theorem [17–19], which shows that the internal field arises from dipole radiation that simultaneously cancels the incident wave in free space and generates a propagating field inside the medium. This directly implies intrinsic first-order relations among incident, reflected, and transmitted waves.

The correlations identified between the transmission and reflection coefficients of metamaterials therefore originate from fundamental scattering physics encoded in the Green's function framework. While such effects are typically negligible in conventional materials, metamaterials amplify them to an observable level, making their physical significance explicit and requiring a systematic theoretical formulation.

https://doi.org/10.3390/mi17040493

3. Derivation of Field Relations and Physical Picture of Modulation

3.1. Relation Derivation Based on Green's Function

Fundamental equations governing the observed relations in metamaterials can be derived from the Green's function. The basic solution expression for the Green's function is given in [19]:

$$U(\boldsymbol{r}) = U^{(i)}(\boldsymbol{r}) + \int U(\boldsymbol{r}')F(\boldsymbol{r}')G(\boldsymbol{r}-\boldsymbol{r}')d^3\boldsymbol{r}' \tag{1}$$

where U denotes a specific component of the total electric field, $U^{(i)}$ denotes a specific component of the incident electric field, $\boldsymbol{r}$ represents the observation position in the observer's reference frame, $\boldsymbol{r}'$ denotes the source position, F is the scattering potential of the material, and G is the Green's function.

Expanding Equation (1) with the Liouville–Neumann series and retaining its integral representation [19], one obtains:

$$\begin{aligned} U(\boldsymbol{r}) = {} & U^{(i)}(\boldsymbol{r}) + \int U^{(i)}(\boldsymbol{r}')F(\boldsymbol{r}')G(\boldsymbol{r}-\boldsymbol{r}')d^3\boldsymbol{r}' \\ & + \int\int U^{(i)}(\boldsymbol{r}')F(\boldsymbol{r}')G(\boldsymbol{r}'-\boldsymbol{r}'')F(\boldsymbol{r}'')G(\boldsymbol{r}-\boldsymbol{r}'')d^3\boldsymbol{r}'d^3\boldsymbol{r}'' + \cdots \end{aligned} \tag{2}$$

This expression explicitly decomposes multiple scattering processes into successive orders: the first term represents zeroth-order scattering, the second term is first-order scattering, the third term is second-order scattering, and so forth.

When Equation (2) is applied to a single-layer metamaterial, each scattering order can be attributed to contributions of individual meta-atoms. Treating each meta-atom as a scattering unit and assuming a lattice period p, Equation (2) can be rewritten as:

$$\begin{aligned} U(\boldsymbol{r}) = {} & U^{(i)}(\boldsymbol{r}) + \sum U^{(i)}(np)F_{meta}(np)G_{meta}(\boldsymbol{r},np) \\ & + \sum\sum U^{(i)}(np)F_{meta}(np)G_{meta}(np,mp)F_{meta}(mp)G_{meta}(\boldsymbol{r},mp) + \cdots \end{aligned} \tag{3}$$

In Equation (3), the second term represents the first-order response of the meta-atoms, the third term represents the second-order response, and so forth.

By explicitly isolating the radiation emitted by the meta-atoms, the total radiated field can be expressed as the superposition of radiation from meta-atoms located at positions np:

$$U(\boldsymbol{r}) = U^{(i)}(\boldsymbol{r}) + \sum A_n F_{meta}(np)G_{meta}(\boldsymbol{r},np) \tag{4}$$

Here:

$$A_n = U^{(i)}(np) + \sum U^{(i)}(mp)F_{meta}(np)G_{meta}(np,\ mp) + \cdots \tag{5}$$

The first term of A_n represents the direct excitation of the n-th meta-atom by the incident wave, while the second term represents the first-order excitation of the n-th meta-atom by the m-th meta-atom, and so forth. Therefore, A_n represents the total excitation source of the n-th meta-atom.

If all meta-atoms are identical, the product $A_n F_{meta}$ is the same for each meta-atom and can be treated as a constant. For oblique incidence, additional spatial phase factors associated with the incident angle must be included. For simplicity, normal incidence is assumed, and Equation (4) can be rewritten as:

$$U(\boldsymbol{r}) = U^{(i)}(\boldsymbol{r}) + A_n F_{meta} \sum G_{meta}(\boldsymbol{r},np) \tag{6}$$

https://doi.org/10.3390/mi17040493

Applying the Floquet's theorem to the periodic sources [20], the field can be expressed as:

$$\sum G_{meta}(\boldsymbol{r}, np) = \sum B_m e^{j\vec{k}_{\perp,\,m}\cdot\vec{r}_{\perp}} e^{j\vec{k}_{\|,\,m}\cdot\vec{r}_{\|}} \tag{7}$$

Here:

$$B_m = \frac{1}{p}\int G_{meta}\left(\vec{r}_{\|}, \vec{r}_{\perp}\right)e^{-j\vec{k}_{\|}\cdot\vec{r}_{\|}}d\vec{r}_{\|} \tag{8}$$

$$\left|\vec{k}_{\|,\,m}\right| = \frac{2\pi m}{p} \tag{9}$$

$$\vec{k}_{\perp,\,m} = \pm\sqrt{\left|\vec{k}\right|^2 - \left|\vec{k}_{\|,\,m}\right|^2} \tag{10}$$

where $\vec{k}$ is the incident wave vector. The directions parallel and perpendicular to the metamaterial array are denoted by $\|$ and $\perp$, respectively. B_m denotes the amplitude of the m-th order Floquet plane wave determined by meta-atom radiation, and m is an integer.

We assume the incident wavelength is larger than the lattice period. In the case where only the fundamental resonance is considered and sidelobe radiation is negligent, then only the zeroth-order ($m = 0$) Floquet mode has contribution in the far field, and Equation (6) reduces to:

$$U(\boldsymbol{r}) = U^{(i)}(\boldsymbol{r}) + A_n B_0 F_{meta} e^{\pm j\vec{k}\cdot\vec{r}} \tag{11}$$

We assume that all meta-atoms are identical single-layer, thin metallic structures. Under this condition, the effective scattering response F_{meta} is identical for all meta-atoms and is symmetric in the two propagation directions. If the incident wave propagation direction is positive (+) (the opposite direction denoted as negative (−)), Equation (11) can be written as:

$$U^{+} = U^{(i)} + U_0^{(s)} e^{j\vec{k}_{+}\cdot\vec{r}_{+}} \tag{12}$$

$$U^{-} = U_0^{(s)} e^{j\vec{k}_{-}\cdot\vec{r}_{-}} \tag{13}$$

Here, $U_0^{(s)}$ denotes the total radiation strength of each meta-atom after accounting for multiple excitation, which is given by:

$$U_0^{(s)} = A_n B_0 F_{meta} \tag{14}$$

Recognizing that the positive propagating electromagnetic field is the transmitted wave, while that along the negative direction is the reflected wave, the governing equations describing electromagnetic wave modulation by the metamaterial can be obtained:

$$t = i + s_{+} \tag{15}$$

$$r = s_{-} \tag{16}$$

where t, r, and i denote the transmitted, reflected, and incident electric fields, respectively, and s_{+} and s_{-} represent the radiation fields emitted by the metamaterial into the opposite directions.

According to Equations (11)–(13), it is evident that resonant radiation must be symmetric in opposite directions, implying opposite momenta and equal energy in the positive

https://doi.org/10.3390/mi17040493

and negative direction. Thus, the constraint conditions associated with the modulation can be derived as:

$$|s_-| = |s_+| = s \tag{17}$$

$$s_+(r) = s_-(-r) \tag{18}$$

Rewrite Equation (15) as:

$$s_+ = t + (-i) \tag{19}$$

It shows that the electromagnetic radiation emitted by the metamaterial consists of two components: $-i$, which exactly cancels the incident field, and t, which propagates forward as the transmitted wave. This result is fully consistent with the classical extinction theorem, which is here reformulated with metamaterials as the scattering object.

3.2. Reconstructed Description of Electromagnetic Wave Modulation by Metamaterials

Based on the modulation relations Equations (15)–(18), this section reconstructs the description of the electromagnetic wave modulation process in metamaterials. As illustrated in Figure 6, we assume a single-layer metamaterial divided the whole space into an incident half-space and a transmission half-space. The half-space toward which the incident wave propagates to metamaterials is defined as the incident half-space (upper half-space in Figure 6), while the opposite region is defined as the transmission half-space. Figure 6a is the conventional description of electromagnetic modulation by metamaterials, and Figure 6b is the reconstructed description from the perspective of the scattering theorem.

In the traditional description, an incident field i impinges on the metamaterial, which splits the incident wave into a transmitted wave t and a reflected wave r. Consequently, the field in the incident half-space consists of the incident wave i and the reflected wave r propagating in opposite directions, while the field in the transmission half-space contains only the transmitted wave t. Within this picture, the amplitudes and phases of the three waves imply a possible design freedom that may allow for independent control through structural engineering. However, as shown above, the incident, reflected, and transmitted waves are mutually constrained, with their amplitudes and phases subject to intrinsic limitations. Therefore, the conventional description requires refinement to incorporate these constraints.

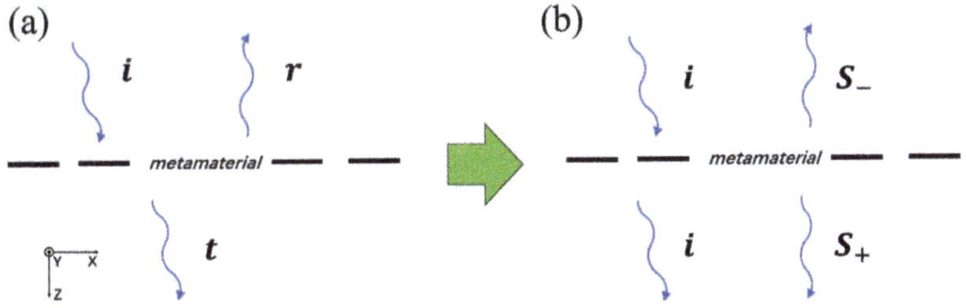

Figure 6. Descriptive diagram of electromagnetic wave modulation by metamaterials. (**a**) Traditional description. (**b**) Reconstructed description.

According to the governing equations describing metamaterial modulation (Equations (15) and (16)) and the associated constraint conditions (Equations (17) and (18)), the modulation process can be reformulated as shown in Figure 6b. It can be interpreted as the incident field i being distributed throughout the entire space. Under the incident field,

https://doi.org/10.3390/mi17040493

the metamaterial radiates, via its induced resonant dipoles, equal-amplitude, equal-phase fields into the two half-spaces in opposite directions. These radiated fields are denoted as s_+ and s_-. As a result, the total field in the incident half-space is the superposition of the incident field i and the radiated field s_-, while that in the transmission half-space is the superposition of the incident field i and the radiated field s_+.

Compared with the conventional picture, the reconstructed description retains it as a special case while rendering the intrinsic constraints explicit. The conventional representation follows directly from Equation (16), where the correlation of the radiated fields encodes the mutual relations among i, r, and t.

Furthermore, from the perspective of energy conservation, the difference between the sum of the individual energies of the incident field i and the radiated field s_+ and the energy of their coherently superposed field must be equal to the energy carried by the radiated field s_- (or r) plus the thermal loss associated with the resonant dipoles. This condition imposes strict constraints on the relative amplitude and phase of the radiated field s_+ with respect to the incident field i. These constraints ultimately manifest as the amplitude–phase relations between the transmitted wave t and the reflected wave r.

The present description is established for electrically resonant metamaterials. In addition to electric resonances, metamaterials can also support magnetic resonances. For magnetically resonant metamaterials, a possible approach is to decompose the structure into coupled multilayer units and then apply the present description to each individual layer. This strategy provides a potential route toward constructing a corresponding description for magnetically resonant metamaterials.

The present description is established for single-layer metallic resonant metamaterials. In addition to this class, there also exist single-layer dielectric resonant metamaterials. For the latter, the two interfaces of the dielectric layer can be regarded as a coupled bilayer system. Accordingly, such metamaterials can be decomposed into multiple coupled layers through a multilayer coupling expansion, after which the present description may be applied to each individual layer. This approach provides a feasible route toward constructing a corresponding description for dielectric resonant metamaterials.

4. Two Corollaries Based on the Reconstructed Description

4.1. Coupling Relation Between the Transmitted and Reflected Waves

From the phenomena discussed in Section 2 and the reconstructed description presented in Section 3, it is evident that the transmitted and reflected electric fields satisfy a definite and stringent relationship. In this subsection, this relationship is explicitly derived.

Starting from Equations (15) and (16), one obtains:

$$t(\infty) = i(\infty) + s_+(\infty) \tag{20}$$

$$r(-\infty) = s_-(-\infty) \tag{21}$$

Further combining Equations (17) and (18) yields:

$$t(\infty) = i(\infty) + r(-\infty) \tag{22}$$

By expanding the position variables into spatial phase factors, the expression can be rewritten as:

$$|t|e^{j\angle t}e^{jk\infty} = |i|e^{jk\infty} + |r|e^{j\angle r}e^{-jk(-\infty)} \tag{23}$$

https://doi.org/10.3390/mi17040493

where $\angle t$ and $\angle r$ denote the phase shifts introduced by the metamaterial, while $k\infty$ and $k(-\infty)$ represent the phase accumulation associated with spatial propagation. The reference phase is defined such that the incident field i has zero phase at the position of the metamaterial. Equation (23) can be rewritten as:

$$|t|e^{j\angle t} = |i| + |r|e^{j\angle r} \tag{24}$$

By normalizing the amplitude of the incident wave, t and r can be identified as the transmission and reflection coefficients, respectively:

$$t = 1 + r \tag{25}$$

This expression shows that, for a single-layer metamaterial, the transmission coefficient differs from the reflection coefficient only by a unit offset in the real part. In other words, the transmission and reflection coefficients are not independent quantities but are rigidly coupled through this relation. This result demonstrates that, at the field level, transmitted and reflected waves in a single-layer metamaterial obey a strict relation that does not depend on the specific geometry or morphology of the meta-atoms.

In contrast, in conventional approaches such as S-parameter constraints, transmission-line models, and transfer-matrix methods, the transmission and reflection coefficients are typically related indirectly through intermediate quantities (e.g., effective impedance) and are simultaneously constrained by general principles such as energy conservation. However, these approaches do not provide an explicit and direct intrinsic coupling relation between the transmission and reflection coefficients.

4.2. Amplitude–Phase Relations of the Reflected and Transmitted Waves

Let the absorption capability of the metamaterial be characterized by the absorption rate A. From energy conservation, one obtains:

$$tt^* + rr^* = 1 - A \tag{26}$$

Substituting Equation (25) into Equation (26) yields:

$$r + r^* + 2\,rr^* + A = 0 \tag{27}$$

By expressing the reflection coefficient as $r = |r|e^{j\angle r}$ and substituting it into Equation (27), the expression can be simplified to:

$$|r| = \frac{-cos(\angle r) \pm \sqrt{cos^2(\angle r) - 2A}}{2} \tag{28}$$

Here, both signs $\pm$ correspond to mathematically valid solutions. The physically admissible branch can be determined by the scattering characteristics of the system. Specifically, the negative branch applies when no side lobes are present in the scattering pattern, whereas the positive branch is selected when side lobes appear.

Therefore, the branch selection is not arbitrary but is directly linked to the presence or absence of higher-order scattering features. This provides a clear and practical criterion for identifying the physically relevant solution. The detailed physical mechanism underlying this correspondence will be investigated in future work.

https://doi.org/10.3390/mi17040493

When $A = 0$, only one nonzero solution remains, and the other branch can be unambiguously excluded as a spurious solution. In this case, one obtains:

$$|r| = -cos(\angle r) \tag{29}$$

Following the same procedure, the corresponding relation for the transmitted wave can be derived:

$$|t| = \frac{cos(\angle t) \pm \sqrt{cos^2(\angle t) - 2A}}{2} \tag{30}$$

For the transmission coefficient, the branch selection does not need to be specified independently. Once the physically admissible branch of the reflection coefficient is determined (e.g., based on the presence or absence of side lobes), the corresponding transmission branch is uniquely fixed by the intrinsic transmission–reflection coupling relation derived in this study.

Physically, this reflects the different roles of the two channels: the reflected wave is directly governed by the resonant response of the meta-atoms, leading to a clear branch selection criterion, whereas the transmitted wave results from the coherent superposition of the incident field and the resonant scattering field. As a result, its branch is not independently chosen but is constrained by the overall coupling relation.

When $A = 0$, this relation reduces to:

$$|t| = cos(\angle t) \tag{31}$$

The above results demonstrate that, in single-layer metamaterials, the reflected and transmitted waves strictly obey Equations (28) and (30). In the absence of absorption, both relations reduce to cosine-type dependencies (Equations (29) and (31)), which are fully consistent with the analysis presented in Section 2. This consistency indicates that, for both transmission and reflection, the phase and amplitude are in one-to-one correspondence and can only take restricted values. Importantly, this correspondence is independent of the specific geometry of the meta-atoms.

In contrast, in conventional approaches such as S-parameter constraints, transmission-line models, and transfer-matrix methods, the amplitude and phase of the transmission and reflection waves are typically determined through intermediate variables or structural parameters. Although phase information can be obtained in transfer-matrix formulations, it mainly reflects inter-layer coupling in multilayer systems and does not provide an explicit amplitude–phase coupling relation for a single-layer metamaterial.

Therefore, these methods do not establish a direct and intrinsic correspondence between amplitude and phase. By comparison, the present results reveal a one-to-one amplitude–phase relation with restricted values, representing a more explicit and geometry-independent constraint at the field level.

4.3. Complex-Plane Representation and Potential Implications

To clearly illustrate the relationship between the reflected and transmitted waves, this section adopts a complex-plane representation to visualize their intrinsic correlation. For the reflection coefficient, using the phase $\angle r$ and the amplitude relation $|r| = -cos(\angle r)$, the vector representation of the reflection coefficient r in the complex plane is constructed and referred to as the reflection vector. Similarly, based on $\angle t$ and $|t| = cos(\angle t)$, the vector representation of the transmission coefficient t is defined as the transmission vector. The resulting geometric construction is shown in Figure 7.

https://doi.org/10.3390/mi17040493

In Figure 7, $\vec{OA}$ denotes the reflection vector, $\vec{OB}$ denotes the transmission vector, and $\vec{OC}$ is the unit vector along the real axis. The line l_3 passes through points A and B, l_2 passes through points C and B, and l_1 passes through points A and O. The red circle represents the admissible values of the reflection vector, while the blue circle represents the admissible values of the transmission vector.

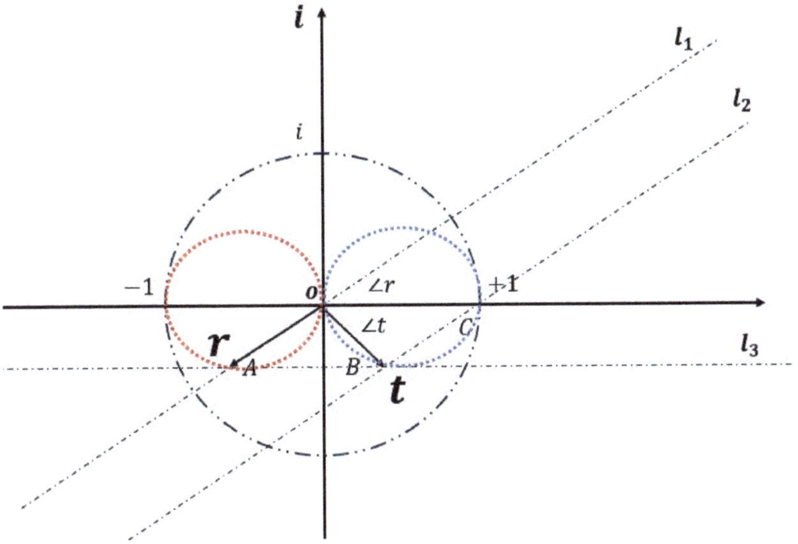

Figure 7. The complex plane distribution diagram of the relationship between the transmission and reflection coefficients.

From Equation (25), one obtains:

$$\vec{OB} = \vec{OA} + \vec{OC} \tag{32}$$

According to the parallelogram law of vector addition, the quadrilateral OABC is a parallelogram. Consequently, $\vec{AB} = 1$ and $\angle r = \angle OAB$. Considering the relation $|r| = -cos(\angle r)$, one finds:

$$\vec{OA} = cos(\angle OAB) = \vec{AB}cos(\angle OAB) \tag{33}$$

Ecidently, $\angle AOB$ is a right angle, indicating that the reflection and transmission vectors are orthogonal to each other. In other words, the reflection and transmission coefficients differ in phase by 90°. This result is fully consistent with the phenomenon revealed in Section 2.

In summary, the reflection coefficient is confined to a circle in the complex plane whose diameter lies on the negative real axis from 0 to −1 (the red circle in Figure 7), while the transmission coefficient lies on a circle whose diameter spans from 0 to 1 on the real axis (the blue circle in Figure 7). Moreover, the two vectors are strictly orthogonal.

This distribution indicates that, for single-layer metamaterials, the accessible phase range becomes limited when the amplitude approaches high values. Breaking this constraint requires the introduction of multiple resonant couplings, such as interlayer coupling in multilayer metamaterials, where more complex behaviors are expected and remain to be explored.

https://doi.org/10.3390/mi17040493

Based on these intrinsic relations between reflection and transmission, several potential research directions can be envisaged. Specifically, the deterministic coupling between the two channels implies that they are not independently tunable but intrinsically constrained by the same underlying relations. This feature suggests that one channel (e.g., transmission) can be designed to achieve a target functionality, while the other (e.g., reflection), being inherently linked, may serve as a correlated signal for monitoring or feedback.

From this perspective, the metamaterial system can be interpreted as a self-constrained system, where the output channels are intrinsically coupled rather than independently controlled. This property may provide a new perspective for designing information metamaterials and intelligent metamaterials, in which internal correlations between channels could be exploited for feedback or regulation. A schematic illustration is provided in Figure 8a, where an incident wave interacts with the metamaterial; the reflected wave can be utilized as a modulation channel, while the transmitted wave, due to its intrinsic relation with the reflection, may serve as a feedback signal to guide the system response.

Alternatively, since the phase and amplitude are constrained by trigonometric relations, the output fields are not independently variable but follow a coupled mapping determined by these relations. This intrinsic coupling introduces a structured dependence between the input and output, which may effectively act as a built-in nonlinear mapping.

From this viewpoint, such amplitude–phase coupling could be utilized in nonlinear operations. In particular, it may provide a possible mechanism for realizing activation-like responses in physical neural networks (also known as diffractive neural networks), or for nonlinear information encoding in information metamaterials. A schematic diagram is shown in Figure 8b, where the incident wave first passes through an input plane and undergoes linear transformations within a diffractive neural network. Subsequently, a properly designed metamaterial layer, governed by the intrinsic amplitude–phase coupling, may function as an activation-like element to introduce nonlinearity, and the resulting electromagnetic field is then collected at the output plane.

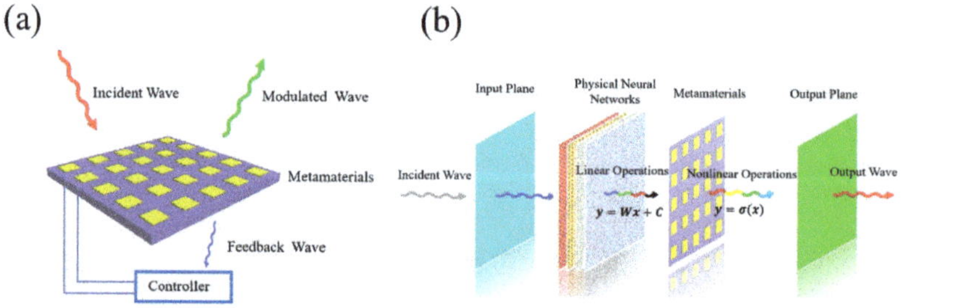

Figure 8. Schematics illustrating the potential applications of the coupling relationships. (**a**) Intelligent metamaterials and information metamaterials. (**b**) Diffractive neural networks.

5. Case Studies for Verifying the Proposed Description

To verify the newly proposed description of the electromagnetic-wave modulation process in metamaterials, this section examines several consequences derived from the preceding analysis and validates them through specific examples.

The verification strategy is as follows: For a given metamaterial, the transmission–reflection relation expressed in Equation (25) is used to calculate the transmittance from the reflection coefficient, while the amplitude–phase relation in Equation (30) is used to calculate the transmittance from the phase. The transmittance is defined as $T = tt^*$.

From Equation (25), one obtains:

$$T = 1 + r + r^* + rr^* \quad (34)$$

https://doi.org/10.3390/mi17040493

From Equation (30), one obtains:

$$T = \left(\frac{cos(\angle t) \pm \sqrt{cos^2(\angle t) - 2A}}{2} \right)^2 \quad (35)$$

To avoid ambiguity associated with the elimination of spurious solutions, only the case of $A = 0$ is considered. In this situation, the alternative solution reduces to zero and can be naturally excluded by selecting the nonzero solution:

$$T = cos^2(\angle t) \quad (36)$$

Four representative single-layer metamaterials are selected for discussion. Each metamaterial features a distinct meta-atom configuration: one is a simple strip-shaped structure (Figure 9a); the second is a more complex I-shaped structure (Figure 10a); the third is a split ring structure with an additional current-loop response (Figure 11a); and the last is a slit-type structure that differs fundamentally from the previous three (Figure 12a). These four configurations can represent metallic inclusion structures and their complementary aperture structures, simple and complex geometries, and straight and split ring structures. For the simple strip-shaped metamaterial, an oblique incidence of 30° is considered to extend the analysis beyond the limitation of normal incidence.

The transmission spectra of the four structures are shown in Figures 9b, 10b, 11b, and 12b. In these figures, the red curve represents reflectance, the blue curve represents transmittance, and the green curve represents absorptance, which is calculated as $1 - R - T$ and characterizes the energy not distributed into either the reflected or transmitted directions.

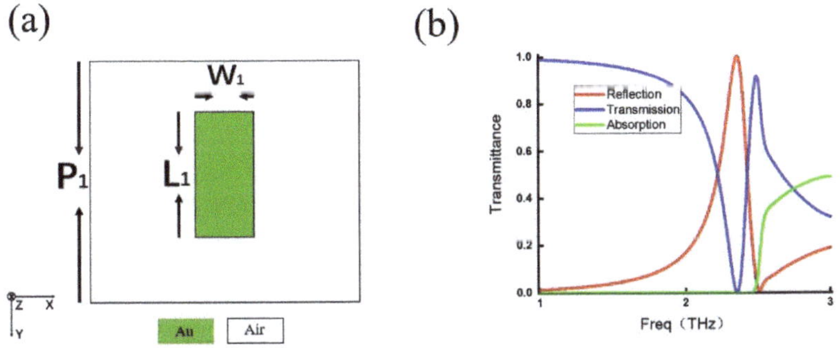

Figure 9. (**a**) Strip-shaped meta-atom with $P_1 = 80$ μm, $L_1 = 50$ μm, and $W_1 = 8$ μm. (**b**) Reflection efficiency curve (red line), transmission efficiency curve (blue line), and absorption efficiency curve (green line) with incidence of 30°.

For each metamaterial, Figure 13 compares three transmittance curves: the transmittance derived from the reflection coefficient (Equation (34)), that calculated from the phases (Equation (36)), and the directly simulated transmission.

It can be seen that the analytically calculated transmittance curves (blue and red lines) agree well with the simulated transmission spectra (green line) commonly for the four different metamaterials; however, small discrepancies appear in limited frequency intervals, such as the higher-frequency portion beyond 2.5 THz in Figure 13a and beyond 3.5 THz in Figure 13b. In these regions, the discrepancy arises from the transmittance calculated using the phase relation. This deviation originates from the assumption $A = 0$ underlying Equation (36), but the absorptance in these frequency ranges is nonzero actually as demonstrated evidently in Figures 9b and 10b, indicating that part of the energy is neither reflected nor transmitted. This energy may be attributed to scattering or thermal

https://doi.org/10.3390/mi17040493

loss. So, for such absorption frequency bands, Equation (35) can be used to correct the result after excluding spurious solutions. However, as discussed earlier, apart from empirical validation, no general criterion for eliminating spurious solutions is provided in the present study. This issue requires further investigation, and the discussion here is therefore limited.

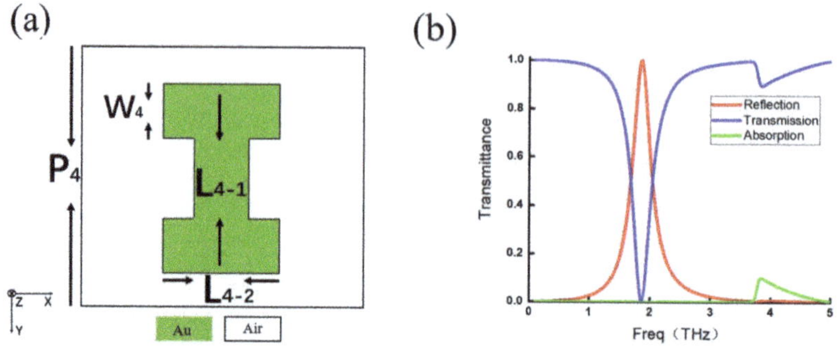

Figure 10. (**a**) I-shaped meta-atom with $P_4 = 80$ μm, $L_{4\text{-}1} = 40$ μm, $L_{4\text{-}2} = 40$ μm, and $W_4 = 6.4$ μm. (**b**) Reflection efficiency curve (red line), transmission efficiency curve (blue line), and absorption efficiency curve (green line).

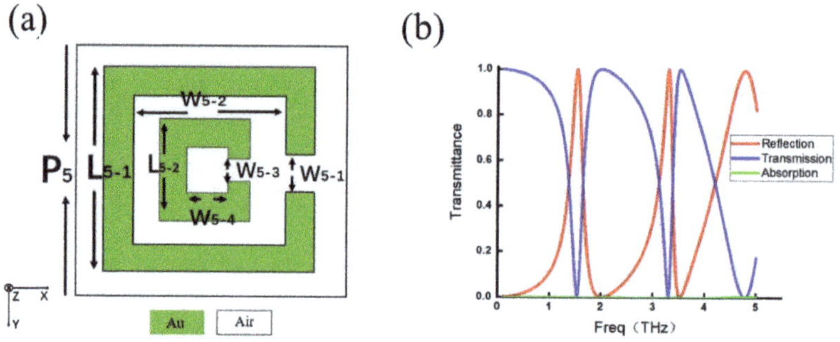

Figure 11. (**a**) Complex split ring meta-atom with $P_5 = 40$ μm, $L_{5\text{-}1} = 30$ μm, $L_{5\text{-}2} = 15$ μm, $W_{5\text{-}1} = 4.8$ μm, $W_{5\text{-}2} = 20.4$ μm, $W_{5\text{-}3} = 2.4$ μm, and $W_{5\text{-}4} = 10.2$ μm. (**b**) Reflection efficiency curve (red line), transmission efficiency curve (blue line), and absorption efficiency curve (green line).

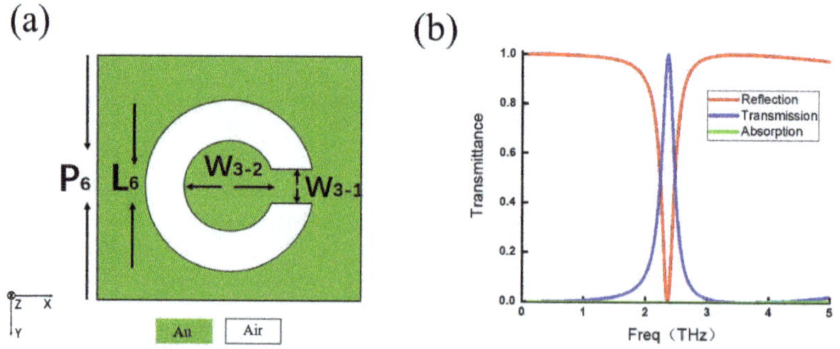

Figure 12. (**a**) Slit split ring meta-atom with $P_6 = 40$ μm, $L_6 = 25$ μm, $W_{5\text{-}1} = 4$ μm, and $W_{5\text{-}2} = 1.7$ μm. (**b**) Reflection efficiency curve (red line), transmission efficiency curve (blue line), and absorption efficiency curve (green line).

https://doi.org/10.3390/mi17040493

Apart from these cases, the calculated transmittance agrees well with the actual transmittance for all structures. Moreover, the transmittance derived from the reflection coefficient using Equation (34) remains valid even in frequency regions where absorption is nonzero. These observations confirm the validity of the two corollaries derived earlier and, to some extent, support the credibility of the reconstructed description of the electromagnetic-wave modulation process in metamaterials proposed in this study.

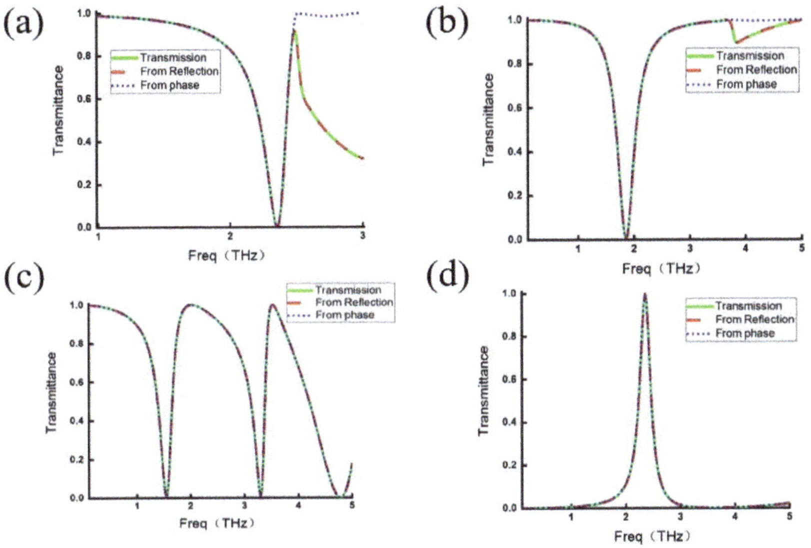

Figure 13. Transmission efficiency curve (green line) calculated from the reflection coefficient (red line) and calculated from the transmission coefficient phase (blue line). (**a**) Strip-shaped metamaterial. (**b**) I-shaped metamaterial. (**c**) Complex split ring metamaterial. (**d**) Slit split ring metamaterial.

6. Summary and Conclusions

This study begins with an analysis of the frequency-dependent amplitude and phase of the transmission and reflection coefficients of three representative metamaterials. By employing a complex-plane representation, a general and geometry-independent rule governing the transmission and reflection waves—independent of the specific meta-atom geometry—is revealed.

Based on the scattering theory described by Green's functions, the mathematical expression of this rule is then derived, and the conventional description of the electromagnetic wave modulation process in metamaterials is reformulated in a more general form. In this context, the manifestation of the extinction theorem in metamaterials is also discussed.

Subsequently, two corollaries are derived from the reformulated description: the coupling relation between the transmission and reflection coefficients, and the amplitude–phase coupling relations of the transmission and reflection waves. The distributions and correlations of these waves are further summarized using a complex-plane representation. Potential applications of these coupling relations are briefly discussed as illustrative implications of the proposed theoretical results, including intelligent metamaterials, information metamaterials, and diffractive neural networks. The revealed coupling relations are of significant importance for the fundamental understanding and rational design of micro/nano devices.

Finally, several specific metamaterial structures are examined, and their transmittance calculated from the derived relations is compared with the actual results, demonstrating

https://doi.org/10.3390/mi17040493

the validity of both the corollaries and the reformulated description. This study provides a general theoretical framework for understanding electromagnetic wave modulation in metamaterials and establishes intrinsic constraints and coupling relations that may serve as a foundation for further investigations of multilayer coupling and multi-resonant coupling in metamaterial systems.

Author Contributions: B.X. completed the research and writing of the entire article, and R.Z. provided guidance for the research and writing of the article. All authors have read and agreed to the published version of the manuscript.

Funding: This paper was supported by the Sichuan Science and Technology Program (No. 2024NSFSC0465), the National Natural Science Foundation of China (No. 62471119), and the National Key Research and Development Program of China (Nos. 2018YFF01013001 and 2017YFA0701000).

Institutional Review Board Statement: Not applicable.

Informed Consent Statement: Not applicable.

Data Availability Statement: The data that support the findings of this study are available from the corresponding author upon reasonable request.

Conflicts of Interest: The authors declare no conflicts of interest.

References

1. Kuznetsov, A.I.; Brongersma, M.L.; Yao, J.; Chen, M.K.; Levy, U.; Tsai, D.P.; Zheludev, N.I.; Faraon, A.; Arbabi, A.; Yu, N.; et al. Roadmap for optical metasurfaces. *ACS Photonics* **2024**, *11*, 816–865. [CrossRef] [PubMed]
2. You, J.; Ma, Q.; Zhang, L.; Liu, C.; Zhang, J.; Liu, S.; Cui, T. Electromagnetic metamaterials: From classical to quantum. *Electromagn. Sci.* **2023**, *1*, 0010051 . [CrossRef]
3. Cheng, Y.Z.; Withayachumnankul, W.; Upadhyay, A.; Headland, D.; Nie, Y.; Gong, R.Z.; Bhaskaran, M.; Sriram, S.; Abbott, D. Ultrabroadband plasmonic absorber for terahertz waves. *Adv. Opt. Mater.* **2015**, *3*, 376–380. [CrossRef]
4. Zhong, R.; Yang, L.; Liang, Z.; Wu, Z.; Wang, Y.; Ma, A.; Fang, Z.; Liu, S. Ultrawideband terahertz absorber with a graphene-loaded dielectric hemi-ellipsoid. *Opt. Express* **2020**, *28*, 28773–28781. [CrossRef] [PubMed]
5. Schurig, D.; Mock, J.J.; Justice, B.J.; Cummer, S.A.; Pendry, J.B.; Starr, A.F.; Smith, D.R. Metamaterial electromagnetic cloak at microwave frequencies. *Science* **2006**, *314*, 977–980. [CrossRef] [PubMed]
6. Chen, W.T.; Zhu, A.Y.; Sanjeev, V.; Khorasaninejad, M.; Shi, Z.; Lee, E.; Capasso, F. A broadband achromatic metalens for focusing and imaging in the visible. *Nat. Nanotechnol.* **2018**, *13*, 220–226. [CrossRef] [PubMed]
7. Cui, T.J.; Liu, S.; Zhang, L. Information metamaterials and metasurfaces. *J. Mater. Chem. C* **2017**, *5*, 3644–3668. [CrossRef]
8. Ma, Q.; Hong, Q.R.; Gao, X.X.; Jing, H.B.; Liu, C.; Bai, G.D.; Cheng, Q.; Cui, T.J. Smart sensing metasurface with self-defined functions in dual polarizations. *Nanophotonics* **2020**, *9*, 3271–3278. [CrossRef]
9. Liu, C.; Ma, Q.; Luo, Z.J.; Hong, Q.R.; Xiao, Q.; Zhang, H.C.; Miao, L.; Yu, W.M.; Cheng, Q.; Li, L.; et al. A programmable diffractive deep neural network based on a digital-coding metasurface array. *Nat. Electron.* **2022**, *5*, 113–122. [CrossRef]
10. Erdemli, Y.E.; Sertel, K.; Gilbert, R.A.; Wright, D.; Volakis, J. Frequency-selective surfaces to enhance performance of broad-band reconfigurable arrays. *IEEE Trans. Antennas Propag.* **2003**, *50*, 1716–1724. [CrossRef]
11. Decker, M.; Staude, I.; Falkner, M.; Dominguez, J.; Neshev, D.N.; Brener, I.; Pertsch, T.; Kivshar, Y.S. High-efficiency dielectric Huygens' surfaces. *Adv. Opt. Mater.* **2015**, *3*, 813–820. [CrossRef]
12. Yu, N.; Genevet, P.; Kats, M.A.; Aieta, F.; Tetienne, J.-P.; Capasso, F.; Gaburro, Z. Light propagation with phase discontinuities: Generalized laws of reflection and refraction. *Science* **2011**, *334*, 333–337. [CrossRef]
13. Farahani-Farahani, M.; Mosallaei, H. Birefringent reflectarray metasurface for beam engineering in infrared. *Opt. Lett.* **2013**, *38*, 462–464. [CrossRef] [PubMed]
14. Chen, H.T.; Taylor, A.J.; Yu, N. A review of metasurfaces: Physics and applications. *Rep. Prog. Phys.* **2016**, *79*, 076401. [CrossRef]
15. Ewald, P.P. Zur begründung der kristalloptik. *Ann. Phys.* **1916**, *354*, 1–38. [CrossRef]
16. Sabaruddin, N.R.; Tan, Y.M.; Chen, S.; Chao, C.C.; Lim, C.M.; Thotagamuge, R.; Kooh, M.R.R.; Chau, Y.C. Designing a broadband terahertz metamaterial absorber through bi-layer hybridization of metal and graphene. *Plasmonics* **2024**, *19*, 3259–3272. [CrossRef]
17. Bozzi, M.; Perregrini, L.; Weinzierl, J.; Winnewisser, C. Design, fabrication, and measurement of frequency-selective surfaces. *Opt. Eng.* **2000**, *39*, 2263. [CrossRef]
18. Oseen, C.W. Über die Wechselwirkung zwischen zwei elektrischen Dipolen und über die Drehung der Polarisationsebene in Kristallen und Flüssigkeiten. *Ann. Phys.* **1915**, *353*, 1–56. [CrossRef]

19. Born, B.M.; Wolf, E. *Principles of Optics*, 7th ed.; Cambridge University Press: Cambridge, UK, 1999; pp. 581–592.
20. Zhang, K.Q. *Electromagnetic Theory in Microwaves and Optoelectronics*, 1st ed.; Springer: Berlin/Heidelberg, Germany, 1994; pp. 405–409.

https://doi.org/10.3390/mi17040493

micromachines

MDPI

Article

MUTC-PD-Based High-Efficiency Photonic Terahertz Generation and Radiation in the 275–296 GHz Band

Yun Wang, Xiaorui Liu and Jianguo Yu *

School of Electronic Engineering, Beijing University of Posts and Telecommunications, Beijing 100876, China; wyun@bupt.edu.cn (Y.W.); lxr1100@bupt.edu.cn (X.L.)
* Correspondence: yujg@bupt.edu.cn

Abstract

We design a photonic terahertz (THz) antenna operating in the 275–296 GHz band, integrating two identical modified uni-traveling-carrier photodiodes (MUTC-PDs), two impedance matching networks, a Wilkinson power combiner and a Vivaldi antenna. Simulation results show a saturated output power of 1.58 dBm at 280 GHz, achieving a 4.13× power enhancement through optimized impedance and power combining compared to a standalone device. The integrated antenna achieves a peak gain of 7.93 dBi, with reflection coefficients below −10 dB. The system demonstrates a power combining efficiency exceeding 95% for phase imbalances up to 20°, and the Wilkinson combiner exhibits only 0.76 dB insertion loss at 285 GHz, demonstrating high radiation and combining efficiency in the 275–296 GHz band.

Keywords: MUTC-PD; impedance matching; power combiner; photonic antenna

1. Introduction

Building upon our previous study on bandwidth-enhanced MUTC-PDs [1], this work further integrates passive and radiating components to realize a compact photonic THz antenna. The terahertz (THz, 0.1–10 THz) frequency range has emerged as a promising frontier for next-generation communication systems, offering unparalleled bandwidth, low latency, and high spatial resolution [2–6]. Particularly, the sub-range spanning 275–450 GHz is gaining significant attention as a key resource for future 6G wireless networks and ultra-high-resolution applications, owing to the availability of vast contiguous bandwidth capable of supporting terabit-per-second data rates and millimeter-scale imaging resolution. Recent experimental breakthroughs in this frequency window have successfully demonstrated 3D terahertz imaging utilizing chromatic metasurfaces, providing a powerful tool for high-precision spatial detection [7]. Beyond 6G communications, this band has also exhibited immense potential for non-contact sensing, such as the real-time quantification of plant leaf water status for sustainable agriculture [8]. To fully exploit these capabilities, ongoing research continues to explore broadband wave manipulation through advanced metasurface designs based on circuit-model optimization, facilitating more versatile THz systems [9].

The uni-traveling-carrier photodiode (UTC-PD), known for its wide bandwidth, fast response time, and high output power, plays a pivotal role in THz signal generation and photonic communication systems [10,11]. The modified UTC-PD (MUTC-PD) further enhances responsivity without compromising bandwidth; however, its output power remains limited beyond 200 GHz. To address this limitation, integrating MUTC-PDs with impedance-matching networks, power combiners, and broadband antennas is crucial for improving efficiency at higher frequencies.

While electronic THz emitters such as resonant tunneling diodes (RTDs) have achieved notable progress in compact source design, their output power and conversion efficiency rapidly deteriorate at higher frequencies [12,13]. Recent studies have demonstrated oscillation frequencies beyond 2 THz and milliwatt-level power at 1 THz [14], yet the DC-to-RF efficiency remains below 1%, and parasitic resistance, carrier transit-time delay, and thermal limitations restrict scalability [15]. Moreover, RTD oscillators typically require resonant cavities for impedance matching, complicating integration with photonic platforms. In contrast, photomixing emitters based on UTC-PDs offer broadband, tunable, and phase-stable THz generation without electronic oscillation.

Several integrated approaches have been reported to enhance the efficiency of photonic THz emitters. Song et al. demonstrated a WR3 waveguide module integrating two UTC-PDs with a T-junction power combiner, achieving 1.2 mW at 300 GHz [16]. Sun et al. realized an MUTC-PD integrated with a Vivaldi antenna and matching network, yielding 5 dBm EIRP at 100 GHz [17]. Ohara et al. and Nagatsuma et al. further developed integrated UTC-PD emitters, achieving >2 mW at 300 GHz and 200 Gbit/s wireless transmission [18,19]. Ssali et al. provided a scalable solution for THz wave generation by integrating a 4 × 1 UTC-PD array with a T-junction combiner and antenna, achieving a 10.9 dB increase in detected power at 300 GHz [20]. Despite these achievements, issues such as low radiation efficiency and fabrication complexity persist, particularly for SiC substrates with high dielectric constants.

To achieve efficient terahertz emission and lay the groundwork for future broadband systems, we design a photonic terahertz (THz) antenna operating within the 275–296 GHz band. This system integrates two identical modified uni-traveling-carrier photodiodes (MUTC-PDs), two impedance matching networks, a Wilkinson power combiner, and a Vivaldi antenna. This frequency band serves as a critical stepping stone towards the 275–450 GHz range, allowing for the validation of key device optimizations and integration concepts under current technological constraints. Initial experimental testing revealed that a first-generation MUTC-PD with a 130 nm depletion region suffered from velocity overshoot failure, which underscored the necessity of optimizing both the epitaxial structure and contact resistance. Informed by these findings, we refined the device design and performed full-system integration and simulation. The simulated results show a saturated output power of 1.58 dBm at 280 GHz, corresponding to a 4.13× power enhancement through efficient impedance matching and power combining compared to a standalone device. The integrated antenna exhibits a peak gain of 7.93 dBi and maintains reflection coefficients below −10 dB across the operating band.

2. Modeling and Analysis of the MUTC-PD

2.1. Design of the Epitaxial Structure

The epitaxial design of MUTC-PD focuses on minimizing carrier transit time and the RC time constant to enhance high-frequency performance. Carrier transit time is influenced by diffusion, drift, and velocity overshoot effects, determined by the thicknesses of the undepleted absorption layer (W_a), depleted absorption layer (W_{ad}), and collection layer (W_c), respectively [21]. The RC time constant is mainly governed by the junction area, collection layer thickness, and the contact resistance of the P- and N-type electrodes.

To optimize performance, the absorption layer thickness must be carefully balanced to ensure sufficient optical absorption while minimizing diffusion and drift delays. The collection layer thickness influences the junction capacitance. The electric field in the depletion region, which controls velocity overshoot [22], is adjusted by modifying the doping concentration in the cliff layer and the conduction band offset of the barrier material relative to InGaAs. Additionally, minimizing the junction area and contact resistance helps reduce

https://doi.org/10.3390/mi17020196

the RC-limited bandwidth. The MUTC-PD design flow is illustrated in Figure 1a, and the corresponding epitaxial structure is shown in Figure 1b.

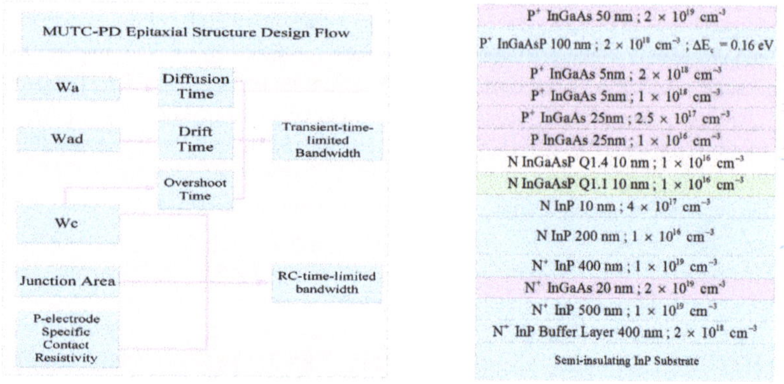

Figure 1. (**a**) Schematic workflow and (**b**) corresponding epitaxial layer configuration of the proposed MUTC-PD. In (**b**), each layer is labeled with its thickness (nm) and carrier doping concentration (cm^{-3}), separated by a semicolon (e.g., 50 nm; 2×10^{19} cm^{-3}).

To achieve an intrinsic 3−dB bandwidth exceeding 100 GHz, the absorption layer must be reduced in thickness. As shown in Figure 2a, the responsivity decreases as the absorption layer becomes thinner, reaching a minimum of 0.062 A/W at 60 nm. Reducing the thickness further causes the responsivity to drop below 0.05 A/W, so a thickness of 60 nm is selected as the optimal balance between bandwidth and responsivity. The overall 3−dB bandwidth is jointly determined by carrier transit time and RC limitations. Simulation results in Figure 2b show that a maximum 3−dB bandwidth of 271 GHz can be achieved with a 10 nm undepleted absorption layer and a 230 nm collection layer, with the active region diameter set to 3 μm.

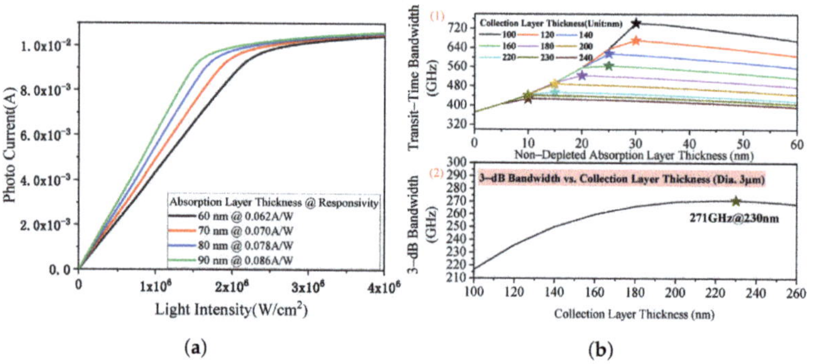

Figure 2. (**a**) Simulated responsivity dependence on absorption layer thicknesses and injection light intensity. A thickness of 60 nm yields a responsivity of 0.062 A/W, balancing bandwidth and optical-to-electrical conversion efficiency. (**b**) Analysis of bandwidth performance: (1) transit-time-limited bandwidth as a function of non−depleted absorption layer thickness, with the collection layer thickness varying from 100 nm to 240 nm; (2) 3-dB bandwidth as a function of collection layer thickness. The asterisk (*) indicates the maximum 3-dB bandwidth of 271 GHz achieved at a collection layer thickness of 230 nm.

Velocity overshoot in the collection region is critical for achieving this bandwidth, requiring the electric field to be maintained within 20–40 kV/cm [23,24]. As shown in Figure 3a–d, the field distribution is influenced by the conduction band offset (ΔE_c), cliff

https://doi.org/10.3390/mi17020196

layer doping concentration, and applied bias. With optimized conditions—$\Delta E_c = 0.16$ eV, doping concentration of 4×10^{17} cm^{-3}, and a reverse bias of -2 V—the electric field remains within the optimal range up to a photocurrent of 8.6 mA. Beyond this point, space-charge effects cause the electric field to collapse. Based on this analysis, Figure 4a presents the simulated output power as a function of photocurrent at 280 GHz. The discrete MUTC-PD reaches a saturated output power of 1.58 dBm at a photocurrent of 7.86 mA.

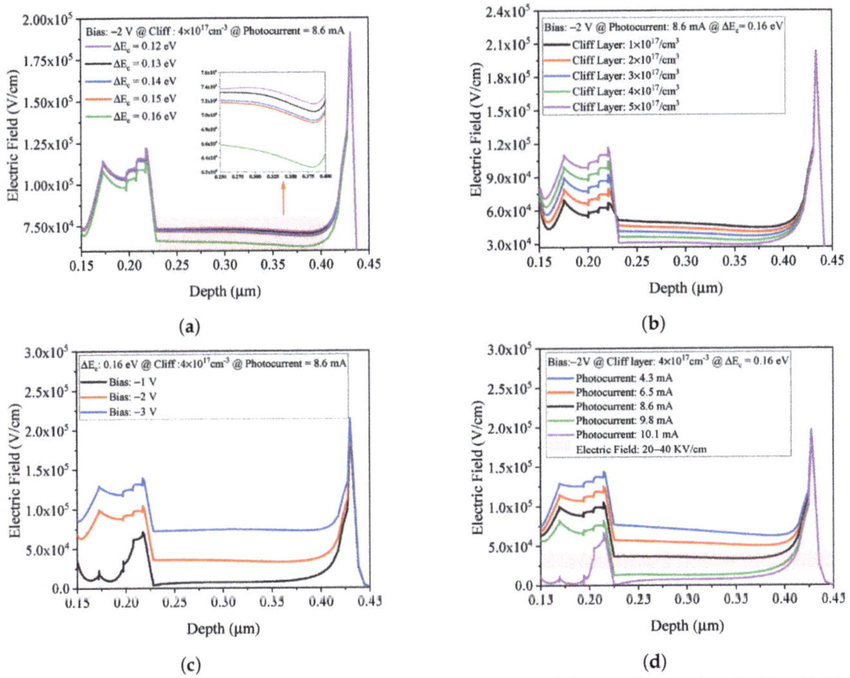

Figure 3. (**a**) Electric field in the depletion region as a function of the conduction band offset (ΔE_c), with larger ΔE_c leading to a lower electric field in the collection region under fixed reverse bias and incident light intensity. (**b**) Electric field versus cliff layer doping concentration at a fixed ΔE_c, where increasing doping concentration results in a lower electric field in the collection region under fixed reverse bias and light intensity. (**c**) Electric field profile across the intrinsic region under different reverse bias voltages, showing that increasing reverse bias increases the electric field in the collection region. (**d**) Electric field in the collection region versus output photocurrent, highlighting the velocity overshoot window (20–40 kV/cm).

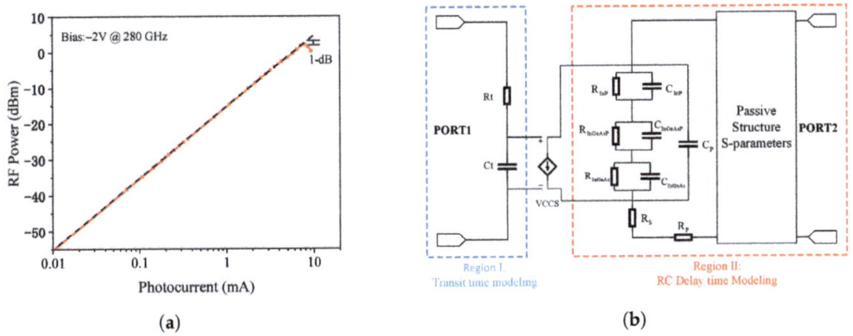

Figure 4. (**a**) RF power versus photocurrent at 280 GHz. (**b**) Detailed equivalent-circuit model of the MUTC-PD including transit-time and RC delay submodules.

https://doi.org/10.3390/mi17020196

2.2. Accurate Modeling of the MUTC-PD

In the previous section, a simplified model was adopted [25]; the equivalent resistance of the MUTC PD included only the contact resistances of the P- and N-type electrodes and the load resistance, assuming a specific contact resistivity of $1 \times 10^{-6}\ \Omega \cdot \text{cm}^2$ for the P electrode. However, this approach lacks the accuracy required to reflect the actual behavior of the device. A more refined model must account for the junction capacitances and series resistances of both the collection region and the depleted absorption region. In addition, the equivalent resistance and capacitance of the InGaAsP (Q1.4) and InGaAsP (Q1.1) buffer layers must also be considered [26].

Figure 4b illustrates the detailed equivalent circuit model of the MUTC-PD, which consists of two main parts: the transit-time model and the RC delay model. In this model, the injected photocurrent is represented by an ideal voltage-controlled current source. Region 1 comprises a series resistance R_t and a capacitance C_t, where the voltage across C_t governs the operating state of Region 2. The values of R_t and C_t are determined by the transit-time bandwidth. Region 2 includes not only the equivalent resistances and capacitances of individual layers but also parasitic resistances and capacitances, as well as the series resistances introduced by the P and N electrodes. Furthermore, the S-parameters of passive circuits, as discussed in the next section, are also incorporated into the model.

Table 1 lists the specific parameter values used in the equivalent circuit, with their derivations detailed in our previous work [1]. Using the parameters from Table 1, the intrinsic bandwidth is calculated to be 124 GHz.

Table 1. Equivalent circuit parameters of the MUTC-PD.

Parameters	Physical Meaning	Value
R_{InP}	Junction resistance of InP	311 Ω
$R_{InGaAsP}$	Junction resistance of InGaAsP	228 Ω
R_{InGaAs}	Junction resistance of InGaAs	5.5×10^4 Ω
C_{InP}	Junction capacitance of InP	2.61 fF
$C_{InGaAsP}$	Junction capacitance of InGaAsP	33.2 fF
C_{InGaAs}	Junction capacitance of InGaAs	13.6 fF

3. Design of Passive Components

3.1. Design of Photonic Terahertz Generator

As demonstrated in the previous section, even reducing the absorption layer thickness to 60 nm fails to significantly improve the frequency response of the MUTC-PD beyond 200 GHz without applying bandwidth enhancement techniques such as inductive peaking at the output electrode. To enable photonic antenna operation in the 275–296 GHz band, improving the MUTC-PD's output power within this frequency band is crucial.

In this work, we propose a passive integration approach—designated as the Photonic Terahertz Generator-A (PTG-A)—which incorporates an impedance matching circuit and a Wilkinson power combiner to address this challenge. The impedance matching circuit is optimized to match the MUTC-PD's output impedance to 50 Ω at 285 GHz [27]. The Wilkinson power combiner, comprising three well-matched ports, is employed to coherently combine the outputs of two identical MUTC-PDs. A schematic top view of PTG-A is shown in Figure 5a. The matching network consists of a series coplanar waveguide (CPW) of length L_1, followed by a short-circuited CPW of length L_2, whose terminal is connected to the top electrode of a metal–insulator–metal (MIM) capacitor functioning as a DC block.

The Wilkinson power combiner in PTG-A employs CPWs with a characteristic impedance of 100 Ω at its input ports [28,29]. The isolation resistor is implemented using a material with an aspect ratio (L_r/W_r) of 4, with its thickness and conductivity carefully op-

https://doi.org/10.3390/mi17020196

timized to achieve a resistance of 200 Ω. Detailed dimensional parameters are provided in Table 2. All passive components are designed on a quartz substrate (ϵ_r = 3.78). The MUTC-PD, which is fabricated on an InP substrate, is flip-chip bonded onto the quartz-based passive circuit. This hybrid integration approach not only ensures reliable high-frequency performance but also significantly lowers the overall manufacturing cost, making it suitable for scalable production.

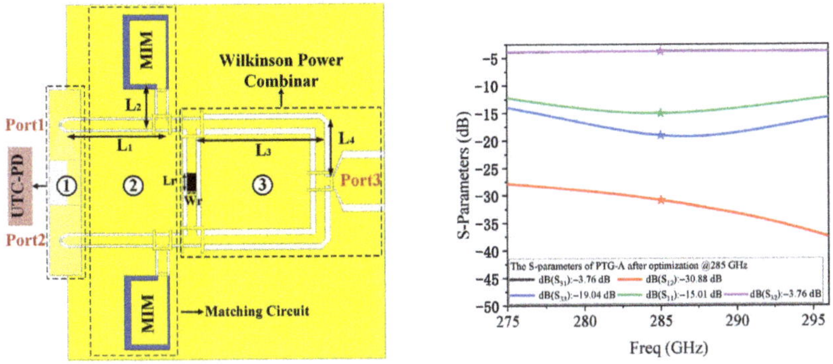

(a) (b)

Figure 5. (**a**) Schematic top view of the proposed Photonic Terahertz Generator-A (PTG-A). (**b**) Simulated S-parameters of PTG-A.

Table 2. Detailed dimensional parameters of PTG-A.

Parameter	L_1	L_2	L_3	L_4	Lr	Wr
Value	68 μm	38 μm	66 μm	67.5 μm	12 μm	3 μm

The simulated performance of PTG-A, obtained using High Frequency Structure Simulator (HFSS), is presented in Figure 5b. The transmission loss from the two input ports to the output port is approximately 3.76 dB, indicating that the actual loss, after accounting for the ideal 3 dB power split, is only about 0.76 dB. The reflection coefficients at ports 1 and 2 remain below −10 dB across the 275–296 GHz band, and reach a minimum of −15.01 dB at 285 GHz. Furthermore, the isolation between the two input ports is below −25 dB, indicating excellent port isolation and minimal crosstalk.

For comparison, the top-view schematic of PTG-B is shown in Figure 6a. PTG-B replaces the Wilkinson combiner with a traditional T-junction power combiner. As shown in Figure 6b, this change leads to a degradation in performance. The transmission loss from the two input ports to the output port is approximately 5.8 dB, indicating that the actual loss, after accounting for the ideal 3 dB power split, is about 2.8 dB. The reflection coefficients at ports 1 and 2 are approximately −19.31 dB, and the isolation between the two ports is around −13.95 dB. These results demonstrate that the Wilkinson-based PTG-A offers significantly improved impedance matching and isolation characteristics compared to the T-junction implementation, thereby providing a more effective solution for high-frequency photonic THz signal generation.

By importing the S-parameters of PTG-A into the equivalent circuit model shown in Figure 4b—where ports 1 and 2 are connected to the outputs of two identical MUTC-PDs, and port 3 is terminated with a 50 Ω load—we compared the relative output power across three configurations in Figure 7: a discrete MUTC-PD, an MUTC-PD with only the impedance matching circuit, and PTG-A. The results reveal that incorporating the impedance matching circuit significantly enhances the output power at the resonant frequency. Specifically, at 280 GHz, the relative output power increases from 0.15 (single

https://doi.org/10.3390/mi17020196

MUTC-PD) to 0.352. However, this improvement is still limited. When the Wilkinson power combiner is added to PTG-A, the relative output power at 280 GHz rises to 0.623, an increase of 4.13 times compared to the single MUTC-PD. Building upon the PTG-A configuration that achieves optimized impedance and power combining, a broadband radiating structure is subsequently designed to realize free-space THz emission.

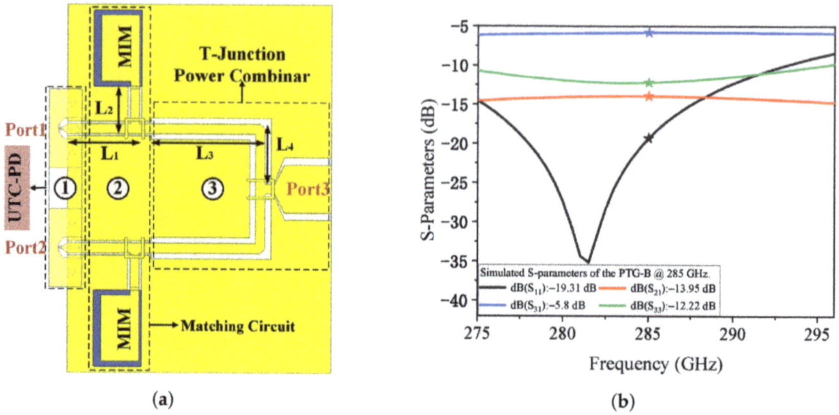

Figure 6. (**a**) Schematic top view of the proposed Photonic Terahertz Generator-B (PTG-B). (**b**) Simulated S-parameters of PTG-B.

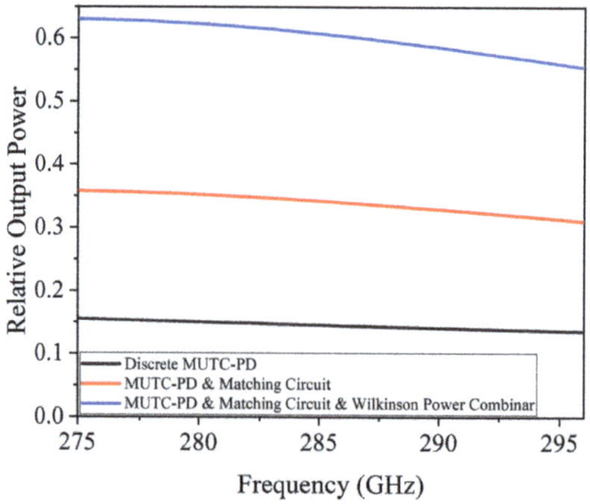

Figure 7. Relative output power across three configurations: a discrete MUTC-PD, an MUTC-PD with only the impedance matching circuit, and PTG-A.

3.2. Fabrication Tolerance Analysis

To rigorously address the fabrication challenges at terahertz frequencies, a systematic tolerance analysis was performed for the integrated PTG-A. We evaluated the sensitivity of the S-parameters (S_{11}, S_{21}, S_{31}, and S_{33}) by introducing deliberate dimensional deviations to three critical geometric parameters: the CPW gap width, the signal line width, and the MIM capacitor width.

The fabrication robustness of the proposed PTG-A was further investigated by introducing dimensional deviations. When the gap width between the signal line and ground plane is varied by ±10%, the impedance matching remains stable. As shown in Figure 8a–d, both input reflection (S_{11}) and output reflection (S_{33}) stay well below −10 dB, confirming the design's resilience to over-etching or under-etching. Figure 8e–h

https://doi.org/10.3390/mi17020196

shows the impact of a ±1 μm deviation in the signal line width. The transmission coefficients S_{21} and S_{31} are nearly identical to the optimized case, ensuring stable power combining efficiency. Furthermore, the sensitivity to MIM capacitor dimensions was evaluated in Figure 9. Adjusting the MIM capacitor width by ±1 μm results in negligible frequency shifts and magnitude variations as depicted in Figure 9a–d. Additionally, the influence of the isolation resistor width on the port-to-port isolation was analyzed. As shown in Figure 10a, the isolation between Port 1 and Port 2 remains better than 20 dB across the target band despite width variations, ensuring effective decoupling between the two MUTC-PD sources. Finally, considering potential synchronization errors in the optical feeding network, the combining efficiency as a function of phase mismatch between the two PDs was investigated. Figure 10b demonstrates that the PTG-A maintains a combining efficiency of over 95% for phase imbalances up to 20°, further validating the robustness of the proposed architecture. These results confirm that the proposed PTG-A possesses a generous design margin. The stability under these variations significantly relaxes the fabrication precision requirement from the initially estimated 0.5 μm to a more practical range, ensuring a high manufacturing yield.

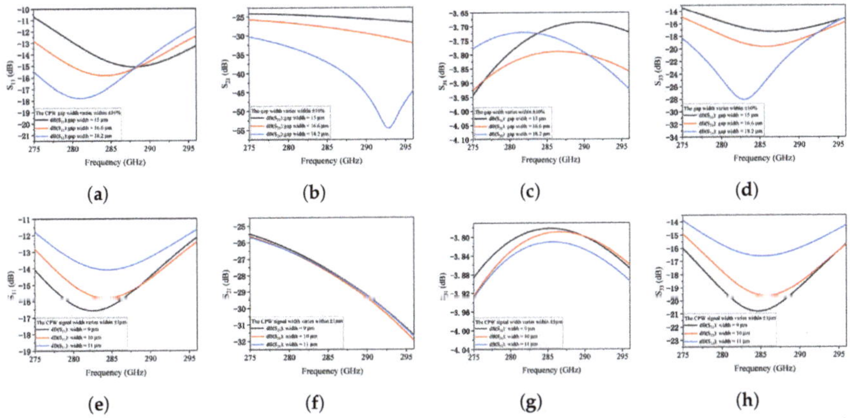

Figure 8. Fabrication tolerance analysis of the proposed PTG-A S-parameters under CPW dimensional variations: (**a**–**d**) ±10% deviation in CPW gap width, and (**e**–**h**) ±1 μm deviation in CPW signal line width.

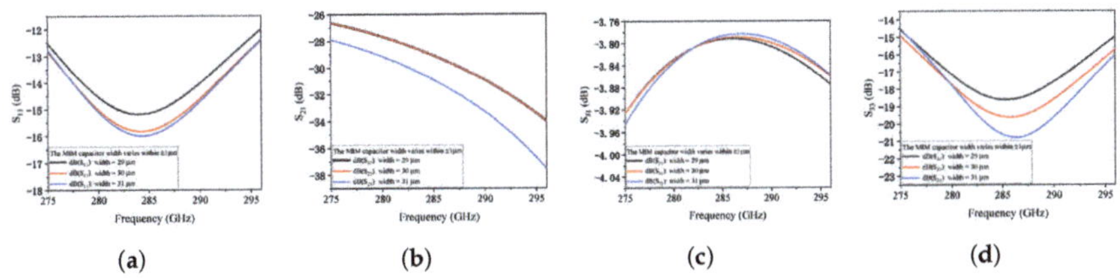

Figure 9. Fabrication tolerance analysis of the proposed PTG-A S-parameters under MIM capacitor width variations: (**a**–**d**) ±1 μm deviation in MIM capacitor width.

https://doi.org/10.3390/mi17020196

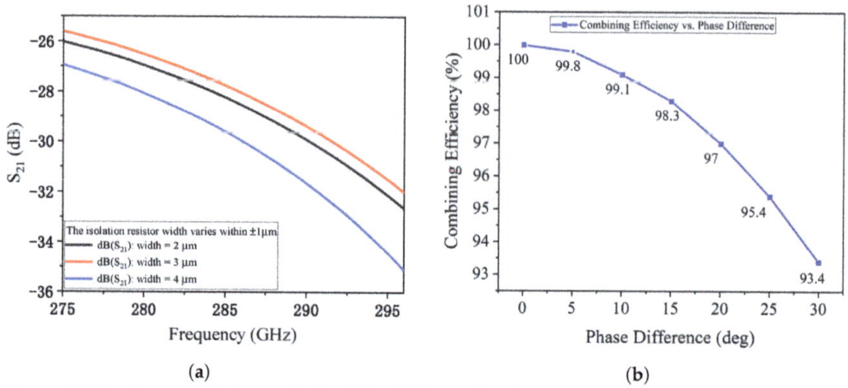

Figure 10. Further sensitivity analysis of the PTG-A: (**a**) Isolation between Port 1 and Port 2 with respect to isolation resistor width variations. (**b**) Power combining efficiency as a function of phase mismatch between the two MUTC-PDs.

4. Design of Photonic Antenna

4.1. Design of Vivaldi Antenna

To achieve efficient radiation and broad bandwidth in high-frequency applications, the Vivaldi antenna is often chosen due to its wideband performance and compact design [30,31]. The Vivaldi antenna consists of a feed port, a radiating structure, and a reflecting structure on the backside. The feed port uses a coplanar waveguide, matching the third port of the Wilkinson power combiner. The schematic diagram of the Vivaldi antenna is shown in Figure 11a, and its reflection performance is illustrated in Figure 11b. The radiating structure features exponentially tapered slots designed for broadband radiation. To further enhance gain, multiple slots are introduced into the radiating element, and several metal blocks are embedded on the center of the dielectric substrate to improve signal guidance. Each slot is 30 μm wide, with a spacing of 100 μm between adjacent slots. The slot height starts at 200 μm and decreases stepwise by 50 μm along the taper.

On the backside of the dielectric substrate, an arc-shaped reflecting structure is designed to suppress backlobe reflection. For operation within the 275–296 GHz frequency range, the antenna's physical dimensions are optimized to meet efficient radiation conditions. Specifically, the maximum aperture must be at least half the wavelength at the lowest frequency (275 GHz), which corresponds to approximately 0.55 mm. The length of the exponential taper section, L, is chosen as 1.1 mm to ensure strong radiation performance and higher gain. The opening width, W, at the end of the exponential taper section is set to 0.65 mm, which determines the low-frequency cutoff and is constrained by the overall antenna size. The high-frequency cutoff depends on the width at the taper's starting end and is optimized to balance impedance matching.

To ensure a wide bandwidth, the relative dielectric constant of the substrate should be moderate, and its thickness should not exceed one-tenth of the wavelength (i.e., $\leq$0.11 mm) to minimize surface waves and parasitic radiation. The metal layer thickness is 2 μm. After full-wave simulation and optimization, the antenna achieves a backlobe gain of −2.1 dBi and a peak gain of 8.77 dBi in the main radiation direction. The E-plane radiation pattern and the 3D radiation pattern are shown in Figure 12a and Figure 12b, respectively.

https://doi.org/10.3390/mi17020196

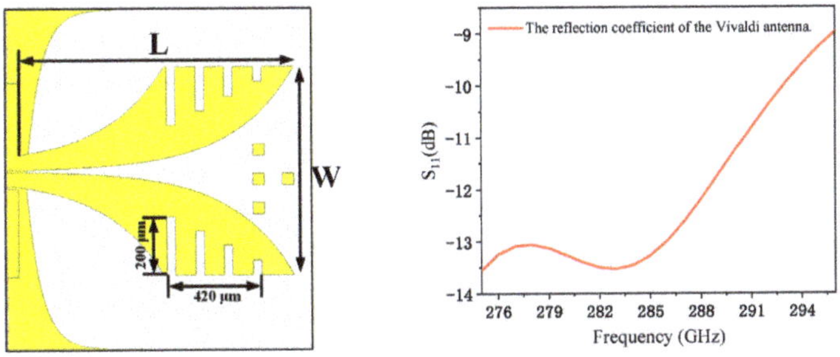

(a) (b)

Figure 11. (**a**) Schematic diagram of the Vivaldi antenna; (**b**) simulated reflection coefficient across 275–296 GHz.

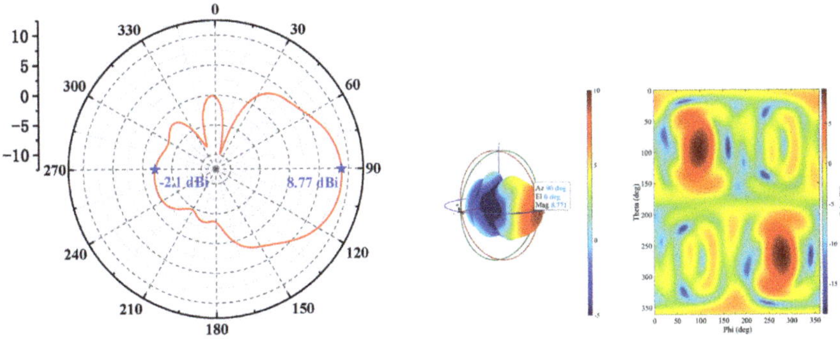

(a) (b)

Figure 12. (**a**) E-plane radiation pattern and (**b**) 3D radiation pattern of the Vivaldi antenna at 285 GHz.

4.2. Photonic Antenna Performance

Figure 13 shows the 3D structure of the photonic antenna. Figure 14a presents the top view of the photonic antenna, while Figure 14b shows the S-parameter curves of the antenna. The MUTC-PD body of the photonic antenna absorbs optical signals to generate electrical signals. The gradient-doped absorption layer of the MUTC-PD introduces a built-in electric field, accelerating the movement of electrons within the absorption layer and enhancing the photoelectric conversion efficiency. The cliff layer between the absorption layer and the collection layer strengthens the electric field in the absorption layer while reducing the electric field in the collection layer, thereby promoting the flow of electrons in the collection region. The signal output from the PD passes through an impedance matching circuit, ensuring that the output impedance of the MUTC-PD is matched to 50 Ω in the 275–296 GHz frequency range, thereby maximizing power transfer and minimizing reflection loss.

Subsequently, the Wilkinson power combiner combines the output signals from two MUTC-PD sources (each with an impedance matching circuit) at the feed port of the Vivaldi antenna, enhancing the overall system output power. Finally, the signal is radiated by the Vivaldi antenna in the form of electromagnetic waves within the 275–296 GHz frequency range, achieving efficient signal transmission.

Simulation results show that, within the 275–296 GHz frequency band, the isolation between the two input ports of the photonic antenna is approximately −20 dB, and the reflection coefficient at the ports is below −10 dB. At 285 GHz, the antenna gain reaches

https://doi.org/10.3390/mi17020196

7.93 dBi. The E-plane and 3D radiation patterns of the photonic antenna are shown in Figure 15a and Figure 15b, respectively.

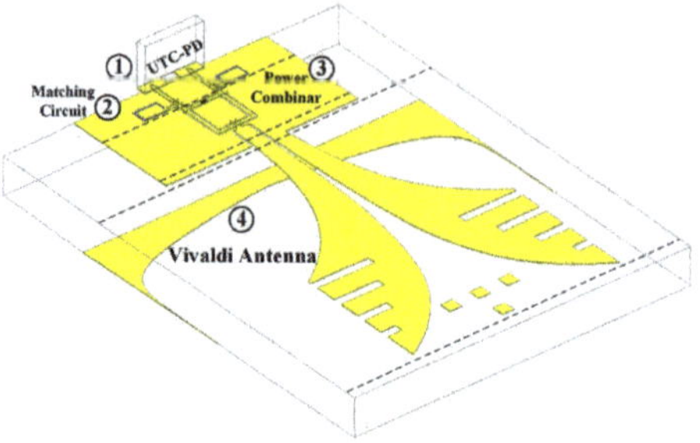

Figure 13. 3D structure of the photonic terahertz antenna module.

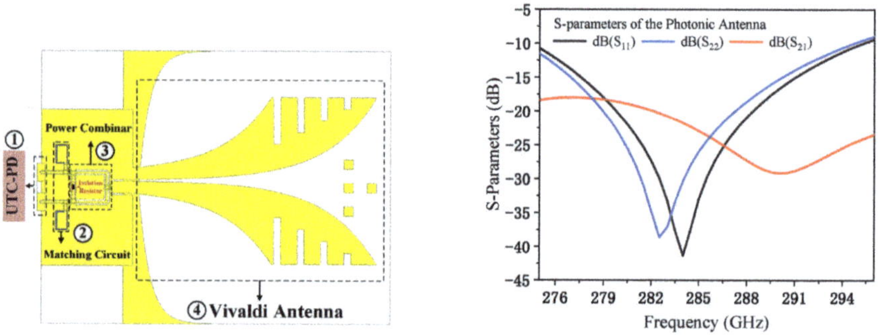

(**a**) (**b**)

Figure 14. (**a**) Top view of the photonic antenna structure; (**b**) simulated S-parameters.

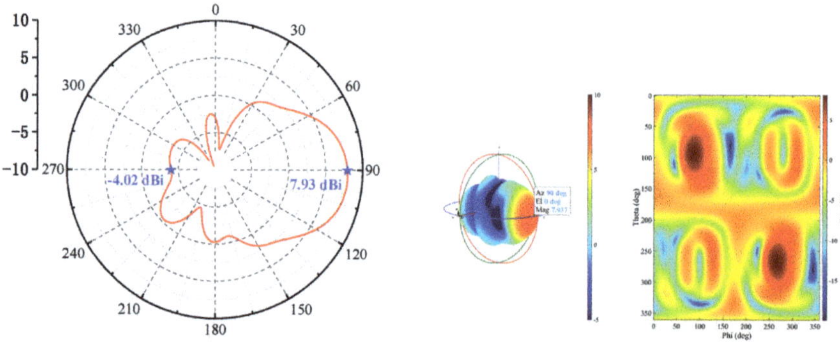

(**a**) (**b**)

Figure 15. (**a**) E-plane radiation pattern and (**b**) 3D radiation pattern of the complete photonic antenna system at 285 GHz.

5. Device Fabrication and Characterization

It should be clarified that, at the current stage, only the baseline MUTC-PD prototype has been fabricated and characterized to verify the device-level modeling and identify the physical bottlenecks limiting high-frequency performance. The full integrated photonic terahertz antenna system is validated through full-wave simulations informed by these experimental findings.

The fabrication process detailed below pertains to the baseline MUTC-PD, which was characterized to identify the performance limitations outlined in the Introduction. It is

https://doi.org/10.3390/mi17020196

important to note that these fabricated devices do not correspond to the optimized epitaxial structure presented in the simulated antenna design; rather, their experimental analysis provided the critical insights that informed the subsequent device optimization. For clarity, only the absorption and collection layer thicknesses were modified in the optimized design; all other structural and material parameters remained identical between the baseline and optimized versions.

The devices were fabricated on InP-based wafers using photolithography, metal deposition, and etching. After cleaning and adhesion baking, P-type electrodes were formed with Ti/Pt/Au (20/20/200 nm) by image reversal lithography and evaporation, followed by lift-off and rapid thermal annealing at 350 °C in N_2 for improved contact. The mesa structure was defined through two-step dry etching and selective wet etching in HCl:H_3PO_4, while N-type contacts were made with Ni/Au (20/100 nm) deposition. Device isolation was achieved with mesa etching to a depth of 13 μm. A SiO_2 dielectric layer (800 nm) was deposited by PECVD, and via holes were created using two lithography and etch cycles. A Ti/Au (300 Å/800 Å) seed layer was sputtered, followed by gold electroplating for 9.5 min, resulting in ∼1.6 μm thick gold with good coverage. After seed layer removal, the Au electrodes were smooth and uniform (1.65 ± 0.2 μm), confirming reliable fabrication and suitable quality for further characterization.

Device Performance

Figure 16 presents the optical microscope image of a fabricated MUTC-PD with a 3 μm P-mesa diameter. Figure 17a shows the scanning electron microscope (SEM) image of the fabricated MUTC-PD, where the P-mesa, second mesa, N-mesa, and air-bridge structures can be clearly observed, while Figure 17b shows the measured I–V characteristics of devices with 3 μm and 5 μm P-mesa diameters. Under a reverse bias of −3 V, the dark currents of the 3 μm and 5 μm devices are 2 nA and 3.2 nA, respectively.

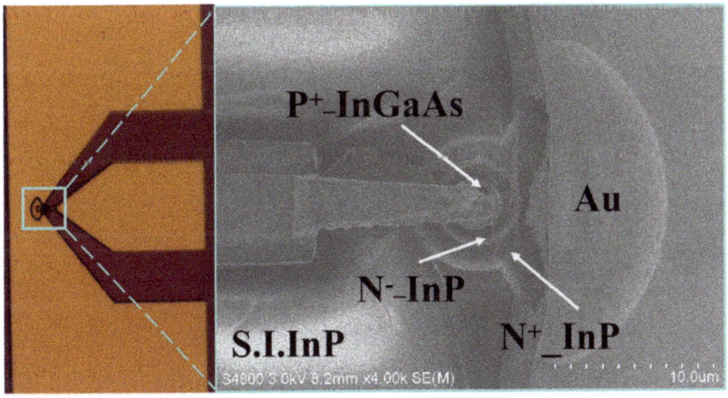

Figure 16. Optical microscope image of the fabricated MUTC-PD. The notations N^+-InP and N^--InP refer to heavily doped (2×10^{18} cm^{-3}) and lightly doped (1×10^{16} cm^{-3}) n-type Indium Phosphide layers, respectively, used for ohmic contacts and electric field management.

The relative frequency response of the device was characterized using a standard optical heterodyne setup. A 1.55 μm continuous-wave laser was amplified by an EDFA and modulated by a Mach–Zehnder modulator driven by an RF signal from a vector network analyzer (VNA). The modulated optical signal was again amplified and directed onto the photodiode under test. The generated electrical signal was separated by a bias-T into DC and RF components, with the RF output fed back to the VNA. By sweeping the RF frequency, the VNA recorded the amplitude and phase response of the photodiode, from which the frequency response curve was obtained.

https://doi.org/10.3390/mi17020196

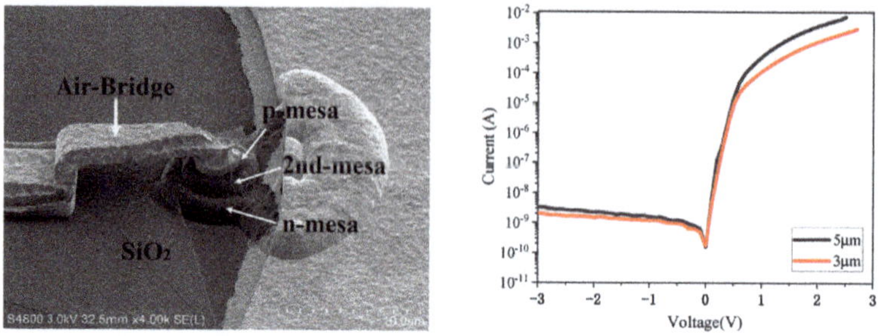

(a) (b)

Figure 17. (**a**) Scanning electron microscope (SEM) image of the fabricated MUTC-PD and (**b**) the measured I–V characteristics of devices with 3 μm and 5 μm P-mesa diameters.

The frequency response tested under a fixed reverse bias of 4 V are plotted in Figure 18. The 3 μm diameter PD exhibits a 3dB bandwidth of 27.6 GHz at a 4 mA photocurrent, while the 5 μm diameter PD shows a 3 dB bandwidth of 13.6 GHz at the same photocurrent. The saturation characteristics at −4 V bias are shown in Figure 19 at 75 GHz. The 1-dB compression points for the 3 μm and 5 μm PDs occur at photocurrents of 7.15 mA and 10.75 mA, respectively, with corresponding RF saturation powers of −3.73 dBm and −3.44 dBm. These values show significant deviation from the simulation results, which will be analyzed in detail in Section 6.

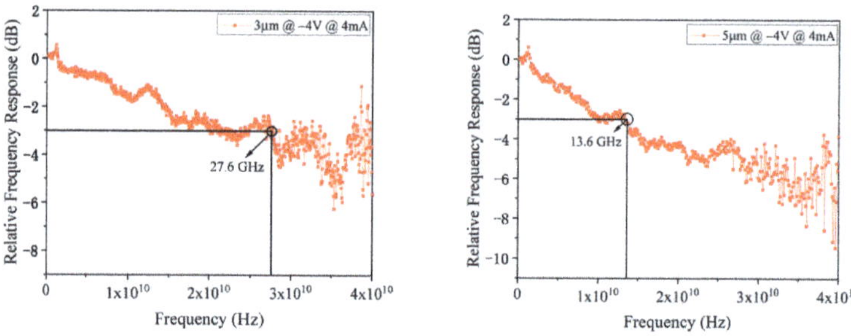

(a) (b)

Figure 18. (**a**) The frequency response tested under a fixed reverse bias of 4 V for 3 μm-diameter PD and (**b**) 5 μm-diameter PD.

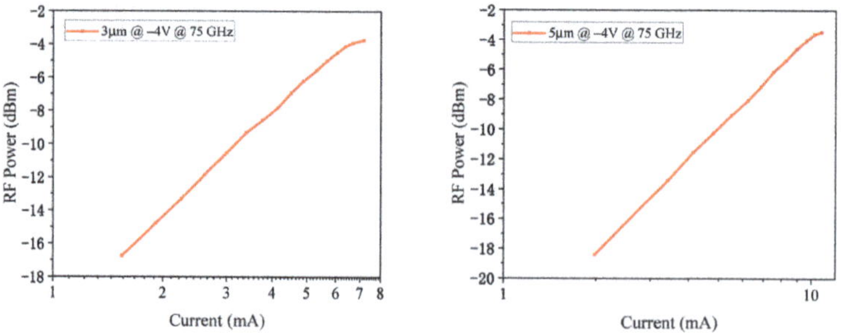

(a) (b)

Figure 19. (**a**) The saturation characteristics at −4V for 3 μm-diameter PD and (**b**) 5 μm-diameter PD.

6. Discussion

In this experiment, the epitaxial structure used differs slightly from the design illustrated in Figure 1b. Specifically, the depletion layer thickness is 130 nm and the absorption

https://doi.org/10.3390/mi17020196

layer thickness is 150 nm; detailed epitaxial parameters can be found in our previous publication [1]. The epitaxial layers were grown by metal–organic chemical vapor deposition (MOCVD). One known issue with MOCVD growth is dopant diffusion from highly doped regions into adjacent low-doped regions by approximately 100 nm, which leads to a significant deviation between the actual and the ideal epitaxial profiles. This deviation was further confirmed by electrochemical capacitance–voltage (ECV) measurements.

To evaluate its impact, we simulated and compared the electric field distributions of the actual and ideal epitaxial structures under a reverse bias of 1.5 V, as shown in Figure 20a. The results show that when the doping concentration in the depletion region increases by one order of magnitude (from 1×10^{16} to $1.5 \times 10^{17}/\text{cm}^{-3}$), the electric field decreases correspondingly, with the minimum value approaching zero under the same bias condition.

Experimental characterization revealed that the device bandwidth increased as the reverse bias rose from 1.5 V to 4 V. Under ideal conditions, the electric field at 1.5 V should lie within the optimal range for electron velocity overshoot. However, due to the epitaxial deviation introduced during fabrication, the electric field varies drastically from nearly 0 to over 10 kV/cm as the bias increases, as shown in Figure 20b. Consequently, the velocity overshoot effect is suppressed, reducing the average electron drift velocity from the ideal 3×10^7 cm/s to 2×10^7 cm/s. According to the transit-time model and simulation methods validated in our previous study [1], this reduction in velocity specifically constrains the carrier transit-time-limited bandwidth, resulting in the observed degradation from 325 GHz to 252 GHz.

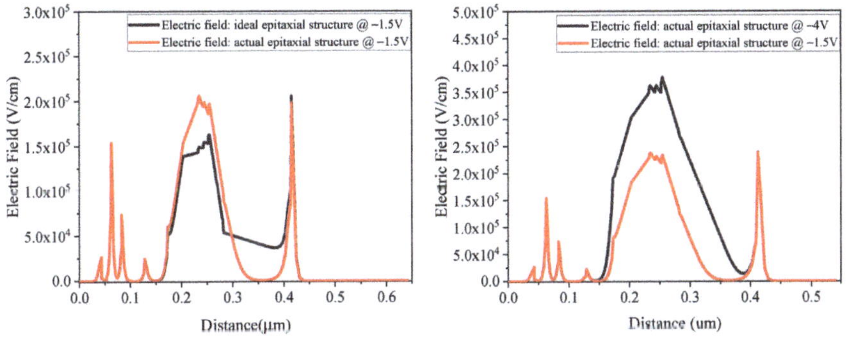

(a) (b)

Figure 20. (**a**) Electric field distributions of the actual and ideal epitaxial structures under a reverse bias of 1.5 V. (**b**) Electric field distributions of the actual epitaxial structure under 1.5 V and 4 V.

In addition to transit-time effects, the RC time constant also contributes to bandwidth degradation, primarily due to high contact resistance. The contact resistance is highly sensitive to annealing conditions such as temperature and duration. After rapid thermal annealing, the specific contact resistivity measured by the CTL-M method was $9.48 \times 10^{-6} \Omega \cdot \text{cm}^2$. Based on this value, the contact resistances for photodiodes with 3 μm and 5 μm P-mesa diameters were calculated to be 301 Ω and 134 Ω, respectively. Incorporating these parameters into the equivalent circuit model shown in Figure 4b, the simulated frequency responses matched well with the experimental results as shown in Figure 21, confirming that suboptimal annealing parameters led to excessive contact resistance, which in turn decreased the RC-limited bandwidth and ultimately reduced the device's 3 dB bandwidth. These insights provide valuable feedback for epitaxial process control and device-level integration in future photonic THz systems.

https://doi.org/10.3390/mi17020196

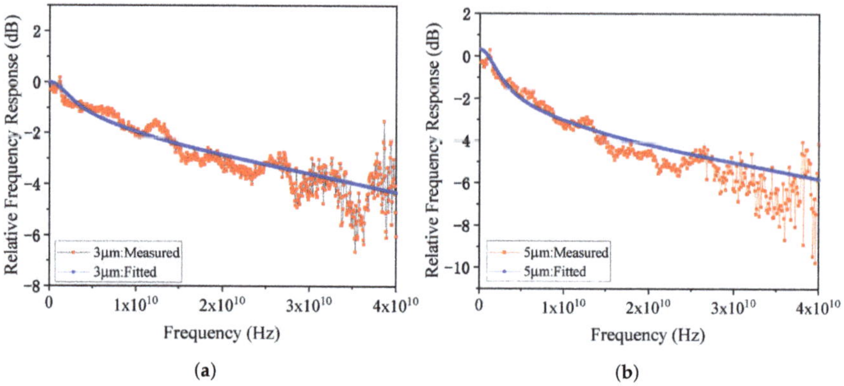

Figure 21. (**a**) The measured and fitted S_{21} parameters for PD with diameter of 3 μm and (**b**) 5 μm under 4 V reverse bias.

To further evaluate the performance of the proposed Photonic Terahertz Generator-A (PTG-A), a comprehensive comparison with other recently reported photonic THz radiation sources is summarized in Table 3. As discussed in the introduction, various strategies have been employed to enhance THz output. For instance, Song et al. utilized a T-junction combiner in a waveguide module to achieve 1.2 mW at 300 GHz [16]. Ohara et al. reported a significant output power of 2.53 mW by focusing on thermal management through the heterogeneous integration of UTC-PDs on SiC substrates [18].

Table 3. Performance comparison of state-of-the-art photonic THz radiation sources.

Ref.	Technology	Freq. (GHz)	Power (mW)	Key Features
[16]	2 × UTC-PD + T-junction	300	1.2	Waveguide module
[18]	UTC-PD on SiC	300	2.53	Thermal management
[20]	4 × 1 UTC-PD Array	300	N/A	T-junction array
[17]	MUTC-PD + Vivaldi	100	3.16	W-band integration
This Work	2 × MUTC-PD + Wilkinson	280	1.44	Integrated Antenna

As shown in Table 3, the proposed PTG-A achieves a competitive saturated power of 1.58 dBm (1.44 mW) at 280 GHz. This represents a 4.13× power enhancement compared to a standalone device, validating the effectiveness of the integrated Wilkinson power combiner and impedance matching networks. Our design provides superior port isolation and robust reflection performance without the specialized fabrication complexity required for SiC bonding.

7. Conclusions

We present a photonic THz antenna for the 275–296 GHz band, featuring an MUTC-PD optimized based on experimental findings. The simulated design integrates the improved PD with matching networks, a Wilkinson combiner, and a Vivaldi antenna on quartz. It achieves 1.58 dBm saturated output at 280 GHz with 4.13× power enhancement, 7.93 dBi peak gain, and reflection coefficients below −10 dB across the band.

Experiments on earlier photodiodes revealed two key limitations: epitaxial deviations reduced transit-time bandwidth from 325 GHz to 252 GHz, while elevated contact resistance from annealing constraints restricted RC-limited bandwidth. These results guided the MUTC-PD redesign, focusing on depletion layer thickness and contact optimization. Future work will focus on the full-system fabrication and experimental validation. To bridge the gap toward practical implementation, we will employ high-precision Electron-Beam Lithography (EBL) and AuSn flip-chip bonding to minimize parasitic effects and phase imbalances in the combining network. A quasi-optical (QO) measurement platform will be established to prioritize the characterization of Effective Isotropic Radiated Power (EIRP) and far-field radiation patterns. This comprehensive roadmap ensures that the proposed architecture serves as a robust and scalable path toward high-power integrated THz transmitters for next-generation 6G wireless systems.

Author Contributions: Methodology, Y.W. and J.Y.; Software, Y.W.; Validation, X.L.; Investigation, Y.W.; Writing—original draft, Y.W.; Writing—review and editing, Y.W. All authors have read and agreed to the published version of the manuscript.

Funding: This work was supported by National Natural Science Foundation of China (62127802).

Data Availability Statement: Data underlying the results presented in this paper are not publicly available at this time but may be obtained from the authors upon reasonable request.

Conflicts of Interest: The authors declare no conflicts of interest.

References

1. Wang, Y.; Yu, J.; Liu, X. Design of Broadband Modified-Uni-Traveling-Carrier Photodiodes with 3 dB Bandwidth of 171 GHz. *Phys. Status Solidi A* **2025**, *222*, 2400858. [CrossRef]
2. Zhang, Q.; Yu, J.; Zhao, X.; Li, W.; Wei, Y.; Yang, X.; Tan, J.; Zhang, B.; Wang, K.; Zhang, Y.; et al. 4.6-km *D*-Band Photonic-Assisted Terahertz Wireless Communication Employing SIMO and MRC Technology. *IEEE Trans. Microw. Theory Tech.* **2024**, *72*, 6657–6668. [CrossRef]
3. Tian, Y.; Dong, B.; Li, Y.; Xiong, B.; Zhang, J.; Sun, C.; Hao, Z.; Wang, J.; Wang, L.; Han, Y.; et al. Photonics-assisted THz wireless communication enabled by wide-bandwidth packaged back-illuminated modified uni-traveling-carrier photodiode. *Opto-Electron. Sci.* **2024**, *3*, 230051. [CrossRef]
4. Ajayan, J. Recent developments in Terahertz wave detectors for next generation high speed Terahertz wireless communication Systems: A review. *Infrared Phys. Technol.* **2024**, *141*, 105467. [CrossRef]
5. Gezimati, M.; Singh, G. Terahertz imaging technology for localization of cancer tumours: A technical review. *Multimed. Tools Appl.* **2024**, *83*, 33675–33711. [CrossRef]
6. Jiang, W.; Zhou, Q.; He, J.; Habibi, M.A.; Melnyk, S.; El-Absi, M.; Han, B.; Di Renzo, M.; Schotten, H.D.; Luo, F.L.; et al. Terahertz communications and sensing for 6G and beyond: A comprehensive review. *IEEE Commun. Surv. Tutor.* **2024**, *26*, 2326–2381. [CrossRef]
7. You, X.; Ako, R.T.; Sriram, S.; Withayachumnankul, W. Experimental Realization of Terahertz 3D Imaging with Chromatic Metasurface. In Proceedings of the 2024 IEEE Asia-Pacific Microwave Conference (APMC), Bali, Indonesia, 17–20 November 2024; pp. 339–341.
8. You, X.; Pagay, V.; Withayachumnankul, W. Non-Contact Monitoring of Plant Leaf Water Status Using Terahertz Waves. *J. Infrared Millim. Terahertz Waves* **2025**, *46*, 53. [CrossRef]
9. You, X.; Fumeaux, C.; Withayachumnankul, W. Circuit-Based Design and Optimization for Broadband Terahertz Metasurfaces. In Proceedings of the 2021 46th International Conference on Infrared, Millimeter and Terahertz Waves (IRMMW-THz), Chengdu, China, 29 August–3 September 2021; pp. 1–2. [CrossRef]
10. Ishibashi, T.; Shimizu, N.; Kodama, S.; Ito, H.; Nagatsuma, T.; Furuta, T. Uni-Traveling-Carrier Photodiodes. In Proceedings of the Ultrafast Electronics and Optoelectronics, Incline Village, NV, USA, 17–19 March 1997; Optica Publishing Group: Washington, DC, USA, 1997; p. UC3. [CrossRef]
11. Ishibashi, T.; Kodama, S.; Shimizu, N.S.N.; Furuta, T.F.T. High-Speed Response of Uni-Traveling-Carrier Photodiodes. *Jpn. J. Appl. Phys.* **1997**, *36*, 6263. [CrossRef]
12. Suzuki, S.; Asada, M. Fundamentals and recent advances of terahertz resonant tunneling diodes. *Appl. Phys. Express* **2024**, *17*, 070101. [CrossRef]
13. Asada, M.; Suzuki, S. Terahertz Emitter Using Resonant-Tunneling Diode and Applications. *Sensors* **2021**, *21*, 1384. [CrossRef]
14. Meng, F.; Tang, Z.; Ourednik, P.; Hazarika, J.; Feiginov, M.; Suzuki, S.; Roskos, H.G. High-power even- and odd mode emission from linear arrays of resonant-tunneling-diode (RTD) oscillators in the 0.4- to 0.8-THz frequency range. *arXiv* **2024**, arXiv:2404.07394.
15. Spudat, C.; Ourednik, P.; Picco, G.; Nguyen, D.; Feiginov, M. Limitations of Output Power and Efficiency of Simple Resonant-Tunneling-Diode Oscillators. *IEEE Trans. Terahertz Sci. Technol.* **2022**, 13, 82–92. [CrossRef]

https://doi.org/10.3390/mi17020196

16. Song, H.J.; Ajito, K.; Muramoto, Y.; Wakatsuki, A.; Nagatsuma, T.; Kukutsu, N. Uni-travelling-carrier photodiode module generating 300 GHz power greater than 1 mW. *IEEE Microw. Wirel. Components Lett.* **2012**, *22*, 363–365. [CrossRef]
17. Sun, K.; Moody, J.; Li, Q.; Bowers, S.M.; Beling, A. High power integrated photonic W-band emitter. *IEEE Trans. Microw. Theory Tech.* **2017**, *66*, 1668–1677. [CrossRef]
18. Ohara, T.; Ishibashi, T.; Kawamoto, Y.; Tojo, M.; Maekawa, K.; Nagatsuma, T. 2-mW-output power uni-traveling-carrier photodiodes in 300-GHz-band. In Proceedings of the 2023 Asia-Pacific Microwave Conference (APMC), Taipei, Taiwan, 5–8 December 2023; pp. 676–678.
19. Nagatsuma, T.; Ohara, T.; Kawamoto, Y.; Maekawa, K.; Ishibashi, T. SiC-substrate uni-traveling-carrier photodiode modules for 300-GHz-band wireless communications. In Proceedings of the 2024 49th International Conference on Infrared, Millimeter, and Terahertz Waves (IRMMW-THz), Perth, Australia, 1–6 September 2024; pp. 1–2.
20. Ssali, H.; Kamiura, Y.; Li, Y.; Li, B.; Che, M.; Kato, K. Enhanced terahertz wave generation by an antenna-coupled 4×1 UTC-PD array and a T-junction combiner. *Opt. Lett.* **2025**, *50*, 6057–6060. [CrossRef]
21. Chao, E.; Xiong, B.; Sun, C.; Hao, Z.; Wang, J.; Wang, L.; Han, Y.; Li, H.; Yu, J.; Luo, Y. D-Band MUTC Photodiodes with Flat Frequency Response. *IEEE J. Sel. Top. Quantum Electron.* **2022**, *28*, 3802208. [CrossRef]
22. Maloney, T.J.; Frey, J. Transient and steady-state electron transport properties of GaAs and InP. *J. Appl. Phys.* **1977**, *48*, 781–787. [CrossRef]
23. Tian, Y.; Xiong, B.; Sun, C.; Hao, Z.; Wang, J.; Wang, L.; Han, Y.; Li, H.; Gan, L.; Luo, Y. Ultrafast MUTC photodiodes over 200 GHz with high saturation power. *Opt. Express* **2023**, *31*, 23790–23800. [CrossRef]
24. Xu, M.; Xiong, B.; Sun, C.; Hao, Z.; Wang, J.; Wang, L.; Han, Y.; Li, H.; Gan, L.; Luo, Y. Ultrafast and High-Power Resonant Cavity Enhanced MUTC Photodiodes with Aperiodic Distributed Bragg Reflectors. *J. Light. Technol.* **2025**, *43*, 4848–4855. [CrossRef]
25. Riesz, R. High speed semiconductor photodiodes. *Rev. Sci. Instruments* **1962**, *33*, 994–998. [CrossRef]
26. Natrella, M.; Liu, C.P.; Graham, C.; van Dijk, F.; Liu, H.; Renaud, C.C.; Seeds, A.J. Accurate equivalent circuit model for millimetre-wave UTC photodiodes. *Opt. Express* **2016**, *24*, 4698–4713. [CrossRef]
27. Nagatsuma, T.; Ishibashi, T.; Hirata, A.; Hirota, Y.; Minotani, T.; Sasaki, A.; Ito, H. Characterisation of uni-travelling-carrier photodiode monolithically integrated with matching circuit. *Electron. Lett.* **2001**, *37*, 1246–1247. [CrossRef]
28. Wolff, I. *Coplanar Microwave Integrated Circuits*; John Wiley & Sons: Hoboken, NJ, USA, 2006.
29. Simons, R.N. *Coplanar Waveguide Circuits, Components, and Systems*; John Wiley & Sons: Hoboken, NJ, USA, 2004.
30. Wang, Z.; Yin, Y.; Wu, J.; Lian, R. A miniaturized CPW-fed antipodal Vivaldi antenna with enhanced radiation performance for wideband applications. *IEEE Antennas Wirel. Propag. Lett.* **2015**, *15*, 16–19. [CrossRef]
31. Moody, J.; Sun, K.; Li, Q.; Beling, A.; Bowers, S.M. A Vivaldi antenna based W-band MUTC photodiode driven radiator. In Proceedings of the 2016 IEEE International Topical Meeting on Microwave Photonics (MWP), Long Beach, CA, USA, 31 October–3 November 2016; pp. 217–220.

https://doi.org/10.3390/mi17020196

micromachines

MDPI

Article

A Compact 380 GHz Zero-Bias Schottky Diode Detector for High-Sensitivity Radiometer Applications

Huilin Tang [1,2], Yongsheng Deng [1,*] and Dehai Zhang [1]

[1] The CAS Key Laboratory of Microwave Remote Sensing, National Space Science Center, Chinese Academy of Sciences, Beijing 100190, China; tanghuilin23@mails.ucas.ac.cn (H.T.); zhangdehai@mirslab.cn (D.Z.)
[2] University of Chinese Academy of Sciences, Beijing 100049, China
* Correspondence: deng@nssc.ac.cn; Tel.: +86-138-1031-6802

Abstract

Reliable terahertz radiometer receiver systems demand detectors that combine high sensitivity with structural robustness. This paper presents the design, fabrication, and characterization of a 380 GHz zero-bias detector based on an ACST Schottky diode. The detector uses a high-impedance grounding topology to reduce parasitic resonances and enhance output stability. A compact U-shaped waveguide transition is adopted to realize an in-line port configuration. This configuration simplifies system integration. Measurements demonstrate a peak voltage responsivity of 2318 V/W and a linearity of 0.9996 at 380 GHz, validating the effectiveness of the proposed design. This work establishes a practical design approach for zero-bias terahertz detectors and supports future high-frequency radiometer and sensing applications.

Keywords: zero-bias Schottky diode; direct grounding; THz detector; radiometer receiver

1. Introduction

Occupying the spectral bridge between microwave electronics and infrared photonics, the terahertz (THz) regime has unlocked critical advancements in scientific and engineering domains [1]. To fully exploit the unique spectral characteristics and application potential of the THz band, the development of high-efficiency, practical receiver technologies is paramount. Among diverse architectures, direct detection receivers have garnered significant interest. By eliminating the complex local oscillator (LO) chain, these receivers provide compact size and low power consumption, which are particularly advantageous for spaceborne and airborne systems subject to stringent size, weight, and power (SWaP) constraints [2].

As the core component converting incident THz signals into measurable DC output, the detector plays a decisive role in determining overall sensitivity and linearity of the receiver [3]. Schottky diodes have benefited from decades of development in high-frequency electronics and remain a mature technology for extending electronic detection toward the THz regime [4]. Despite advancements in Schottky diode and micro/nano manufacturing technologies [5], designing zero-bias detectors operating above 300 GHz remains a formidable challenge. At THz frequencies, parasitic effects associated with the Schottky junction and surrounding metallization increasingly influence detector performance. Existing grounding strategies, such as direct wire bonding or conductive epoxy attachment at the probe tip, have demonstrated effectiveness across millimeter-wave and lower terahertz bands [6,7]. However, at higher frequencies, these methods tend to exhibit increased sensitivity to fabrication tolerances.

https://doi.org/10.3390/mi17030352

The 380 GHz atmospheric window, characterized by its low propagation loss and favorable dispersion, is of strategic importance [8]. Consequently, this frequency band has emerged as a prime candidate for demanding applications, ranging from 6G communications to deep-space exploration and radiometric remote sensing [9,10]. Detector designs specifically optimized for the 380 GHz atmospheric window while offering high sensitivity and compactness remain relatively limited. Representative state-of-the-art results include the waveguide-packaged zero-bias detectors from Virginia Diodes Inc. (VDI), which achieve responsivities of approximately 3000 V/W at 50 GHz and 100 V/W at 1100 GHz [11]. Within the sub-millimeter regime, the Institute of Microelectronics, Chinese Academy of Sciences, has reported a 270 GHz zero-bias detector based on an InP Schottky diode, achieving a typical responsivity of 1400 V/W across 260–280 GHz [12]. Chen et al. demonstrated zero-bias GaAs Schottky detectors at 89 GHz and 150 GHz with stable responsivity and excellent linearity [13]. These studies demonstrate the feasibility of THz zero-bias detection while also underscoring the importance of detector architectures.

This work presents a compact, fabrication tolerant zero-bias THz detector designed for operation in the 380 GHz atmospheric window. The detector utilizes a low barrier GaAs Schottky diode from ACST, enabling high sensitivity and room temperature operation without external bias circuitry. While the use of a commercial diode chip and conventional waveguide-to-planar transition concepts follows established practice, the originality of this work lies in a tailored co-design and integration strategy optimized for the 380 GHz band. A directly grounded high impedance transmission line topology, together with a hammer-head low pass filter (LPF), is adopted to suppress parasitic resonances and enhance high frequency stability. To improve system integrability, a compact U-shaped waveguide bend is incorporated to realize an inline port configuration. The design methodology provides a practical foundation for THz imaging and remote sensing receivers, establishing a transferable framework for future zero-bias detector implementations at even higher frequencies.

2. Circuit Design and Analysis

The operating principle of a zero-bias terahertz detector is based on the square-law rectification characteristics of the Schottky diode, which converts incident THz signal into a measurable dc output. The diode current–voltage relationship is expressed as [14]

$$\mathrm{I}(V) = I_s\left(e^{\frac{qV}{nkT}} - 1\right) \tag{1}$$

where I_s is the saturation current, q is the electron charge, V is the junction voltage, n is the ideality factor, k is the Boltzmann constant, and T is the absolute temperature.

Under small-signal excitation, a small amplitude terahertz voltage $\delta V = V_p\cos(\omega_c t)$ is superimposed on the DC bias V_0 (which is zero in this case). By expanding the diode current using a Taylor series around the operating point and neglecting higher-order terms, the current can be approximated as:

$$\mathrm{i}(V) \approx i(V_0) + V_p\cos(\omega_c t)\left.\frac{di}{dV}\right|_{V_0} + \frac{V_p^2}{4}\left.\frac{d^2i}{dV^2}\right|_{V_0} + \frac{V_p^2\cos(2\omega_c t)}{4}\left.\frac{d^2i}{dV^2}\right|_{V_0} \tag{2}$$

The resulting signal comprises DC, fundamental, and second-harmonic components. The generated DC current is proportional to the square of the input voltage amplitude, enabling estimation of the RF input power through DC voltage measurement.

$$I_{DC} \propto \delta V^2 \tag{3}$$

https://doi.org/10.3390/mi17030352

The system topology of the 380 GHz zero-bias detector is illustrated in Figure 1. The circuit consists of a waveguide-to-microstrip transition, an impedance matching network, a Schottky diode, a high-impedance grounding path, and an LPF. To streamline the circuit implementation, the conventional output matching section is replaced by a phase-shifting transmission line. This approach reduces circuit complexity and physical size while maintaining effective impedance matching.

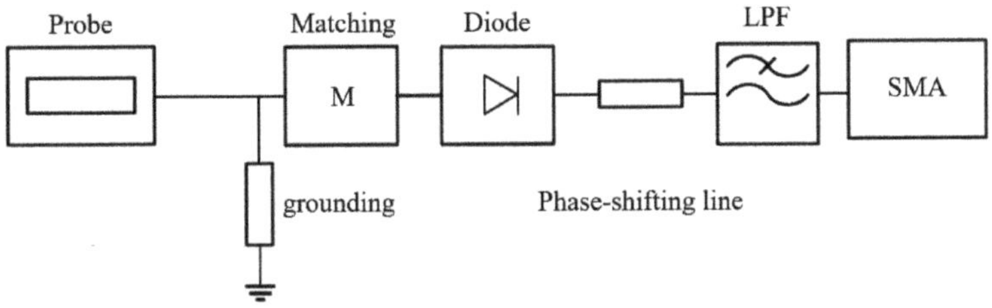

Figure 1. System topology of the 380 GHz zero-bias Schottky detector.

As depicted in Figure 1, the incident THz signal is coupled into the microstrip line via a low loss waveguide-to-microstrip transition. The nonlinear Schottky junction performs the rectification, and the generated DC component is extracted by the LPF. At terahertz frequencies, the detector performance is determined not only by the intrinsic parameters of the Schottky diode but also by the electromagnetic coupling among circuit blocks, parasitic elements, and the physical layout. Optimization of the detector topology is essential for improving detection efficiency, enhancing high frequency stability, and robustness against fabrication tolerances.

2.1. Modelling of Schottky Diodes

Accurately predicting device behavior at 380 GHz requires a rigorous treatment of distributed parasitic effects, which render conventional lumped-element models insufficient. Consequently, a hybrid modeling strategy was adopted to characterize the quasi-vertical GaAs Schottky diode (ACST).

A full-wave electromagnetic analysis was first performed on the three-dimensional (3D) physical structure of the diode, including the GaAs mesa, air-bridge anode, and cathode pads, in order to extract frequency-dependent S-parameters. The cross-sectional schematic is illustrated in Figure 2a. These distributed parameters were then embedded into a nonlinear SPICE model in Figure 2b. This circuit includes the series resistance R_s, junction capacitance C_j, junction resistance R_j, series inductance L_s, shunt parasitic capacitance C_p, together with parasitic contributions from metallization, contacts, and fingers.

Figure 3a shows the established 3D electromagnetic model of the diode. Key parameters, including the saturation current I_s, series resistance R_s, ideality factor n, and zero-bias junction capacitance C_{j0}, were extracted by fitting the model to the measured DC I-V characteristics. As shown in Figure 3b, the co-simulation results show good consistency with the measured data, validating the accuracy of the combined model for the subsequent detector design.

https://doi.org/10.3390/mi17030352

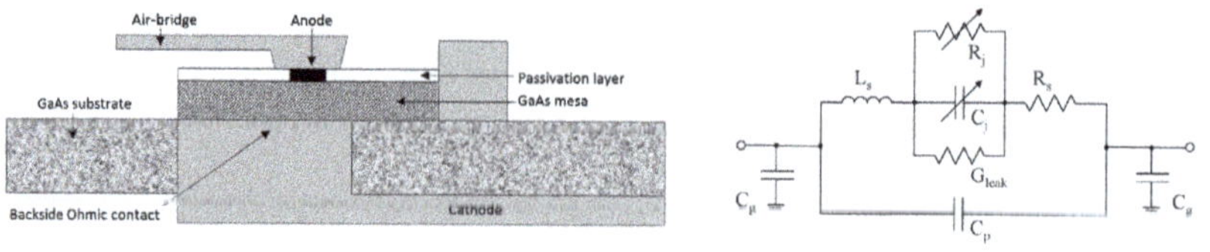

(a) (b)

Figure 2. Structure and equivalent circuit of the Schottky diode: (**a**) cross-sectional schematic of the diode structure; (**b**) equivalent circuit model.

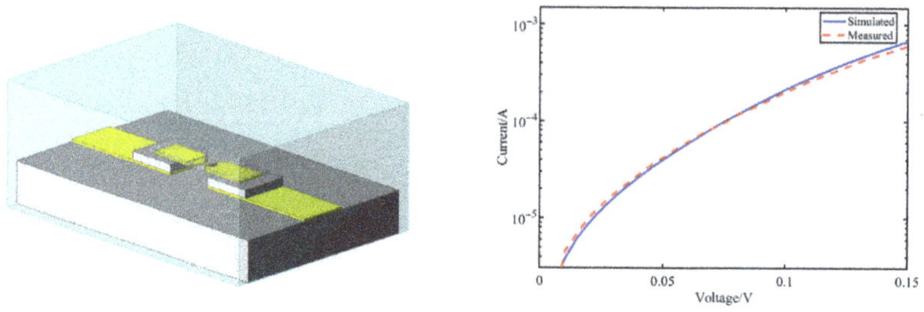

(a) (b)

Figure 3. Schottky diode modeling and validation: (**a**) 3D electromagnetic model of the diode; (**b**) comparison between measured and simulated I–V characteristics.

2.2. *Waveguide-to-Microstrip Transition Structure*

Efficient energy coupling is critical for detector sensitivity. An E-plane probe transition was optimized for the 360–400 GHz band. Key geometric parameters, especially the probe insertion depth and backshort distance, were tuned by parametric sweeps. Figure 4a illustrates the optimized transition structure. As shown in the simulation results in Figure 4b, the transition achieves a return loss better than 25 dB and insertion loss below 0.15 dB across the target frequency band.

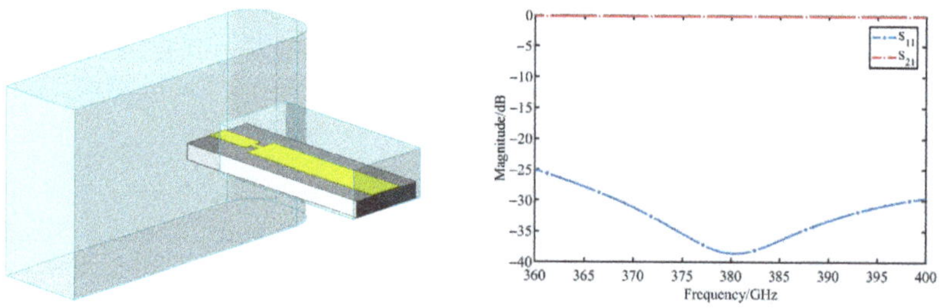

(a) (b)

Figure 4. Waveguide microstrip transition Structure: (**a**) Geometry of the E-plane probe transition; (**b**) Simulated S-parameters of the transition.

However, standard E-plane probes dictate an orthogonal port arrangement, which complicates system stacking. To resolve this, a U-shaped waveguide bend was integrated to realize an inline port configuration, as shown in Figure 5. Unlike conventional mitered bends, which are prone to parasitic reflections and high-order mode excitation at terahertz frequencies, the proposed design features a large-radius continuous curvature. Field distribution simulations confirm that this smoothed profile ensures seamless energy transmission without significant radiation leakage or resonant accumulation, effectively satisfying the

https://doi.org/10.3390/mi17030352

dual requirements of compactness and high-frequency signal integrity. The simulated insertion loss is less than 0.4 dB.

Figure 5. U-shaped waveguide bend structure.

2.3. Grounding Design

In E-plane probe transitions, the DC return path is usually realized by extending the substrate across the waveguide aperture to contact with the cavity wall [15,16]. However, this method extends the substrate length, increasing mechanical fragility for thin quartz materials. Another grounding scheme relies on gold wire bonding [17,18]. At terahertz frequency, thinner gold wires are needed. However, the standard bonding processes typically utilize wires with diameters ranging from 12.5 μm to 25 μm, thinner gold wires pose challenges in terms of process stability and reliability [19].

To mitigate the sensitivity of conventional grounding schemes to fabrication and assembly tolerances at terahertz frequencies, a high-impedance grounding path is adopted in this work. The grounding structure is shown in Figure 6a. Placing the grounding line upstream of the input matching network facilitates a more uniform distribution of grounding points. This configuration reduces the sensitivity of the grounding impedance to position variations, improving robustness against fabrication and assembly tolerances. The grounding wire behaves as a distributed element whose high-frequency response can be approximated by a series inductance and resistance. The equivalent circuit of the grounding path is shown in Figure 6b.

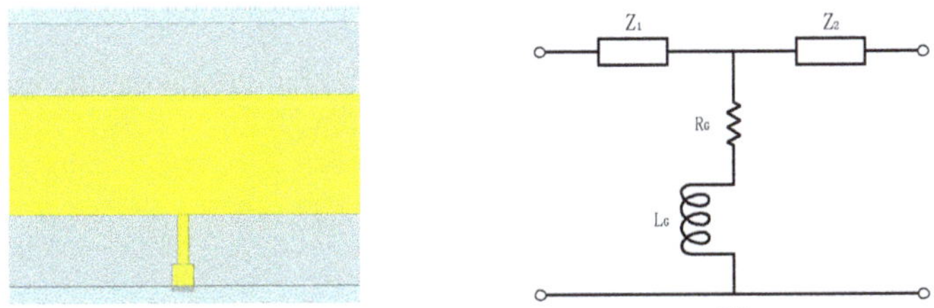

(a) (b)

Figure 6. High-impedance direct-grounding structure: (**a**) physical layout of the grounding line; (**b**) equivalent circuit representation of the grounding path.

The geometrical dimensions of the grounding wire can be precisely controlled during fabrication, and its high frequency behavior can be approximated by a series inductance L_G and resistance R_G, expressed as:

$$L_G = \mu_0 \cdot \frac{1}{2\pi} \cdot \left(\ln\left(\frac{4l}{d}\right) + \frac{\mu_r \tanh\left(\frac{4d_s}{d}\right)}{4} - 1 \right) \quad (4)$$

https://doi.org/10.3390/mi17030352

$$R_G = \begin{cases} \left(\frac{4\rho l}{\pi d^2}\right)\cdot\cosh\left[0.041\left(\frac{d}{d_s}\right)^2\right], & \frac{d}{d_s} \leq 3.394 \\ \left(\frac{4\rho l}{\pi d^2}\right)\cdot\left(\frac{0.25d}{d_s} + 0.2654\right), & \frac{d}{d_s} > 3.394 \end{cases} \quad (5)$$

where μ_0 is the permeability of free space, l is the wire length, d is the wire diameter, d_s is the skin depth at the operating frequency, and ρ denotes the electrical resistivity.

2.4. Low Pass Filter Design

To extract the rectified DC signal while rejecting the terahertz carrier, a compact filter with steep roll-off is required. Traditional stepped-impedance filters are often too lengthy for integrated terahertz modules [20]. To achieve steep stopband rejection within a compact footprint, a Hammer-head topology [21] is adopted. As shown in Figure 7a, this structure introduces capacitive loading via a planar expansion element on the microstrip line, which effectively sharpens the cutoff response and enhances high-frequency attenuation without substantially increasing the circuit length. The equivalent circuit model of the cell structure is shown in Figure 7b.

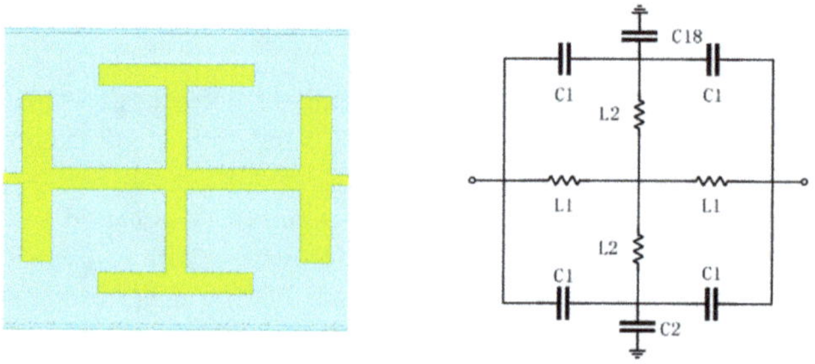

(**a**) (**b**)

Figure 7. Hammer-head low-pass filter unit cell: (**a**) physical layout of the microstrip structure; (**b**) equivalent circuit.

To enhance the stopband attenuation, two Hammer-head cells are cascaded. The resulting dual-stage LPF structure and its simulated frequency response are shown in Figure 8. The simulation results indicate an attenuation level exceeding 42 dB across the 340–420 GHz bandwidth, while maintaining low insertion loss at low frequency, ensuring that the DC signal generated by the detector can pass through the filter smoothly. The proposed LPF simultaneously offers structural compactness and superior stopband performance.

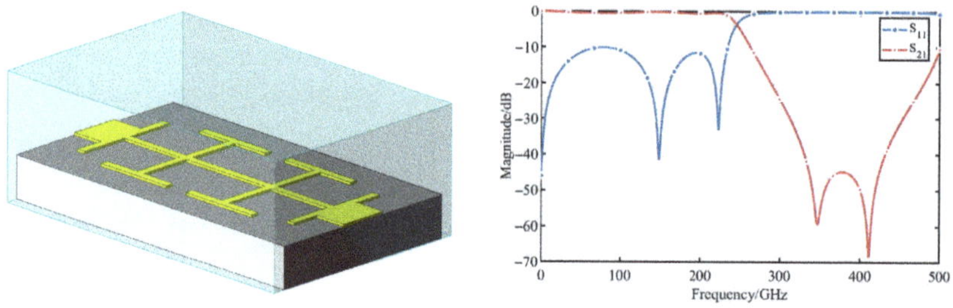

(**a**) (**b**)

Figure 8. Cascaded hammerhead LPF: (**a**) 3D model of the LPF; (**b**) Simulated S-parameters of the LPF.

https://doi.org/10.3390/mi17030352

2.5. Overall Simulation and Analysis

The individual sub-circuits were integrated into a system-level model for final optimization, as shown in Figure 9a. The lengths of the phase-shifting line and the diode input transmission lines were adjusted to compensate for the imaginary part of the input impedance of the diode-LPF cascade at 380 GHz. Considering the compactness requirements of the detector circuit, a single stub matching network is adopted to realize high responsivity detection over the operating bandwidth.

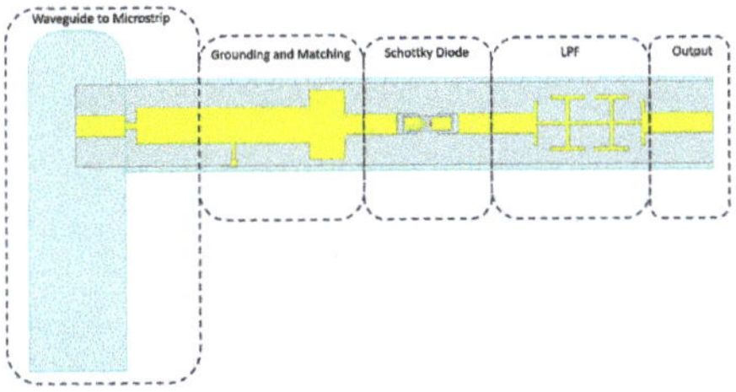

(**a**)

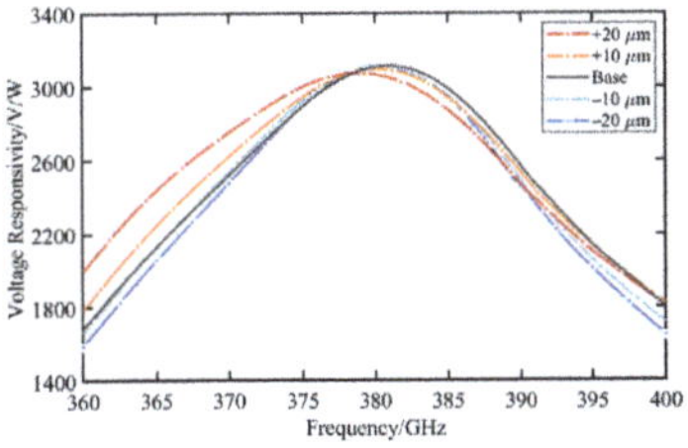

(**b**)

Figure 9. Simulation results of the 380 GHz zero-bias detector: (**a**) Integrated circuit model; (**b**) Voltage responsivity for different grounding points.

To validate the manufacturing robustness, a tolerance-aware parametric analysis was conducted by introducing controlled perturbations to the grounding position. Grounding offsets of ±10 μm and ±20 μm are incorporated into the co-simulation model, and the corresponding results are presented in Figure 9b. Simulation results show that the detector achieves a peak voltage responsivity of about 3000 V/W near 380 GHz and maintains stable performance within the range of ±20 μm offset, with only a slight responsivity variation and no observable frequency offset. These results demonstrate that the proposed grounding topology effectively mitigates performance variations caused by fabrication and provides robust support for stable and reproducible high sensitivity detector performance.

3. Fabrication and Measurements

The detector is implemented in a split-block waveguide architecture fabricated from gold-plated copper, ensuring high conductivity and minimizing surface losses. As shown in Figure 10a, the input port adopts standard WR-2.2 waveguide interface, and the volume of the whole detection module is 20 × 20 × 15 mm^3. The detector circuit was fabricated on a quartz substrate with a relative permittivity of 3.78 and a thickness of 50 μm. As shown in Figure 10b, the substrate is mounted into the split-block channel using conductive silver epoxy, and direct grounding is established. The DC output port was connected to an SMA probe via gold wire bonding. During this practical assembly process, the manual dispensing of conductive silver epoxy and the placement of the quartz substrate inherently introduce microscopic alignment variations, typically within the range of 10 to 20 μm.

The experimental setup is shown in Figure 11. A PSG Analog Signal Generator was used as the small-signal source, and the terahertz signal was generated through a frequency-multiplier chain provided by ACST. The signal is then delivered to the detector under test via a variable attenuator. To characterize the detector voltage responsivity, the input power at the detector was calibrated using a VDI PM5B power meter (Virginia Diodes Inc., Charlottesville, VA, USA), and the input level was maintained at approximately −30 dBm. For the linearity measurement, the carrier frequency is kept constant while the input power is swept by adjusting the attenuator, and the resulting DC voltage response is recorded using a FLUKE digital multimeter(Fluke Corporation, Everett, WA, USA).

https://doi.org/10.3390/mi17030352

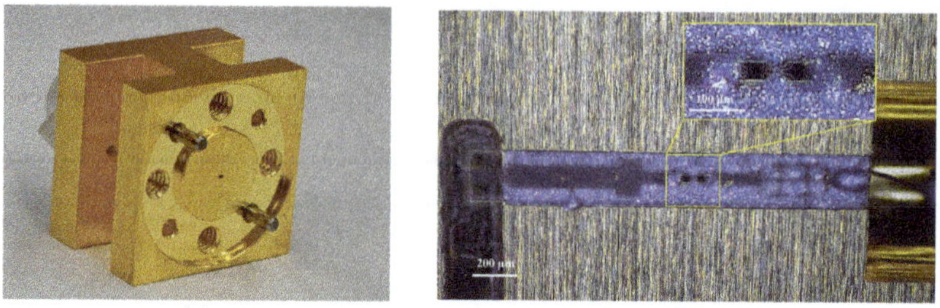

(**a**) (**b**)

Figure 10. Fabricated 380 GHz zero-bias detector: (**a**) detector module; (**b**) Optical image of the detector circuit.

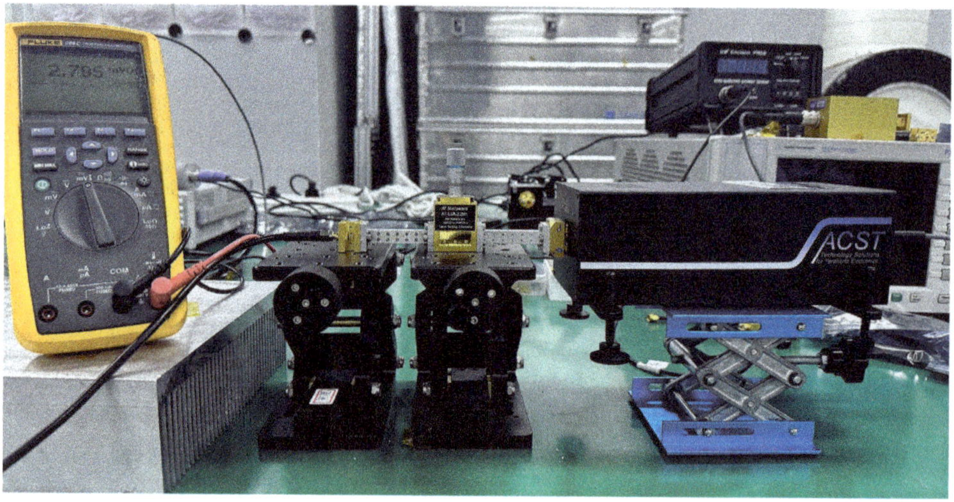

Figure 11. Testing environment.

The relationship between the measured voltage responsivity and frequency is shown in Figure 12a. Within the frequency range of 360–400 GHz, the detector achieves a peak voltage responsivity of 2318 V/W at 380 GHz, demonstrating that the zero-bias Schottky detector exhibits favorable detection performance in this frequency band. For a zero-bias Schottky diode, the total noise is dominated by thermal noise. Using the typical junction resistance R_j of 2500 Ω at room temperature (T = 290 K), the theoretical thermal noise voltage $V_n = \sqrt{4k_B T R_j}$ is calculated to be approximately 6.32 nV/$\sqrt{\text{Hz}}$. Given the measured peak voltage responsivity of 2318 V/W at 380 GHz, the theoretical noise limited equivalent power (NEP) is estimated to be roughly 2.73 pW/$\sqrt{\text{Hz}}$.

Linearity, a critical metric for accurate radiometric brightness temperature retrieval, was evaluated by sweeping the input power at 380 GHz. As shown in Figure 12b, the detector exhibits a clear linear response over the measured input power range.

Compared with the simulated results, the measured responsivity exhibits a certain degree of degradation across the entire frequency band. This discrepancy is primarily attributed to several practical non-idealities at sub-millimeter wavelengths. First, the nonlinear diode model employs nominal parameters, whereas the effective series resistance and junction capacitance of the assembled Schottky diode can deviate from the modeled values under practical DC and high frequency operating conditions, which directly affects the RF-to-DC conversion efficiency and the peak responsivity. Furthermore, parasitic effects

https://doi.org/10.3390/mi17030352

introduced by the conductive silver epoxy assembly, microscopic substrate placement inaccuracies, and the output gold wire bonding all collectively contribute to the overall responsivity degradation. Despite these differences, the measured results are consistent with the simulation in terms of the overall trend and the location of the peak responsivity, which verifies the validity of the proposed design approach. Table 1 presents a comparison of the voltage responsivity of the proposed detector with other reported terahertz detectors.

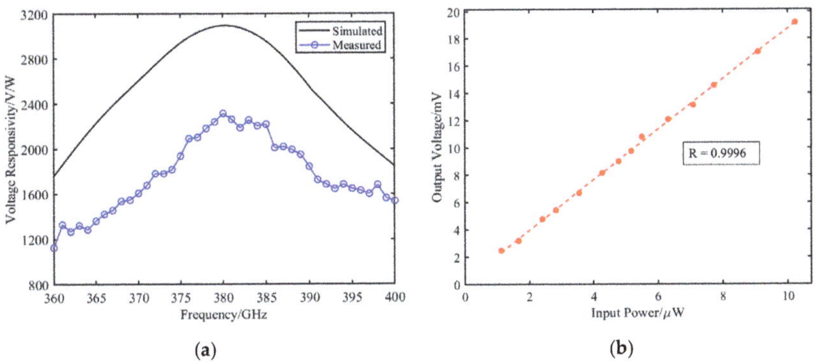

Figure 12. Measured performance: (**a**) voltage responsivity; (**b**) linearity.

Table 1. Comparison with related work.

Ref.	Frequency	Voltage Responsivity	Technology
[11]	330–500 GHz	Typ: 1250 V/W	VDI WR2.2 ZBD
[13]	85–95 GHz 145–155 GHz	Typ: 2500 V/W (85–95 GHz) Typ: 1600 V/W (145–155 GHz)	ACST SBD
[17]	140–165 GHz	1211 V/W (Max @147 GHz)	Schottky diodes
[18]	210–220 GHz	Typ: 1800 V/W	VDI G-band SBD
[22]	0.1–0.8 THz	320 V/W (@300 GHz)	InGaAs/InP SBD
[23]	330–500 GHz	2123 V/W (Max @340 GHz)	InP TB-RTD
This work	360–400 GHz	2318 V/W (Max @380 GHz)	ACST SBD

The results demonstrate that the detector is specifically optimized for the 380 GHz atmospheric window, achieving a highly competitive peak voltage responsivity of 2318 V/W compared to the commercial VDI WR2.2 detector. The proposed design achieves competitive detection performance within a comparable frequency range.

4. Conclusions

A 380 GHz zero-bias Schottky diode waveguide detector is demonstrated, featuring an integrated inline U-shaped waveguide interface, a broadband waveguide-to-microstrip transition, and a high impedance direct grounding topology. The proposed grounding strategy significantly reduces sensitivity to assembly misalignment. Simulations indicate that a ±20 μm lateral grounding offset causes only a minor degradation in peak responsivity and no pronounced center frequency shift, which is beneficial for repeatable manufacturing. A fabricated split-block prototype was characterized from 360 to 400 GHz, achieving a measured peak voltage responsivity of 2318 V/W at 380 GHz, and maintains linearity of 0.9996. These results validate that the proposed transition and grounding co-design enables compact, tolerance resilient THz detector modules suitable for radiometer receivers.

https://doi.org/10.3390/mi17030352

Author Contributions: Writing—original draft, conceptualization, validation and editing, H.T.; Writing—review and editing, Y.D.; supervision, D.Z. All authors have read and agreed to the published version of the manuscript.

Funding: This research was funded by the HY-2E/F Satellite Calibration Radiometer (E41Z200201).

Institutional Review Board Statement: Not applicable.

Informed Consent Statement: Not applicable.

Data Availability Statement: The data that support the findings of this study are available from the corresponding author upon reasonable request.

Acknowledgments: The authors are grateful to Jin Meng for assisting with the measurement.

Conflicts of Interest: The authors declare no conflicts of interest.

References

1. Han, R.; Hu, Z.; Wang, C.; Holloway, J.; Yi, X.; Kim, M.; Mawdsley, J. Filling the Gap: Silicon Terahertz Integrated Circuits Offer Our Best Bet. *IEEE Microw. Mag.* **2019**, *20*, 80–93. [CrossRef]
2. Cooke, C.M.; Leong, K.M.K.H.; Zamora, A.; Gorospe, B.S.; Mei, X.B.; Kangaslahti, P.; Schlecht, E.; Ogut, M.; Goncharenko, Y.; Reising, S.C.; et al. A 670 GHz Integrated InP HEMT Direct-Detection Receiver for the Tropospheric Water and Cloud Ice Instrument. *IEEE Trans. Terahertz Sci. Technol.* **2021**, *11*, 566–576. [CrossRef]
3. Ajayan, J. Recent Developments in Terahertz Wave Detectors for Next Generation High Speed Terahertz Wireless Communication Systems: A Review. *Infrared Phys. Technol.* **2024**, *141*, 105467. [CrossRef]
4. Lewis, R.A. A review of terahertz detectors. *J. Phys. D Appl. Phys.* **2019**, *52*, 433001. [CrossRef]
5. Yadav, R.; Ludwig, F.; Faridi, F.R.; Klopf, J.M.; Roskos, H.G.; Preu, S.; Penirschke, A. State-of-the-Art Room Temperature Operable Zero-Bias Schottky Diode-Based Terahertz Detector Up to 5.56 THz. *Sensors* **2023**, *23*, 3469. [CrossRef] [PubMed]
6. Wu, C.; Zhang, Y.; Xu, Y.; Yan, B.; Xu, R. Millimeter-Wave Waveguide-to-Microstrip Transition with a Built-In DC/IF Return Path. *IEEE Trans. Microw. Theory Tech.* **2021**, *69*, 1295–1304. [CrossRef]
7. Cui, J.; Zhang, Y.; Xia, D.; Xu, Y.; Xiao, F.; Yan, B.; Xu, R. A 220 GHz Broadband Sub-Harmonic Mixer Based on Global Design Method. *IEEE Access* **2019**, *7*, 30067–30078. [CrossRef]
8. Millán, L.F.; Lebsock, M.D.; Kurowski, M.J. Retrieval simulations of a spaceborne differential absorption radar near the 380 GHz water vapor line. *Atmos. Meas. Tech.* **2025**, *18*, 4483–4495. [CrossRef]
9. Serghiou, D.; Khalily, M.; Brown, T.W.; Tafazolli, R. Terahertz channel propagation phenomena, measurement techniques and modeling for 6G wireless communication applications: A survey, open challenges and future research directions. *IEEE Commun. Surv. Tutor.* **2022**, *24*, 1957–1996. [CrossRef]
10. Xiao, X.; Weng, F. A Comparison of Information Content at Microwave to Millimeter Wave Bands for Atmospheric Sounding. *Remote Sens.* **2022**, *14*, 6124. [CrossRef]
11. Virginia Diodes, Inc. Detectors (ZBD). Available online: https://vadiodes.com/detectors-zbd/ (accessed on 30 January 2026).
12. Zhang, J.-J.; Zhou, J.-T.; Yang, C.-Y.; Tian, Z.; Jin, Z. A 270 GHz High Performance Waveguide Detector Utilizing a Zero-Bias Schottky Diode. *J. Infrared Millim. Waves.* **2015**, *34*, 1–5.
13. Chen, Y.; Zhang, S.-W. Design of Millimeter-Wave Detectors Based on Zero-Bias Schottky Diode for Direct Detection System of CubeSat Radiometer. *J. Infrared Millim. Waves.* **2022**, *41*, 285–293.
14. Singh, Y. *Semiconductor Devices*; I.K. International Pvt Ltd.: New Delhi, India, 2013.
15. Ren, T.-H.; Zhang, Y.; Yan, B.; Xu, R.-M.; Yang, C.-Y.; Zhou, J.-T.; Jin, Z. A High Performance Terahertz Waveguide Detector Based on a Low-Barrier Diode. *Chin. Phys. Lett.* **2016**, *33*, 060701. [CrossRef]
16. Liu, X.-Y.; Zhang, Y.; Xia, D.-J.; Ren, T.-H.; Zhou, J.-T.; Guo, D.; Jin, Z. A High-Sensitivity Terahertz Detector Based on a Low-Barrier Schottky Diode. *Chin. Phys. Lett.* **2017**, *34*, 070701. [CrossRef]
17. Liu, S.-Y.; Zhang, D.-H.; Meng, J.; Zhu, H.-T. The Design of a High Voltage Responsivity Terahertz Detector Based on Improved Grounding Model. *Microw. Opt. Technol. Lett.* **2021**, *63*, 2326–2332. [CrossRef]
18. He, W.; Yu, W.; Lv, X. Design and Implementation of Terahertz Detector. In *Proceedings of the 2022 IEEE 10th Asia-Pacific Conference on Antennas and Propagation (APCAP), Xiamen, China, 4–7 November 2022*; IEEE: New York, NY, USA, 2022; pp. 1–2.
19. *GB/T 8750-2022*; Gold-Based Bonding Wire and Bandlet for Semiconductor Package. Standards Press of China: Beijing, China, 2022.
20. Singh, J.; Jha, S.K.; Raturi, S.; Singh, V.; Awasthi, Y.K. Design of Low Pass Filter using Microstrip Line—A Comparative Study. In *Proceedings of the 2021 8th International Conference on Signal Processing and Integrated Networks (SPIN), Noida, India, 26–27 August 2021*; IEEE: New York, NY, USA, 2021; pp. 868–873.

https://doi.org/10.3390/mi17030352

21. He, Y.; Tian, Y.; Miao, L.; Jiang, J.; Deng, X. A Broadband 630–720 GHz Schottky Based Sub-Harmonic Mixer Using Intrinsic Resonances of Hammer-Head Filter. *China Commun.* **2019**, *16*, 76–84.
22. Shin, J.-H.; Park, D.W.; Lee, E.S.; Kim, M.; Lee, D.H.; Lee, I.-M.; Park, K.H. Highly reliable THz hermetic detector based on InGaAs/InP Schottky barrier diode. *Infrared Phys. Technol.* **2021**, *115*, 103736. [CrossRef]
23. Clochiatti, S.; Grygoriev, A.; Kress, R.; Mutlu, E.; Possberg, A.; Vogelsang, F.; van Delden, M.; Pohl, N.; Weimann, N.G. Low-Noise Resonant Tunneling Diode Terahertz Detector. *IEEE Trans. Terahertz Sci. Technol.* **2025**, *15*, 107–119. [CrossRef]

https://doi.org/10.3390/mi17030352

micromachines

MDPI

Article

2 km Uncompressed HD Video Wireless Transmission at 100 GHz Based on All-Optical Frequency Up- and Down-Conversion

Shuang Gao [1], Yutong Jiang [1], Zhuoxin Li [1], Qing Zhong [1], Min Zhu [1,2,*] and Jiao Zhang [2,*]

1 National Mobile Communications Research Laboratory, Southeast University, Nanjing 210096, China; shuang_gao@seu.edu.cn (S.G.); qingzhong@seu.edu.cn (Q.Z.)
2 Purple Mountain Laboratories, Nanjing 211111, China
* Correspondence: minzhu@seu.edu.cn (M.Z.); zhangjiao1@pmlabs.com.cn (J.Z.)

Abstract: The millimeter-wave wireless transmission system is widely regarded as a promising solution for applications of future 6G communication. This paper presents an experimental comparison between all-optical and all-electric receivers for millimeter-wave communication systems over a 15 m wireless link and demonstrates 200 m and 2 km real-time uncompressed HD video transmission using an all-optical transceiver at 100 GHz. The systems leverage photonics-assisted heterodyne beating techniques at the transmitter, while the receivers employ either an avalanche photodiode (APD)-based all-optical approach or an envelope detection-based all-electric approach. Experimental results show that the all-optical transceiver supports significantly higher transmission rates, achieving error-free transmission at up to 11.318 Gbps over a 200 m wireless link without clock recovery, compared to the all-electric receiver, which is limited to only 3.125 Gbps error-free 15 m transmission. This work proves that the proposed system based on the all-optical receiver is more promising for supporting future 6G scenarios requiring ultra-wideband, high capacity, and wide coverage high-speed wireless communications.

Keywords: millimeter-wave; all-optical transceiver; uncompressed HD video transmission

1. Introduction

The millimeter-wave (mmWave) band has garnered significant attention for its high bandwidth efficiency, low latency, and ability to meet the demands of emerging bandwidth-hungry applications [1,2]. At the transmitting side of mmWave wireless communication systems, to convert the baseband data signal into the mmWave frequency band, there are typically two technical approaches to achieve mmWave up-conversion. One is the all-solid-state electronic mixing technique, and the other is the photonics-assisted heterodyne beating technique. The photonics-assisted technique leverages the advantages of optical devices, such as their wide bandwidth and high-frequency response, which has demonstrated potential in overcoming the limitations of conventional electronic devices, enabling ultra-wideband communication with higher data rates and better transmission performance [3]. In addition, mmWave up-conversion based on the photonics-assisted technique has been widely used due to its convenience in facilitating fiber-wireless-fiber (FiWiFi) convergence. At the receiving side of mmWave wireless communication systems, all-electric receivers based on envelope detection and all-optical receivers are two main categories for mmWave down-conversion approaches. Unlike all-electric receivers based on monolithic microwave integrated circuit (MMIC) technology, the all-electric receiver based on envelope detection does not require the generation of a local oscillator signal related to the mmWave carrier frequency. The core task of the envelope detector is to determine the amplitude variations of the intensity-modulated signal. Meanwhile, this approach avoids complex frequency synthesis or mixing operations, making it suitable for low-complexity scenarios. In contrast, the all-optical receiver employs an ultra-wideband mmWave electro–optic modulator to directly modulate the received mmWave signal onto the optical carrier. Subsequently, an optical band-pass filter filters out one of the sidebands as the optical baseband signal [4,5].

Under the IMT-2030 framework, particularly within the context of the newly defined usage scenarios, such as immersive communication, massive communication, and ubiquitous connectivity, the demand for emerging applications like uncompressed 4K/8K ultrahigh-definition (UHD) video has surged, leading to an exponential increase in data traffic [6]. Optical-to-millimeter-wave (O-mmW) and millimeter-wave-to-optical (mmW-O) conversions enable the seamless integration of millimeter-wave wireless communication with existing optical fiber networks. As shown in Figure 1, this approach allows for high-performance and cost-effective transmission in scenarios such as low-altitude unmanned aerial vehicle (UAV) wireless relays, locations where laying optical fibers is impractical (e.g., across rivers), and emergency communications during fiber-optic disruptions caused by disasters.

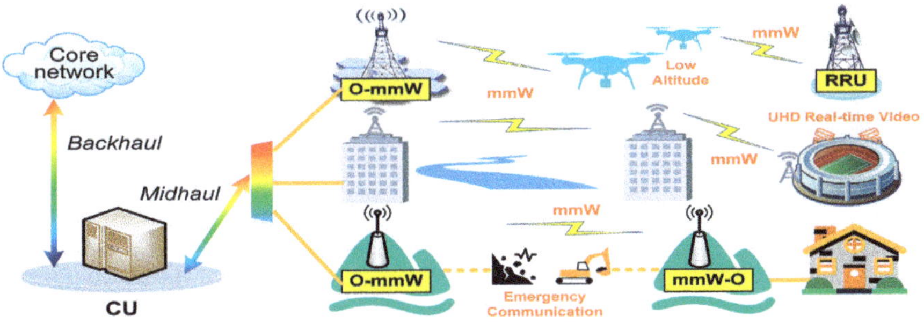

Figure 1. Scenarios for mmWave wireless communication under IMT-2030 framework.

To achieve the aforementioned objectives, one trend in the evolution of 5G and future 6G communications is the use of high-order modulation formats and offline digital signal processing [7,8]. However, this scheme increases receiver complexity and latency, thereby hindering real-time wireless transmission applications. To resolve this issue, at the transmitting side, employing photonics-assisted heterodyne beating techniques is an effective technical solution for generating wideband, high-quality mmWave and terahertz signals. On the receiving side, leveraging advancements in detectors based on Schottky barrier diodes (SBD) for mmWave and terahertz signal detection [9], one practical approach is to utilize the simple architecture of the all-electric receiver based on envelope detection. Another viable approach is to select a suitable high-speed electro–optic modulator, utilizing an all-optical receiver to improve transmission performance while balancing system complexity with the benefits of reduced insertion loss and spurious noise. The key devices in the all-electric receivers and all-optical receivers are envelope detectors and photodetectors, respectively. They are commercially available and have been continuously upgraded, enabling the demonstration of mmWave wireless communication systems with increasingly higher bandwidth and data rates. For instance, a balanced Schottky diode envelope detector, capable of demodulating amplitude shift keying (ASK) signals, is used to demonstrate mmWave transmission at data rates up to 14 Gbit/s, with a tunable carrier frequency between 78 GHz and 92 GHz [10]. On the other hand, although the all-optical receiver is more complex in structure and higher in cost compared to the all-electric receiver, it can achieve significantly higher data rates. As the key component of the all-optical receiver, avalanche photodiodes (APDs) are increasingly being used in wireless communication systems at the mmWave and Terahertz bands, particularly in scenarios where their ability to handle weak signal detection and high-speed data is crucial [11]. For example, a system using APD on the receiving side achieves error-free transmission at data rates between 30 Gbit/s and 50 Gbit/s over a distance of 10 m at a carrier frequency of 300 GHz [12]. However, there has been no simultaneous demonstration and comparison of the transmission performance of an all-electric receiver based on envelope detection and an all-optical receiver based on APD, which is crucial for providing valuable guidance on balancing cost and transmission performance, as well as on practical applications like transmitting uncompressed HD or UHD 4K/8K video.

This paper experimentally demonstrates the mmWave transmission system with an all-electric receiver based on envelope detection and an all-optical receiver based on APD. Firstly, we investigate the transmission performance of the two receiving schemes at a 100 GHz carrier frequency and a 15 m wireless link. The all-optical receiver scheme achieves error-free transmission of an 11.318 Gbit/s OOK signal, while the all-electric receiver based on envelope detection can only achieve error-free transmission up to 3.125 Gbit/s for the OOK signals. Furthermore, in the exploration of long-distance transmission using the all-optical receiver scheme, we achieve 200 m of error-free transmission at 10 Gbit/s. Additionally, to verify the feasibility of this scheme in practical applications, we accomplished real-time transmission of uncompressed HD video over a distance of 200 m and 2 km. The demonstrated systems reveal that the all-optical receiver is a more promising millimeter-wave-receiving approach for supporting future 6G scenarios requiring ultra-wideband, high-capacity, and wide-coverage high-speed wireless communications.

2. Link Budget Analysis

To provide guidance for subsequent wireless transmission experiments, it is necessary to conduct a theoretical analysis of transmission loss and perform the link budget. As is well known, the free-space path loss in wireless transmission is proportional to the square of the frequency. Therefore, the free-space loss in the transmission at the mmWave band is significantly higher than that at the lower frequencies currently used by operators. Thus, conducting theoretical analysis and calculations of transmission loss and power budget before performing specific experiments is of great importance in guiding the selection of experimental parameters.

According to the Friis transmission equation [13,14], the relationship between the transmitted power P_T, received power P_R, transmitter antenna gain G_R, receiver antenna gain G_T, and the path loss is given as follows:

$$P_R = P_T + G_T + G_R - \overline{L_P}(d) - L_f - L_a \times d, \quad (1)$$

$$\overline{L_P}(d) = L_{FS}(d_0) + 10n\log\left(\frac{d}{d_0}\right), \quad (2)$$

where d is the distance of wireless transmission, $\overline{L_P}(d)$ is the average path loss at a specific distance, $L_{FS}(d_0)$ is the free-space path loss at the reference distance, expressed as $L_{FS}(d_0) = 20\log\left(\frac{4\pi d_0}{\lambda}\right)$, where λ is the wavelength, and the typical reference distance is $d_0 = 1\ m$; n is the path loss exponent, which represents the rate at which the path loss increases with the distance between the transmitter and receiver, L_f is the line loss, and L_a is the atmospheric attenuation factor, which can be described by the specific attenuation calculated based on atmospheric conditions.

In the aforementioned theory, the ITU-R P.676-12 recommendation provides the method for calculating atmospheric attenuation; this method cumulatively accounts for the individual spectral lines of oxygen and water vapor under any pressure, temperature, and humidity conditions. It also considers factors such as the non-resonant Debye spectrum of oxygen below 10 GHz, nitrogen attenuation above 100 GHz mainly caused by pressure, and the moist continuum band, which results from excessive water vapor absorption observed in computational experiments [15]. Based on the factors considered in this recommendation, we have adjusted the parameters of the ITU atmospheric loss model using meteorological data from the past 30 years, including annual average relative humidity, atmospheric pressure, and temperature in Nanjing, Jiangsu Province, China, where the experiments are conducted. This makes the atmospheric loss model more closely aligned with the experimental environment of this work. The parameters of these two cases are shown in Table 1. Case 1 refers to the meteorological condition parameters provided in the standard example from the ITU-R P.676-12 document, while Case 2 represents the typical values corresponding to the historical annual meteorological data for Nanjing, as described earlier.

Table 1. Different cases of meteorological conditions used in atmospheric attenuation model.

Case	Temperature	Atmospheric Pressure	Relative Humidity	Actual Vapor Pressure	Absolute Humidity	Specific Humidity
Case 1	15 °C	1013.25 hPa	-	-	7.5 g/m^3	-
Case 2	16 °C	1017 hPa	74%	13.45 hPa	10.08 g/m^3	13.85 g/kg

Figure 2 shows the specific attenuation curve under case 1 and case 2, as well as the specific attenuation curve under dry air conditions, where the absolute humidity is set to zero, for reference. Under dry air conditions and Case 1, the specific attenuation at 100 GHz is 0.033 dB/km and 0.516 dB/km, respectively. Under the typical meteorological conditions in Nanjing (Case 2), the specific attenuation at 100 GHz is 0.73 dB/km, while it will increase to over 50 dB/km at frequency bands above 300 GHz under certain weather conditions. Therefore, for short-range communication in the W-band, the atmospheric attenuation can be almost negligible, but, as the frequency increases and the distance extends, the atmospheric attenuation will significantly affect the transmission performance [16].

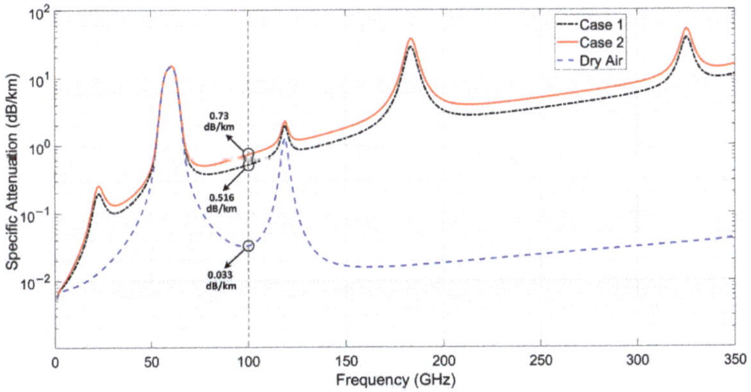

Figure 2. Specific attenuation as a function of frequency under case 1, case 2, and dry air condition.

According to Equation (2), the free-space path loss for transmission over 200 m and 2 km at a carrier frequency of 100 GHz is 118.46 dB and 138.46 dB, respectively. In designing the experimental setup, incorporating the precise atmospheric attenuation and free-space path loss into Equation (1) enables the rational selection of experimental parameters such as transmit power, transmit antenna gain, and receive antenna gain, considering achievable receiver sensitivity. This is particularly crucial for effectively conducting long-distance transmission experiments in the millimeter-wave and terahertz frequency bands. The specific details of the power budget will be provided later in this paper.

3. Scenario Validation

3.1. Experimental Setup

Based on the aforementioned analysis, we construct corresponding experimental platforms for scenario validation. Table 2 summarizes the general experimental setup and relevant parameters of the experiments in this work.

Table 2. Summary of the indoor and outdoor experimental setups.

Scenario	Carrier Frequency (GHz)	Receiver Scheme	Distance (m)	Antenna	PA [3] Used	BERT Test	Video Transmission
Indoor	100	All-optical/All-electric	15	LCA [1]	×	√	Uncompressed HD
Outdoor	100	All-optical	200	RA [2]	×	√	Uncompressed HD
Outdoor	100	All-optical	2000	RA	√	√	Uncompressed HD

[1] LCA: Lens corrected antenna. [2] RA: Reflector antenna. [3] PA: Power amplifier.

The three experiments for scenario validation, as shown in the table, all employ a transmitter architecture based on the photonics-assisted heterodyne beating technique, generating the mmWave with a frequency of 100 GHz. Indoors, we test the transmission performance at a distance of 15 m using the all-optical receiver and the all-electric receiver based on envelope detection at the receiving end. During the measurement process, bit error rate testers (BERTs) are used to generate test signals and evaluate the quality of the transmitted signals. The phase noise is measured by analyzing the frequency stability of the signals using a high-precision signal and spectrum analyzer. Eye diagrams for all experiments are measured using an oscilloscope and an eye diagram analyzer, providing visual assessments of transmission quality along with quantitative jitter analysis, enabling further evaluation and comparison of transmission performance under different schemes and experimental conditions. Outdoors, to further explore the advantages of the all-optical receiver in long-distance transmission and practical application, we test the transmission performance at distances of 200 m and 2 km and demonstrate real-time uncompressed HD video transmission.

Figure 3 shows the architecture of the transmission system in this work. The transmitter employs the photonics-assisted heterodyne beating technique to modulate the OOK signals generated by the BERT or the real-time data stream of a camera onto the optical carrier, thereby generating and transmitting millimeter-wave signals. The receiver is implemented via two approaches: (i) an all-electric receiver based on envelope detection and (ii) an all-optical receiver based on an avalanche photodiode. The measured optical spectrum after the modulator at the transmitting side is shown in Figure 4a; the measured optical spectra of the received signal after the modulator, dense wavelength division multiplexing (DWDM), the tunable optical filter (TOF) in the all-optical receiver, and the passband of TOF are shown in Figure 4b. The parameters of key components used in the experiment are provided in Table 3.

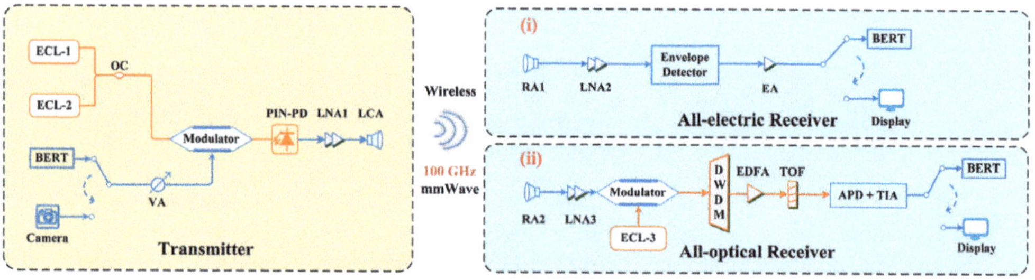

Figure 3. The architecture of the transmission system, including approaches of the all-electric receiver based on envelope detection and all-optical receiver. **(i)** All-electric receiver. **(ii)** All-optical receiver.

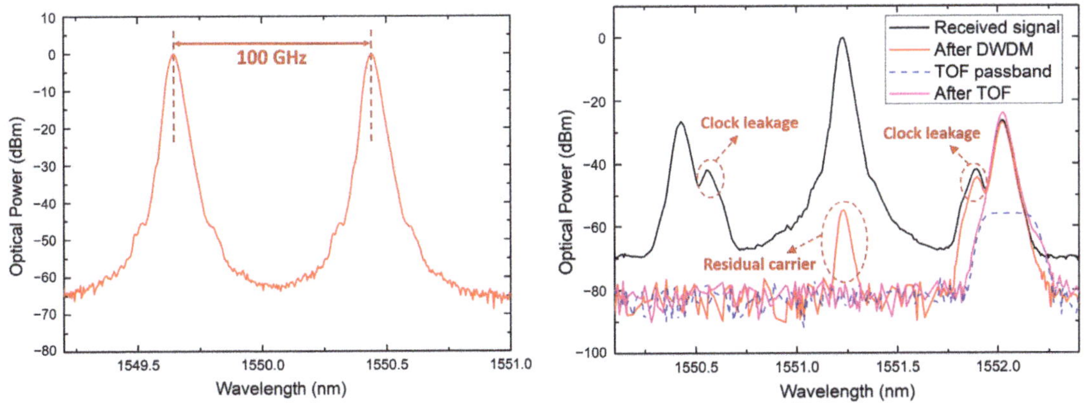

(**a**) (**b**)

Figure 4. The measured optical spectra: (**a**) after the intensity modulator at the transmitting side, (**b**) of the received signal after the phase modulator, DWDM, TOF in the all-optical receiver, and the passband of TOF.

Table 3. Parameters of key components in the experimental system.

Name	Parameter
LNA1	Frequency: 75–110 GHz Gain (Typical): 38 dB Noise Figure (Typical): 4 dB
LNA2, LNA3	Frequency: 75–110 GHz Gain (Typical): 32 dB Noise Figure (Typical): 4 dB
EA	Frequency: 10 MHz–50 GHz Gain (Typical): 35 dB Gain Flatness: ±2 dB
PIN-PD	Typical wavelength: 1550 nm Responsivity: 0.6 A/W 3 dB cut-off frequency: 100 GHz
APD + TIA	Bandwidth: 10 GHz Operating wavelength: 1250~1610 nm Responsivity: >4 A/W Sensitivity: −24.5 dBm Output voltage: 400 mV
Envelope Detector	Typical operating frequency: 75~110 GHz Sensitivity: 3000 V/W

3.2. *Calculation of Power Budget in the Wireless Link*

Table 4 presents the actual weather conditions during the 200 m and 2 km outdoor transmissions. Using the parameters recorded in the table, the ITU P.676-12 model can be applied to further refine the atmospheric loss values, enabling a more accurate power budget. Table 5 lists the power budget in the wireless link of outdoor scenarios based on the aforementioned actual meteorological conditions, the received power reaches −37.835 dBm and −45.352 dBm at transmission distances of 200 m and 2 km, respectively.

Table 4. Actual weather conditions of outdoor transmissions.

Date	Distance (m)	Temperature (°C)	Atmospheric Pressure	Relative Humidity	Absolute Humidity	Actual Vapor Pressure	Specific Attenuation
29 May	200	27	1015 hPa	63%	13.88 g/m^3	16.20 hPa	1.176 dB/km
2 September	2000	30	1010 hPa	78%	20.63 g/m^3	23.64 hPa	1.901 dB/km

Table 5. Power budget of outdoor transmissions.

Parameter	Value	
Operation frequency	100 GHz	100 GHz
Distance	200 m	2 km
P_T	3.66 dBm	20.41 dBm
G_T	39 dBi	39 dBi
G_R	39 dBi	39 dBi
$\overline{L_P}(d)$	118.46 dB	138.46 dB
L_f	0.8 dB	1.5 dB
L_a	1.176 dB/km	1.901 dB/km
P_R	−37.835 dBm	−45.352 dBm

3.3. Indoor 15 m Transmission Using All-Optical and All-Electric Receiver

Figure 5 provides photos of the indoor transmission system; to ensure a fair and reasonable comparison, we use the same structure based on photonics-assisted heterodyne beating techniques at the transmitting side of the mmWave transmission experimental system. Additionally, the transmitting antennas, receiving antennas, and other components are consistently kept identical throughout the indoor experiments.

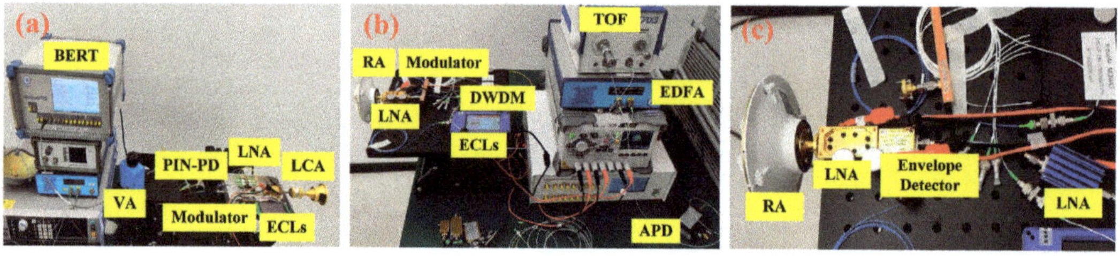

Figure 5. Photos of the indoor transmission system: (**a**) transmitter, (**b**) all-electric receiver, (**c**) all-optical receiver.

On the transmitting side, a BERT is employed to generate OOK signals with modulation frequencies corresponding to data rates ranging from 1.25 Gbit/s to a maximum of 11.318 Gbit/s. The modulation bandwidth is determined by the data rate and spectral efficiency of the OOK signals. For instance, a data rate of 11.318 Gbit/s corresponds to a modulation bandwidth of approximately 11.318 GHz. The BERT is configured to generate signals using a pseudorandom binary sequence of length 31 (PRBS31), with the signal amplitude adjustable between 200 mV and 500 mV. The transmission link is carefully calibrated before each experiment, and the BERT configuration parameters are double-checked to ensure consistency. Two external cavity lasers (ECL-1 and ECL-2), each with a linewidth of less than 100 kHz, are utilized to generate the optical carrier. The wavelengths of the two aforementioned lasers are set to 1549.190 nm and 1549.990 nm, respectively, corresponding to a wavelength interval of 0.8 nm, which means that the frequency of the mmWave signal generated by the heterodyne beating of the two laser beams is 100 GHz. The signal from the BERT is first attenuated using a variable attenuator (VA) with manually adjustable attenuation levels ranging from 0 to 9 dB before being sent to an optical intensity modulator.

This modulated signal, combined with another optical signal generated by ECL-2 using an optical coupler (OC), is transmitted through a single optical fiber. Subsequently, these two signals undergo optical heterodyne beating in a PIN photodiode (PIN-PD), generating a 100 GHz mmWave signal, which is then amplified by a low-noise amplifier (LNA) and transmitted through a lens-corrected antenna.

On the receiving side, we employ experimental system structures based on two different approaches. For the all-electric receiver based on envelope detection, the mmWave signal received by the RA is amplified by an LNA and then input into a W-band envelope detector (AT-PD-75110N1, AT Microwave Limited, in Shanghai, China) for envelope detection. The envelope detector extracts the amplitude information of the modulated signal through nonlinear rectification and removes the carrier and high-frequency components using a low-pass filter. The output is the baseband component, which corresponds to the OOK signal, with a bandwidth identical to that set by the BERT at the transmitter. After being amplified by another LNA, the signal is fed into a BERT at the receiving side to test the BER performance. For the all-optical receiver, the mmWave signal is first received by a reflective antenna (RA), then amplified using a low-noise amplifier (LNA), and subsequently fed into an optical phase modulator. A laser from ECL-3, with a wavelength of 1550.800 nm, is introduced into the other port of the optical phase modulator. This step directly modulates the received mmWave signal onto the optical carrier, generating two optical sidebands. Afterward, a dense wavelength division multiplexing module is used to filter one of the optical sidebands, which acts as the optical baseband signal. This optical baseband signal is then amplified in an erbium-doped fiber amplifier (EDFA), filtered by a tunable optical filter, and sent into an APD for photoelectric conversion. The APD we selected has a bandwidth of 10 GHz, working at the wavelength range of 1250–1610 nm, and achieves a detection sensitivity of −15 dBm. The resulting signal is then amplified by a transimpedance amplifier (TIA) and input into a BERT to measure the transmission performance.

3.4. Outdoor 200 m and 2 km Uncompressed HD Video Transmission

To explore the potential for long-distance transmission, we use the transmission system architecture based on an all-optical receiver to carry out the outdoor experiment. We replace the antennas of the aforementioned indoor demonstration system with larger aperture reflector antennas. A power amplifier (PA) is additionally added before the transmitting antenna for the 2 km transmission. This commercially mature PA operates at frequencies ranging from 75 to 110 GHz, with a typical gain of 16 dB. The rest of the experimental setup is kept unchanged. Then, we move the wireless link outdoors and extend the transmission distance from 15 m to 200 m and 2 km. The experimental setup for the outdoor transmission is shown in Figure 6.

The BER performance is first tested using BERTs. Additionally, simultaneously with long-distance transmission experiments, we also conduct practical application verification by replacing the signal source at the transmitter with a camera and substituting the BERT at the receiver with a display, thereby achieving live streaming of uncompressed HD video. A commercial camera is used in our experiment, and the real-time video captured by the camera is first input into a high-definition multimedia interface (HDMI) to serial digital interface (SDI) converter at the transmitting end. This converter is a high-quality, compact commercial device designed for converting HDMI video signals to professional SDI outputs. The SDI interface, supports a data rate of 1.485 Gbit/s, which is commonly used for uncompressed HD video transmission in broadcast-grade applications and corresponds to a modulation bandwidth of approximately 1.485 GHz in the wireless link when using OOK modulation. This aligns with the requirements for uncompressed HD video transmission. The SDI output interface transmits the video signal via a cable into the intensity modulator for modulation. The subsequent processes at the transmitting side, such as heterodyne beating in the PIN-PD, are identical to the procedures of indoor experiments. Similarly, on the receiving side, the baseband signal output from the APD is first input into an SDI to HDMI converter. The HDMI output interface of this converter is directly connected to a display, allowing the real-time video captured by the camera at the transmitting side to be displayed.

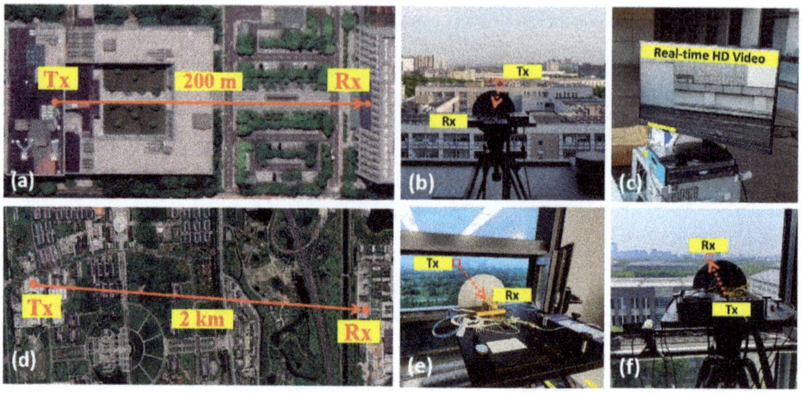

Figure 6. Experimental setup of the outdoor transmission: satellite map of the transmission link over a distance of (**a**) 200 m and (**d**) 2 km. Photos of the receiving side of (**b**) 200 m and (**e**) 2 km transmission. Photos of (**c**) real-time HD video displayed on the screen and (**f**) the transmitting side of 2 km transmission.

4. Experimental Results and Discussion

For both indoor and outdoor transmission experiments, we first measure the bit error rate (BER) performance under different data rates. In the indoor 15 m transmission experiment, by measuring the phase noise and eye diagram, we demonstrate that the all-optical receiver-based architecture offers superior performance for high-rate millimeter-wave wireless communication. Furthermore, in outdoor transmission experiments over distances of 200 m and 2 km, we explore the transmission performance of the proposed all-optical transceiver in long-distance scenarios and successfully demonstrate uncompressed HD video transmission.

4.1. Indoor Transmission Results

To investigate the underlying mechanisms behind the performance differences between the all-electric receiver based on envelope detection and the all-optical receiver based on APD, the phase noise of the two schemes is measured in our experiment. Figure 7 shows the phase noise as a function of frequency offset. This curve is obtained under the BtB configuration of a 1 GHz sine wave generated by an arbitrary waveform generator (AWG). Within the critical frequency offset range below 10^6 Hz, the phase noise of the all-optical receiver is significantly lower than that of the all-electric receiver. Specifically, in the range of from 10 kHz to 100 kHz, the phase noise of the all-optical receiver is approximately 8 dB lower.

The transmission performance of the two schemes shows significant differences under the influence of phase noise, which is caused by the characteristics of the devices and the system principles. Figure 8 shows the BER as a function of signal amplitude input to the transmitting side intensity modulator for 15 m indoor transmission using the all-electric and all-optical receivers, respectively. For the all-electric receiver, due to the bandwidth limitation of the envelope detector, the receiving side BERT fails to recover the data when the transmission rate exceeds 5 Gbit/s, resulting in a transmission link interruption. In contrast, the all-optical receiver is capable of supporting transmission rates up to the maximum of 11.318 Gbit/s generated by the BERT. Note that the system employing the all-electric receiver achieves error-free transmission of OOK signals up to a maximum of 3.125 Gbit/s, whereas the system using the all-optical receiver is capable of achieving error-free transmission at 11.318 Gbit/s with a signal amplitude of 300 mV. It is worth noting that the aforementioned error-free transmissions refer to long-term stable transmission with no bit errors, as indicated by a bit error rate (BER) lower than the measurable limit of 1×10^{-16} on the BERT, and all experiments are conducted without clock recovery.

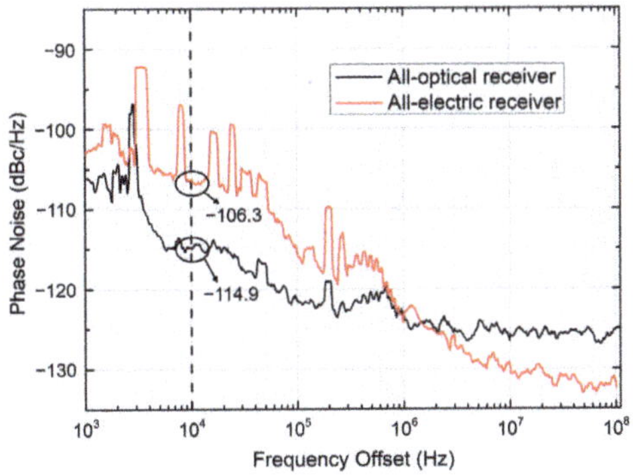

Figure 7. Phase noise as a function of frequency offset.

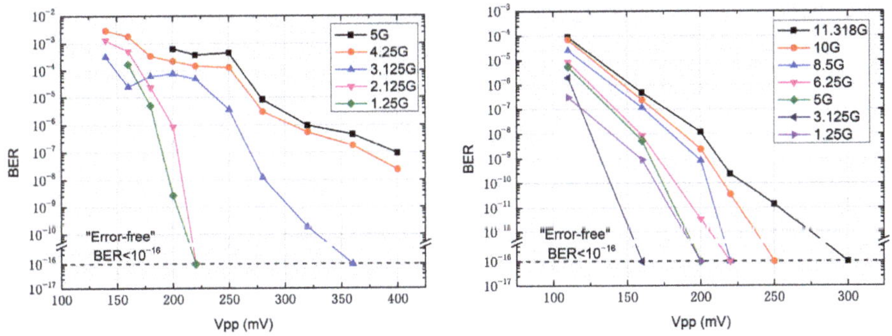

(a) (b)

Figure 8. BER as a function of the signal amplitude into the intensity modulator for indoor transmission over a distance of 15 m based on (**a**) the all-electric receiver and (**b**) the all-optical receive.

To further evaluate the transmission performance, we also compare the quality of the eye diagrams for the two receiving schemes. Figure 9a,b show the waveform, spectrum, and demodulated eye diagrams for the 5 Gbit/s OOK signal, measured under the all-electric receiver scheme and the all-optical receiver scheme, respectively. In Figure 9a, we configured the oscilloscope to use the First Order PLL method for clock recovery in order to obtain a stable eye diagram; the bit error rate corresponding to the eye diagram is 6.8×10^{-7}, making it challenging to achieve live streaming of uncompressed HD videos. Nevertheless, the all-optical receiver can achieve error-free transmission at a rate of 5 Gbit/s, and it can be seen in Figure 9b that the quality of the eye diagram without clock recovery is evidently better than that of the all-electric receiver. However, the rising and falling edges of multiple symbols are also no longer concentrated at a single position in the eye diagrams, appearing more dispersed. Quantitative analysis using an eye diagram analyzer reveals that this phenomenon is closely related to the larger jitter RMS value of 8.88 ps observed in the all-electric receiver, compared to the 7.07 ps jitter RMS in the all-optical receiver. The flicker noise and other factors inherent to the mechanism of the envelope detector are also key factors leading to signal degradation and transmission instability.

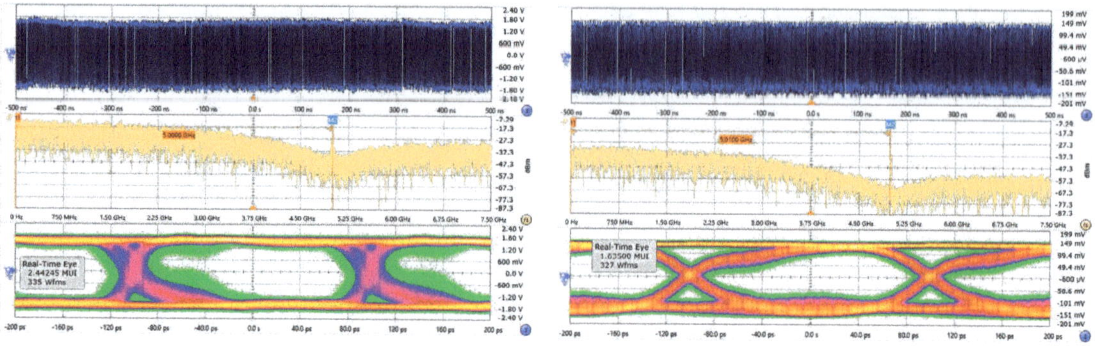

(a) (b)

Figure 9. Signal waveform, spectrum, and demodulated eye diagram for 5 Gbit/s OOK signal of (**a**) all-electric receiver and (**b**) all-optical receiver.

4.2. Outdoor Transmission Results

Furthermore, we explore the capability of the communication system based on the all-optical receiver for long-distance transmission over distances of 200 m and 2 km, both of which successfully demonstrated real-time transmission of uncompressed HD video signals for practical scenarios. In the 200 m transmission, as indicated by the BER curve in Figure 10a, error-free transmission is achieved at data rates of 10 Gbit/s and below. For data rates of 10.52 Gbit/s and 11.318 Gbit/s, the bit error rate was on the order of 10^{-9} when the received optical power (ROP) of the APD on the receiving side approached -5 dBm. On the receiving side, we use an HD display to show the real-time video signal after transmission. The system maintains stable transmission for 24 h, with the bit error rate consistently below 10^{-16} and the video display free of noise. When the distance is further increased to 2 km, the system is still capable of stable transmission of uncompressed HD video. BER performance degrades as shown in Figure 10b; error-free transmission is achieved only at a rate of 1.25 Gbps. BER performance at 3.125 Gbps is superior to that at 2.125 Gbps, remaining on the order of 10^{-8}. At 5 Gbps, BER performance deteriorates to the order of 10^{-5}. These experimental results align with the theoretical estimate that the transmission rate required for uncompressed full HD video is around 3 Gbps, and they are consistent with the system characteristics observed in previous tests.

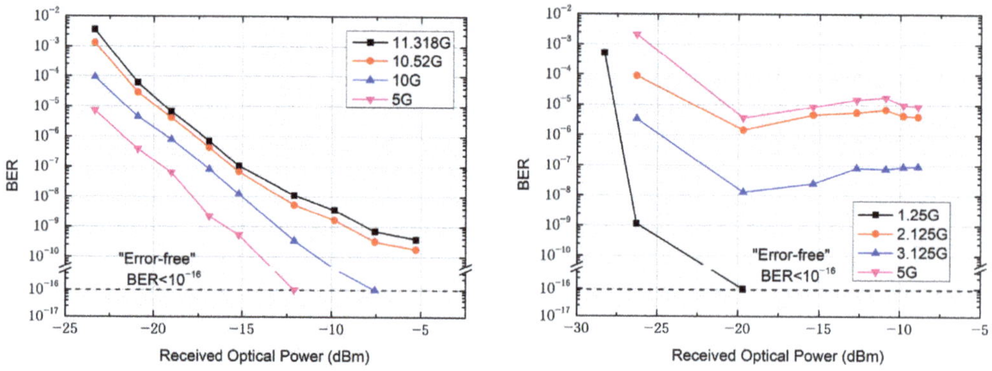

(a) (b)

Figure 10. BER as a function of the received optical power for outdoor transmission over the distance of (**a**) 200 m and (**b**) 2 km.

4.3. Stability and Power Consumption Evaluation

During indoor experiments, even in narrow corridors where multipath effects pose significant challenges, stable transmission is still achieved by carefully selecting appropriate antennas and transmission power. The scheme based on all-optical frequency up- and down-conversion demonstrates minimal jitter, with performance that is robust enough to support emerging 6G indoor applications, eliminating the need for clock recovery. For indoor real-time uncompressed video transmission, we continuously tested for 24 h, and, for outdoor transmission over 200 m and 2 km, we tested for 12 h, during which time the system maintained stable transmission. The experimental results from this work validate that the proposed transmission scheme based on all-optical up- and down-conversion has the potential to support future applications such as the real-time uncompressed transmission of UHD video.

Evaluating the power consumption of the experimental setup in this work is of significant importance in guiding the deployment of the system in practical applications. Based on calculations and measurements, the power consumption of the all-electric receiver and all-optical receiver in indoor experiments is approximately 2.1 W and 5.8 W, respectively. When normalized to the maximum achievable data rates under indoor conditions, the energy efficiency is approximately 0.67 W/Gbps for the all-electric receiver and 0.51 W/Gbps for the all-optical receiver. This indicates that the all-optical receiver is approximately 24% more energy efficient than the all-electric receiver, demonstrating its superior energy efficiency despite its higher absolute power consumption. For outdoor experiments, when adding a PA at the transmitter for long-distance transmission, the total power consumption of the communications system based on all-optical frequency up- and down-conversion is around 23.5 W. These findings highlight the trade-off between power consumption and energy efficiency in selecting the appropriate receiver approach for specific application scenarios.

5. Conclusions

We have experimentally demonstrated and compared the performance of all-optical and all-electric receivers in mmWave communication systems at 100 GHz. The all-optical receiver, utilizing an avalanche photodiode detector, achieves error-free 200 m transmission at data rates up to 11.318 Gbps, while the all-electric receiver, based on envelope detection, is limited to 3.125 Gbps error-free transmission over a 15 m wireless link. In addition to superior bit error rate performance, the mmWave communication system based on an all-optical transceiver demonstrates greater stability, lower phase noise, and a wider bandwidth. Moreover, the aforementioned system enables stable real-time transmission of uncompressed full HD video over distances of both 200 m and 2 km, with the all-optical receiver proving to be more effective for long-distance communication. The results suggest that the all-optical receiver is a more suitable choice for future millimeter-wave and terahertz communication systems that require high data rates and long-range transmission, as well as for applications such as uncompressed UHD video streaming in emerging 6G networks.

Author Contributions: Conceptualization, S.G. and J.Z.; methodology, S.G. and J.Z.; software, S.G., Y.J. and Z.L.; validation, M.Z. and Q.Z.; formal analysis, S.G., Y.J. and Q.Z; investigation, S.G. and Y.J.; resources, M.Z. and J.Z.; data curation, S.G., Y.J. and Z.L.; writing—original draft preparation, S.G.; writing—review and editing, S.G., Y.J., Z.L. and J.Z.; supervision, M.Z.; project administration, M.Z. and J.Z. All authors have read and agreed to the published version of the manuscript.

Funding: This research was funded by the National Key Research and Development Program of China (Grant No. 2023YFB2905600), National Natural Science Foundation of China (Grant Nos. 62101121, 62101126, 62201393, 62201397, 62271135, 62435003, 62405384), Jiangsu Provincial Key Research and Development Program (BE2023001), and the Major Key Project of Peng Cheng Laboratory.

Data Availability Statement: The data presented in this study are available on request from the corresponding author. The data are not publicly available due to privacy restrictions.

Conflicts of Interest: The authors declare no conflicts of interest.

References

1. Zhu, M.; Zhang, J.; Hua, B.; Lei, M.; Cai, Y.; Tian, L. Ultra-wideband fiber-THz-fiber seamless integration communication system toward 6G: Architecture, key techniques, and testbed implementation. *Sci. China Inf. Sci.* **2023**, *66*, 113301. [CrossRef]
2. Liu, X.; Zhang, J.; Zhu, M.; Zhong, Q.; Tong, W.; Xin, Z.; Wang, Y.; Hua, B.; Cai, Y.; Lei, M.; et al. Real-time fifth-generation new radio signal transmission over a transparent fiber-wireless-fiber seamless system in the 75-110 GHz band. *Chin. Opt. Lett.* **2024**, *22*, 12.
3. Nagatsuma, T.; Horiguchi, S.; Minamikata, Y.; Yoshimizu, Y.; Hisatake, S.; Kuwano, S. Terahertz wireless communications based on photonics technologies. *Opt. Express* **2013**, *21*, 23736–23747. [CrossRef] [PubMed]
4. Ummethala, S.; Harter, T.; Koehnle, K.; Li, Z.; Muehlbrandt, S.; Kutuvantavida, Y. THz-to-optical conversion in wireless communications using an ultra-broadband plasmonic modulator. *Nat. Photonics* **2019**, *13*, 519–524. [CrossRef]
5. Dat, P.T.; Yamaguchi, Y.; Motoya, M.; Oikawa, S.; Ichikawa, J.; Kanno, A. Transparent fiber-millimeter-wave-fiber system in 100-GHz band using optical modulator and photonic down-conversion. *J. Light. Technol.* **2022**, *40*, 1483–1493. [CrossRef]
6. Griffiths, P.R.; Homes, C.C. Instrumentation for far-infrared spectroscopy. In *Handbook of Vibrational Spectroscopy*, 1st ed.; Chalmers, J., Griffiths, P.R., Eds.; Wiley: Chichester, UK, 2002; Volume 1.
7. Stöhr, A.; Shih, B.; Abraha, S.T.; Steffan, A.G.; Ng'oma, A. High spectral-efficient 512-QAM-OFDM 60 GHz CRoF system using a coherent photonic mixer (CPX) and an RF envelope detector. In *OSA Technical Digest*; Optica Publishing Group: Washington, DC, USA, 2016; pp. 1–3.
8. Yasuda, Y.; Hisatake, S.; Kuwano, S.; Terada, J.; Otaka, A.; Nagatsuma, T. Terahertz wireless transmission enabled by photonics using binary phase-shift keying at 300 GHz. *URSI Radio Sci. Bull.* **2016**, *356*, 20–23.
9. Liu, L.; Xu, H.; Duan, Y.; Lichtenberger, A.W.; Hesler, J.L.; Weikle, R.M. A 200 GHz Schottky diode quasi-optical detector based on folded dipole antenna. *Int. Symp. Space Terahertz Technol.* **2009**, *145*, 149.
10. Granja, A.B.; Konstantinou, D.; Rommel, S.; Cimoli, B.; Rodriguez, S.; Reese, R. High Data Rate W-Band Balanced Schottky Diode Envelope Detector for Broadband Communications. In Proceedings of the 14th European Microwave Integrated Circuits Conference (EuMIC), Paris, France, 30 September–1 October 2019; pp. 310–313.
11. Du, H.; Xu, G. Optical wireless multiple-input multiple-output system based on avalanche photodiode receiver. *Ann. Telecommun.* **2020**, *75*, 89–99. [CrossRef]
12. Kanno, A.; Kuri, T.; Morohashi, I.; Hosako, I.; Kawanishi, T.; Yoshida, Y.; Kitayama, K.-I. Coherent Terahertz Wireless Signal Transmission Using Advanced Optical Fiber Communication Technology. *J. Infrared Millim. Terahertz Waves* **2015**, *36*, 180–197. [CrossRef]
13. Friis, H.T. A Note on a Simple Transmission Formula. *Proc. IRE* **1946**, *34*, 254–256. [CrossRef]
14. *ITU-R F 2107*; Characteristics and Applications of Fixed Wireless Systems Operating in the Frequency Range from 57 to 134 GHz. International Telecommunication Union: Geneva, Switzerland, 2011.
15. International Telecommunication Union. Attenuation by Atmospheric Gases and Related Effects. In *ITU-R Recommendation P.676-12*, 12th ed.; ITU: Geneva, Switzerland, 2019; pp. 1–24.
16. Xu, Y.M.; Zhang, Z.R.; Xiao, J.N.; Yu, J.J. W-band Millimeter Wave Communication. *J. Terahertz Sci. Electron. Inf. Technol.* **2014**, *12*, 554–558.

micromachines

MDPI

Article

High Spectrum Efficiency and High Security Radio-Over-Fiber Systems with Compressive-Sensing-Based Chaotic Encryption

Zhanhong Wang [1], Lu Zhang [1,*], Jiahao Zhang [1], Oskars Ozolins [2], Xiaodan Pang [1] and Xianbin Yu [1,*]

[1] College of Information Science and Electronic Engineering, Zhejiang University, Hangzhou 310027, China; zhanhongwang@zju.edu.cn (Z.W.); zhangjh914@zju.edu.cn (J.Z.); xipa@zju.edu.cn (X.P.)

[2] Institute of Photonics, Electronics and Telecommunications, Riga Technical University, 1048 Riga, Latvia; oskars.ozolins@rtu.lv

* Correspondence: zhanglu1993@zju.edu.cn (L.Z.); xyu@zju.edu.cn (X.Y.)

Abstract

With the increasing demand for high throughput and ultra-dense small cell deployment in the next-generation communication networks, spectrum resources are becoming increasingly strained. At the same time, the security risks posed by eavesdropping remain a significant concern, particularly due to the broadcast-access property of optical fronthaul networks. To address these challenges, we propose a high-security, high-spectrum efficiency radio-over-fiber (RoF) system in this paper, which leverages compressive sensing (CS)-based algorithms and chaotic encryption. An 8 Gbit/s RoF system is experimentally demonstrated, with 10 km optical fiber transmission and 20 GHz radio frequency (RF) transmission. In our experiment, spectrum efficiency is enhanced by compressing transmission data and reducing the quantization bit requirements, while security is maintained with minimal degradation in signal quality. The system could recover the signal correctly after dequantization with 6-bit fronthaul quantization, achieving a structural similarity index (SSIM) of 0.952 for the legitimate receiver (Bob) at a compression ratio of 0.75. In contrast, the SSIM for the unauthorized receiver (Eve) is only 0.073, highlighting the effectiveness of the proposed security approach.

Keywords: compressive sensing; radio over fiber; chaotic encryption

1. Introduction

As the demand for high capacity continues to grow in the next-generation wireless communication networks, high-frequency electromagnetic waves, such as millimeter waves (mmW) and terahertz waves (THz), have garnered significant attention. These frequencies offer broad bandwidth resources, which are crucial for meeting the capacity requirements of future networks [1]. However, their use in radio access networks (RANs) presents several challenges. On one hand, high-frequency communication can significantly enhance the throughput of radio remote units (RRUs), enabling the future 6G network to achieve peak rates of 1 Tb/s and user-experienced data rates of beyond 1 Gb/s [2]. On the other hand, these high-frequency waves experience greater path loss, necessitating ultra-dense deployment of RRUs to maintain coverage and performance [3]. Therefore, improving spectrum efficiency becomes essential for the optical fronthaul networks. Currently, the common public radio interface (CPRI) is widely used for optical fronthaul networks, particularly in Option 8 of the 3GPP function split options [4]. However, CPRI imposes high communication overhead, as it requires up to 15-bit quantization to maintain signal

 https://doi.org/10.3390/mi17010080

integrity, which results in significant bandwidth consumption and spectrum efficiency degradation [5]. This makes it clear that new methods are needed to improve the spectrum efficiency of optical fronthaul systems.

Several innovative strategies have been proposed to address this challenge. Analog radio-over-fiber (A-RoF) systems, for instance, have been experimentally implemented, as they require less bandwidth for transmission compared to digital signals [6,7]. However, analog signals are more susceptible to the nonlinear effects of optical fibers, necessitating complex nonlinear equalization algorithms that increase the demand for digital signal processing (DSP). Additionally, while data compression algorithms have been suggested to reduce transmission data volume, their effectiveness remains uncertain [8,9]. Digital–analog radio-over-fiber (DA-RoF) systems can enhance the signal-to-noise ratio (SNR) and support high-order modulation formats to improve spectrum efficiency, but they also require digital–analog hybrid transmission and complex signal recovery algorithms, which drive up deployment costs [10]. Alongside these technical challenges, security is a crucial concern for RANs. Due to the broadcast nature of the downlink in optical fronthaul networks, transmitted signals are vulnerable to eavesdropping by unauthorized receivers. In recent years, physical (PHY)-layer encryption techniques have been explored to provide more robust security than traditional media access control (MAC)-layer encryption [11,12]. One promising PHY-layer encryption method is chaotic encryption, which is favored for its high randomness and low implementation cost [13–16]. Recently, a CS-based chaotic encryption algorithm was proposed to improve the security performance of the communication system, embedding the chaotic encryption algorithm into the CS algorithm [17,18]. In addition, the performance of the spectrum efficiency is not discussed in these works.

In this paper, we present an approach that combines high spectrum efficiency with robust security by using the compressive sensing (CS)-based chaotic encryption algorithm in a radio-over-fiber (RoF) system. We generated a 16QAM-DMT signal and mapped it to an 8 Gbit/s BPSK signal via fronthaul quantization and transmitted it over 10 km of standard single-mode fiber (SSMF) at the C-band. A photonics-based method is used to generate the RF signal at 20 GHz. With the CS-based chaotic encryption algorithm, the system achieves quantization-error-free at a 6-bit quantization level, even with compression ratios (CR) of 0.25 and 0.5. Furthermore, the system's security is ensured, as demonstrated by the structural similarity index (SSIM) of 0.952 for the legitimate receiver (Bob) at a compression ratio of 0.75, indicating minimal loss of information. In contrast, the SSIM for the eavesdropper (Eve) is only 0.073 under the same conditions, highlighting the effectiveness of the proposed security method.

The remainder of the paper is organized as follows: Section 2 introduces the selected chaotic sequence and the principle of the algorithms. Section 3 explains the experiment setup for the 20 GHz RoF system, and Section 4 discusses the experiment results and optimizes the parameters of the algorithm. Finally, Section 5 concludes the paper with a summary of the research.

2. Principle

The architecture of the proposed RoF system is shown in Figure 1. The data is processed by the CS-based chaotic encryption algorithm at the BBU and then transmitted to multiple RRUs. At the RRUs, RF signals are generated and sent to the users. First, the amount of transmitted data is reduced, and a lesser quantization bit number will be used for the optical fronthaul network because of the CS algorithm. On top of that, because of the chaotic encryption, the legal receiver Bob can reconstruct the origin signal with minimal quality degradation, and the illegal receiver Eve obtains less effective information due to

https://doi.org/10.3390/mi17010080

the absence of the legal keys. Therefore, high spectrum efficiency and high security can be achieved in the proposed system.

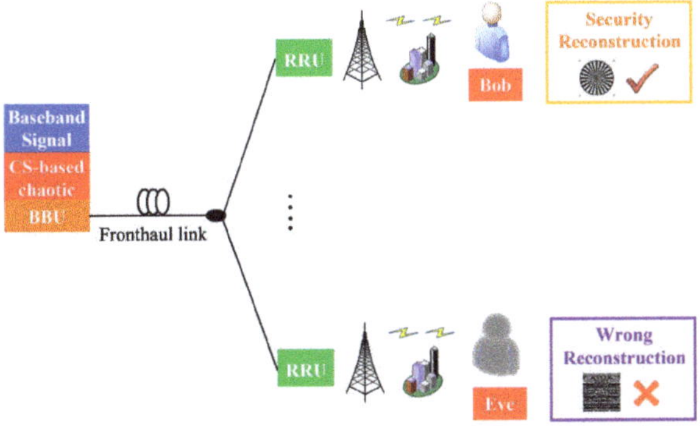

Figure 1. The architecture of the high-spectrum-efficiency and high-security RoF system.

A generalized algorithm block diagram is shown in Figure 2. The algorithm will be discussed in detail. First, the key is prestored in the BBU and the legal receivers. With the key, the chaotic sequence can be generated at the BBU. In the proposed system, the 2-dimensional logistic-sine-coupling map (2D-LSCM) is chosen as the chaotic sequence, which can be expressed as

$$\begin{cases} x_{i+1} = \sin(\pi(4\theta x_i(1-x_i) + (1-\theta)\sin(\pi y_i)) \\ y_{i+1} = \sin(\pi(4\theta x_i(1-x_i) + (1-\theta)\sin(\pi x_{i+1})) \end{cases} . \tag{1}$$

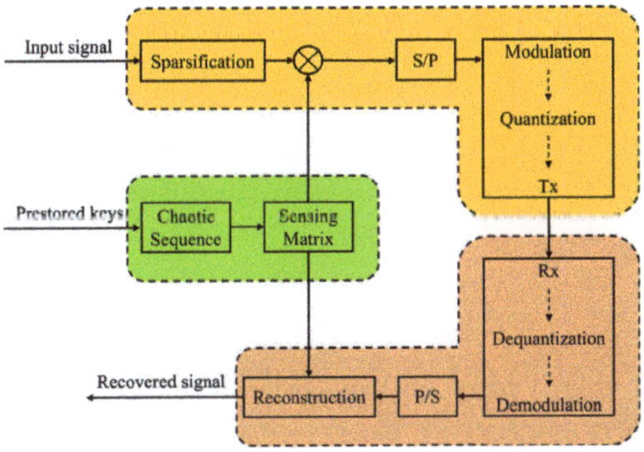

Figure 2. Schematic for the compressive-sensing-based chaotic encryption algorithm.

Here, x_1, y_1 and θ are chosen as the keys, where $\theta \in [0, 1]$, x_1, $y_1 \in (0, 1)$. Compared to other sequences, such as the Logistic map and the Lorenz system, 2D-LSCM exhibits hyperchaotic properties over a larger range, with better chaotic performance and lower computational complexity [19]. Then utilizing the keys, the sensing matrix **B** for the compressive sensing can be obtained. According to mathematical theory, the matrix **B** needs to meet the restricted isometry property (RIP) condition for signal reconstruction at the receiver. The random circular matrix (RCM) is utilized in our system to meet the condition because of its simplicity [20]. The cyclic characteristic of RCM is

$$\begin{aligned} &\mathbf{B}(j,1) = \lambda \cdot \mathbf{B}(j-1,n) \\ &\mathbf{B}(j,2:n) = \lambda \cdot \mathbf{B}(j-1,1:n-1), \end{aligned} \tag{2}$$

https://doi.org/10.3390/mi17010080

where $2 \leq j \leq$ m, $\lambda > 1$. The first line of **B** can be set as the corresponding number of the average of x_n and y_n. For the input transmitted signal, it needs to be sparsified by orthonormal basis decomposition due to the condition for compressive sensing. For example, an $n \times 1$ vector **x** can be decomposed as

$$\mathbf{x} = \sum_{k=1}^{n} \psi_k s_k = \boldsymbol{\psi}\mathbf{s}, \tag{3}$$

where $\boldsymbol{\psi}$ is the orthonormal basis of **x**, and **s** is an $n \times 1$ sparse vector. For multimedia signals such as images, the frequency domain matrix is sparse because of low high-frequency components, and the sparsification can be completed by time-frequency transform directly, such as Fast Fourier Transform (FFT), Discrete Cosine Transform (DCT), and Discrete Wavelet Transform (DWT). Among them, the range of matrix elements obtained after DCT can be represented using only 14 binary digits, which is smaller than the 18 bits of DWT and the 31 bits of FFT [17]. Therefore, in algorithm design, DCT is adopted for sparsification to reduce data volume. Then the vector **x** is multiplied by an $m \times n$ measurement matrix $\boldsymbol{\Phi}$ for low-dimensional mapping, which can be expressed as

$$\mathbf{y} = \boldsymbol{\Phi}\mathbf{x} + \mathbf{e} = \boldsymbol{\Phi}\boldsymbol{\psi}\mathbf{s} + \mathbf{e} = \mathbf{B}\mathbf{s} + \mathbf{e}, \tag{4}$$

where **B** is the $m \times n$ sensing matrix obtained by a chaotic sequence. Here **y** is an $m \times 1$ matrix, and the CR is defined as m/n. Then after parallel serial conversion, the signal is modulated and quantized by the BBU and then transmitted to RRU for frequency conversion.

At the legal receiver, the keys are prestored and the chaotic sensing matrix **B** can be cloned. In fact, the problem of solving **s** can be transformed into an optimization problem [21], which is

$$\begin{aligned} &\min ||\mathbf{s}||_0 \\ &\text{subject to } ||y - Bs||_2 \leq \varepsilon_e, \end{aligned} \tag{5}$$

where ε_e is the energy of the sampling noise. It is difficult to solve the NP-hard problem by an analytical method. The smooth l_0 method can be used in this scene [22]. This method uses a Gaussian function to fit the l_0-norm of vector **s**, converting it into a smooth continuous function for optimization. The Gaussian function is

$$f_\sigma(s_i) = 1 - \exp(-\frac{s_i^2}{2\sigma^2}), \tag{6}$$

where σ is the smoothness factor. Then the l_0-norm of vector **s** is

$$||\mathbf{s}||_0 = \lim_{\sigma \to \infty} F_\sigma(s_i) = \lim_{\sigma \to \infty} (n - \sum_{i=1}^{n} f_\sigma(s_i)). \tag{7}$$

Because of the smooth, continuous characteristic, gradient descent can be utilized to find the minimum point of l_0-norm. The initial vector is the multiplication of the pseudo-inverse of the matrix **B** and the measurement vector **y**. After several iterations, the algorithm convergence is achieved, and the sparsest solution of the problem in Equation (5) is obtained. However, because of a lack of knowledge of the keys, the chaotic sensing matrix cannot be cloned correctly at the illegal receiver Eve. Therefore, Eve can obtain less effective information, and the security of the RoF system is ensured.

https://doi.org/10.3390/mi17010080

3. Experimental Setup

The setup for the proof-of-concept experiment is illustrated in Figure 3. A 64 × 64 image signal, labeled 'testpat1', is chosen as the input signal. First, the image signal undergoes a sparse transformation using DCT, followed by processing with the CS-based chaotic encryption algorithm. The resulting compressed signal is then modulated into a 16-QAM signal. Next, the signal is converted into a DMT signal through subcarrier mapping and IFFT, with 2048 IFFT points. Of these, 800 subcarriers are used to carry the 16-QAM signal, while another 800 subcarriers carry the complex conjugate of the 16-QAM signal. The DMT signal is subsequently quantized and remapped to a BPSK signal for optical fronthaul transmission.

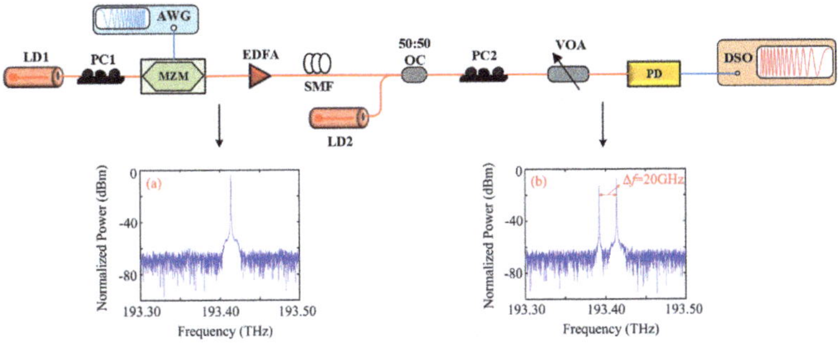

Figure 3. Experimental setup of the proposed RoF system. Inset: (**a**) the normalized optical spectrum after MZM, (**b**) the normalized optical spectrum before PD.

For the optical transmission, intensity modulation is employed. The output wavelength of laser diode LD1 (100 kHz linewidth) is set to 1500.00 nm, and the laser is modulated by a Mach-Zehnder Modulator (MZM, EOSpace, 40 GHz bandwidth). An 8 Gbps total signal is generated by an arbitrary waveform generator (AWG, Keysight M8195A, 64 GSa/s) to drive the MZM. The modulated laser is then amplified by an erbium-doped fiber amplifier (EDFA) to increase the power budget. After traveling through 10 km standard single-mode fiber (SSMF), the signal is coupled with a local oscillator (LO) laser from LD2 (100 kHz linewidth), which operates at a wavelength of 1550.16 nm, using a 50:50 coupler. A variable optical attenuator (VOA) is used to control the input power to the photodiode (PD, 40 GHz bandwidth). In the PD, the 20 GHz RF signal is generated via photo-mixing and transmitted through an RF cable. The RF signal is captured by a digital storage oscilloscope (DSO, Keysight DSOX93204A, 80 GSa/s) for offline DSP.

The signal recovery process follows these steps in the DSP flow. First, the signal is resampled at 64 GSa/s from DSO and converted to baseband through digital downconversion. Due to the free-running characteristics of the two laser diodes, frequency offset (FO) can impact communication performance. The FO is estimated using a digital phase-locked loop [23], and synchronization is achieved via a cross-correlation algorithm. To compensate for dispersion, a 4-tap feedforward equalizer is employed, followed by a decision feedback equalizer with four feedforward taps and three feedback taps [24]. After equalization, the BPSK signal is recovered, and dequantization and demodulation are performed to retrieve the compressed signal. The sparsest solution to the optimization problem is found using the greedy descent algorithm, with 40 iterations and a step size of 1.01. The original signal is then reconstructed through IDCT.

https://doi.org/10.3390/mi17010080

4. Results and Discussion

The transmission performance of the proposed RoF system is shown in Figure 4. The bit error rate (BER) is evaluated by comparing the transmitted and received BPSK signals. Given the non-Gaussian distribution of the actual transmission signal, the robustness of the system is critical. The results show that the BER improves with increasing optical power at the photodiode (PD). The system maintains a BER below the hard-decision forward error correction (HD-FEC) threshold of 3.8×10^{-3} [25] when the optical power exceeds −4 dBm, and error-free transmission is achieved when the optical power exceeds 0 dBm. The BPSK constellation diagram in Figure 3 further demonstrates the performance.

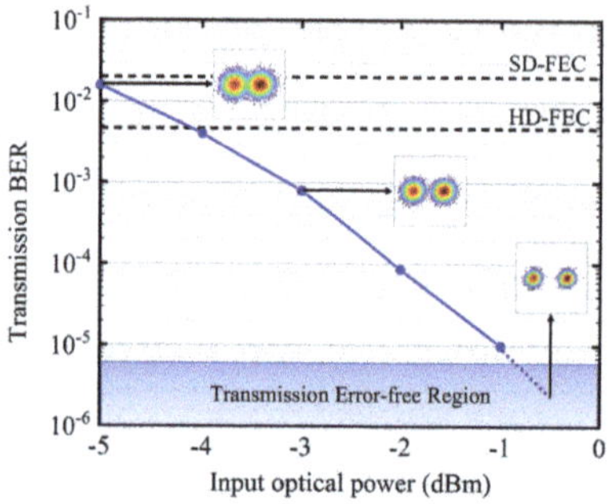

Figure 4. The transmission performance of the proposed RoF system.

The spectrum efficiency of the system is discussed. To ensure a fair comparison, spectral efficiency is evaluated under two conditions: one using the CS-based chaotic encryption algorithm and the other with traditional encryption but without CS. In both cases, the input power to the photodiode (PD) is set to 5 dBm to achieve error-free transmission. With the CS-based algorithm, the amount of transmitted data is significantly reduced. For instance, when the CR is 0.5, the number of symbols per subcarrier in a frame decreases from 20 to 9. This reduction allows the system to transmit twice the amount of data in the same transmission bandwidth. Similarly, when the CR is 0.25, nearly four times the amount of data can be transmitted. However, it is important to note that a lower CR can lead to greater signal degradation during reconstruction. This represents a trade-off between spectrum efficiency and signal quality.

The effect of fronthaul quantization is also examined in our experiment. Figure 5 shows the BER performance of the system with different CR values. Since the system achieves error-free transmission, the focus is on estimating the effect of transmission errors. With the CS algorithm, the quantization BER improves compared to the case without CS, meaning that fewer quantization bits are required in the RoF system. As a result, the system's spectrum efficiency is further enhanced. This improvement is because the CS algorithm reduces signal redundancy and the amplitude of the DMT signal [26], allowing the signal to be represented with fewer quantization bits. However, the quantization BER degrades as the CR increases since lower CR values eliminate more redundant information from the sparse signal. Notably, error-free quantization can be achieved when the CR is 0.25 or 0.5, demonstrating the effectiveness of the CS-based approach in reducing the required quantization bits while maintaining signal integrity.

https://doi.org/10.3390/mi17010080

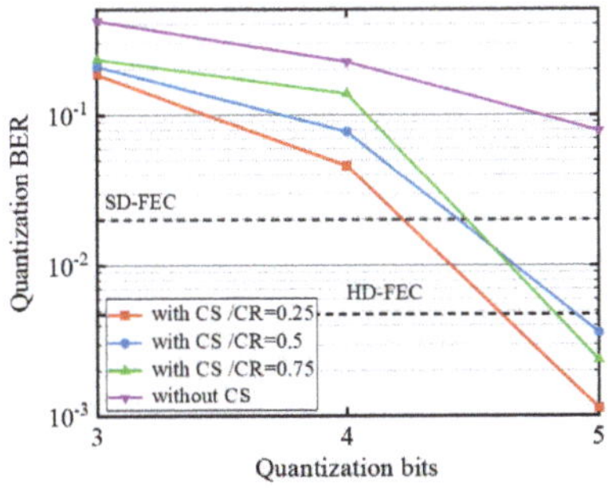

Figure 5. The quantization BER performance under different fronthaul quantization bit numbers when error-free transmission is achieved.

Then the security of the system is verified. The SSIM performance and the quality of the reconstructed images are shown in Figure 6 and Table 1. The legal keys ($x_1 = 0.6$, $y_1 = 0.7$, $\theta = 0.8$) are prestored in Bob ($x_1 = 0.7$, $y_1 = 0.7$, $\theta = 0.8$), and there are only incorrect keys in Eve. As a result, Bob, the legitimate receiver, can successfully reconstruct the image using the correct decryption keys, while Eve, the unauthorized receiver, cannot. When the CR is 0.75, the SSIM between Alice and Bob reaches 0.952, indicating only minimal quality degradation due to the CS-based chaotic encryption algorithm. In contrast, the SSIM between Alice and Eve remains close to 0, meaning that Eve is unable to extract meaningful information from the reconstructed image. Then the situation of perturbation is also considered. Here, the disturbance is set to 10^{-14}, and the SSIM of Eve is 0.033 with CR = 0.5 and 0.028 with CR = 0.75, confirming the sensitivity to the keys. With the sensitivity of the chaotic sequency is 10^{-14}, the key space can be calculated as $(10^{14})^3 = 10^{42}$, which is sufficient to prevent attacks from Eve. This demonstrates that the CS-based chaotic encryption algorithm effectively ensures the security of the RoF system, preventing unauthorized access.

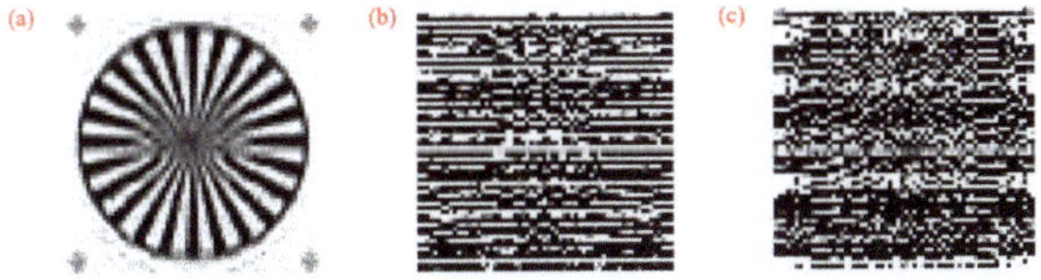

Figure 6. Images recovered by (**a**) Bob and (**b**) Eve with incorrect keys $(x_1, y_1, \theta) = (0.7, 0.7, 0.8)$ and (**c**) Eve with incorrect keys $(x_1, y_1, \theta) = (0.6 + 10^{-14}, 0.7, 0.8)$ when CR = 0.5.

Table 1. The SSIM between the image transmitted and the one recovered with 6-bit quantization.

Receiver	CR = 0.5	CR = 0.75
Bob	0.877	0.952
Eve $[(x_1, y_1, \theta) = (0.7, 0.7, 0.8)]$	0.052	0.073
Eve $[(x_1, y_1, \theta) = (0.6 + 10^{-14}, 0.7, 0.8)]$	0.033	0.028

The optimized parameters for the smooth l_0 method are also discussed. The SSIM performance of Bob with different iteration numbers is illustrated in Figure 7. For iteration numbers, the step size is set as 1.01, and the CR is set as 0.5. From the results, we found that there is not much difference in SSIM performance under different iteration times. When the iteration number is small, SSIM performance has slightly improved with its improvement. Afterwards, the algorithm converges and SSIM performance gradually stabilizes. Since the improvement is not very significant, the iteration number can be

https://doi.org/10.3390/mi17010080

reduced to lower the algorithm complexity and meet real-time communication requirements. In fact, the complexity of the compressive sensing algorithm is approximately $O(N)$, which is completely acceptable for real-time communication [17]. Based on existing deductions, the initial smooth factor for the compressive sensing algorithm may be chosen about two to four times the maximum absolute value of the obtained sources. In the proof-of-concept experiment, we set the factor to two times the maximum absolute value to ensure mathematical interpretability.

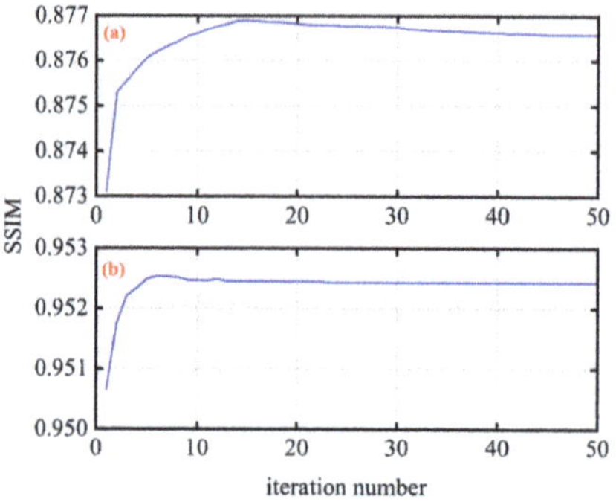

Figure 7. The SSIM performance under different iteration numbers of the smooth l_0 method with (**a**) CR = 0.5 and (**b**) CR = 0.75.

The selection of CR also has a significant impact on the signal reconstruction performance of Bob. The SSIM performance of Bob under different CRs is shown in Figure 8. Clearly, when the CR is less than 0.5, the SSIM performance significantly decreases. There are two main reasons for this result. Firstly, the DCT of the image signal is not a strictly sparse matrix, and there are still small coefficients at high frequencies. When the number of samples is reduced, the effective sparsity of the matrix remains unchanged, making it difficult for the algorithm to distinguish between the main components and noise based on redundant information. On the other hand, the observation process of CS is accompanied by noise. When the CR decreases, the number of elements in the matrix also decreases. Due to the diffusion effect of matrix multiplication, the SNR of each sampling point decreases. Therefore, to ensure the signal quality reconstructed by Bob, it is best to choose a CR greater than 0.5. And adaptive adjustment algorithms can be further adopted to balance the computational complexity and signal reconstruction quality.

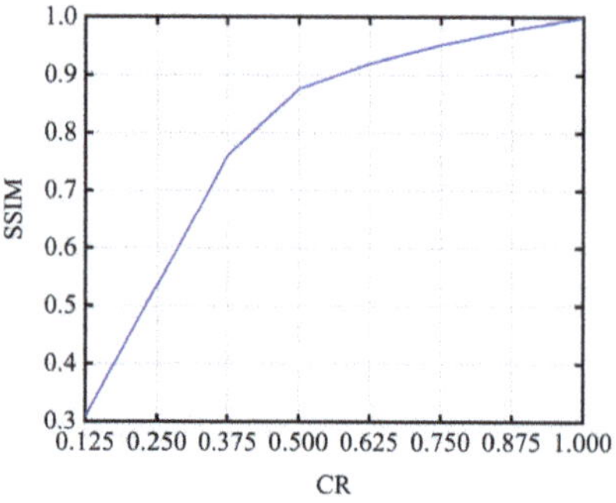

Figure 8. The SSIM between Alice and Bob under different CR conditions.

https://doi.org/10.3390/mi17010080

5. Conclusions

In this paper, we propose a novel method that utilizes a CS-based chaotic encryption algorithm to enhance both spectrum efficiency and security of the RoF system. Regarding spectrum efficiency, the CS algorithm reduces the amount of transmitted data, and the system achieves quantization-error-free at only a 6-bit quantization level with the proposed algorithm. In terms of security, an SSIM of 0.952 for Bob at a CR of 0.75 is achieved, indicating minimal information loss. In contrast, under the same condition, the SSIM for Eve is only 0.073, confirming the effectiveness of the algorithm. On this basis, the parameters are optimized and explored to further reduce computational complexity.

Author Contributions: Conceptualization, Z.W. and L.Z.; methodology, Z.W. and L.Z.; software, Z.W. and J.Z.; validation, Z.W. and J.Z.; formal analysis, Z.W. and L.Z.; investigation, Z.W. and L.Z.; resources, L.Z. and X.Y.; data curation, Z.W.; writing—original draft preparation, Z.W.; writing—review and editing, Z.W., L.Z., X.P., O.O., and X.Y.; supervision, L.Z. and X.Y.; project administration, L.Z., X.P., O.O., and X.Y. All authors have read and agreed to the published version of the manuscript.

Funding: This work is supported by the National Natural Science Foundation of China (62471433); the Sweden's Innovation Agency (VINNOVA) funded project (2024-02451); the strategic innovation program Smarter Electronic Systems—a joint venture by Vinnova, Formas, and the Swedish Energy Agency A-FRONTAHUL project (2023-00659).

Institutional Review Board Statement: Not applicable.

Informed Consent Statement: Not applicable.

Data Availability Statement: Dataset available on request from the authors.

Conflicts of Interest: The authors declare no conflicts of interest.

References

1. Xiao, M.; Mumtaz, S.; Huang, Y.; Dai, L.; Li, Y.; Matthaiou, M.; Karagiannidis, G.K.; Bjornson, E.; Yang, K.; I, C.-L.; et al. Millimeter Wave Communications for Future Mobile Networks. *IEEE J. Select. Areas Commun.* **2017**, *35*, 1909–1935. [CrossRef]
2. Rasti, M.; Taskou, S.K.; Tabassum, H.; Hossain, E. Evolution Toward 6G Multi-Band Wireless Networks: A Resource Management Perspective. *IEEE Wirel. Commun.* **2022**, *29*, 118–125. [CrossRef]
3. Hemadeh, I.A.; Satyanarayana, K.; El-Hajjar, M.; Hanzo, L. Millimeter-Wave Communications: Physical Channel Models, Design Considerations, Antenna Constructions, and Link-Budget. *IEEE Commun. Surv. Tutor.* **2018**, *20*, 870–913. [CrossRef]
4. Common Public Radio Interface, Interface Specification V7.0. 2015. Available online: https://www.cpri.info/downloads/CPRI_v_7_0_2015-10-09.pdf (accessed on 5 January 2026).
5. Liu, X.; Zeng, H.; Chand, N.; Effenberger, F. Efficient Mobile Fronthaul via DSP-Based Channel Aggregation. *J. Light. Technol.* **2016**, *34*, 1556–1564. [CrossRef]
6. Nimura, S.; Tanaka, K.; Ishimura, S.; Nishimura, K.; Inohara, R.; Tsuritani, T.; Suzuki, M. 10.51-Tbit/s Multi-Channel IF-over-Fiber Transmission with SDM/WDM/SCM for beyond-5G Mobile Fronthaul Accommodating Ultra-High-Density Antennas. *J. Opt. Commun. Netw.* **2023**, *15*, C263. [CrossRef]
7. Maximidis, R.; Vagionas, C.; Kalfas, G.; Leiba, Y.; Miliou, A.; Pleros, N. A 51 Gb/s Reconfigurable mmWave Fiber-Wireless C-RAN Supporting 5G/6G MNO Network Sharing. *J. Light. Technol.* **2023**, *41*, 4705–4712. [CrossRef]
8. Qiao, R.; Jiang, T.; Yu, W. Learning-Based Fronthaul Compression for Uplink Cloud Radio Access Networks. In Proceedings of the ICC 2023—IEEE International Conference on Communications, Rome, Italy, 28 May–1 June 2023; pp. 5928–5933.
9. Grönland, A.; Klaiqi, B.; Gelabert, X. Learning-Based Latency-Constrained Fronthaul Compression Optimization in C-RAN. In Proceedings of the 2023 IEEE 28th International Workshop on Computer Aided Modeling and Design of Communication Links and Networks (CAMAD), Edinburgh, UK, 6–8 November 2023; pp. 134–139.
10. Zhang, C.; Zhu, Y.; He, B.; Lin, J.; Liu, R.; Xu, Y.; Yi, L.; Zhuge, Q.; Hu, W.; Hu, W.; et al. Clone-Comb-Enabled High-Capacity Digital-Analogue Fronthaul with High-Order Modulation Formats. *Nat. Photonics* **2023**, *17*, 1000–1008. [CrossRef]
11. Argyris, A.; Syvridis, D.; Larger, L.; Annovazzi-Lodi, V.; Colet, P.; Fischer, I.; García-Ojalvo, J.; Mirasso, C.R.; Pesquera, L.; Shore, K.A. Chaos-Based Communications at High Bit Rates Using Commercial Fibre-Optic Links. *Nature* **2005**, *438*, 343–346. [CrossRef]
12. Kodama, T.; Nakagawa, N.; Kataoka, N.; Wada, N.; Cincotti, G.; Wang, X.; Miyazaki, T.; Kitayama, K.-I. Secure 2.5 Gbit/s, 16-Ary OCDM Block-Ciphering with XOR Using a Single Multi-Port En/Decoder. *J. Light. Technol.* **2010**, *28*, 181–187. [CrossRef]

https://doi.org/10.3390/mi17010080

13. Bi, M.; Fu, X.; Zhou, X.; Zhang, L.; Yang, G.; Yang, X.; Xiao, S.; Hu, W. A Key Space Enhanced Chaotic Encryption Scheme for Physical Layer Security in OFDM-PON. *IEEE Photonics J.* **2017**, *9*, 1–10. [CrossRef]
14. Cui, M.; Chen, Y.; Zhang, C.; Liang, X.; Wu, T.; Liu, S.; Wen, H.; Qiu, K. Chaotic RNA and DNA for Security OFDM-WDM-PON and Dynamic Key Agreement. *Opt. Express* **2021**, *29*, 25552. [CrossRef] [PubMed]
15. Zhang, W.; Zhang, C.; Chen, C.; Zhang, H.; Qiu, K. Brownian Motion Encryption for Physical-Layer Security Improvement in CO-OFDM-PON. *IEEE Photonics Technol. Lett.* **2017**, *29*, 1023–1026. [CrossRef]
16. Hu, Z.; Chan, C.-K. A 7-D Hyperchaotic System-Based Encryption Scheme for Secure Fast-OFDM-PON. *J. Light. Technol.* **2018**, *36*, 3373–3381. [CrossRef]
17. Wu, T.; Zhang, C.; Chen, Y.; Cui, M.; Huang, H.; Zhang, Z.; Wen, H.; Zhao, X.; Qiu, K. Compressive Sensing Chaotic Encryption Algorithms for OFDM-PON Data Transmission. *Opt. Express* **2021**, *29*, 3669. [CrossRef]
18. Zhong, Q.; Guo, Z.; Liu, B.; Ren, J.; Mao, Y.; Wu, X.; Wu, Y.; Zhao, L.; Sun, T.; Ullah, R. Block Compressive Sensing Chaotic Embedded Encryption for MCF-OFDM Transmission System. *Opt. Express* **2022**, *30*, 21774. [CrossRef]
19. Hua, Z.; Jin, F.; Xu, B.; Huang, H. 2D Logistic-Sine-Coupling Map for Image Encryption. *Signal Process.* **2018**, *149*, 148–161. [CrossRef]
20. Rauhut, H.; Romberg, J.; Tropp, J.A. Restricted Isometries for Partial Random Circulant Matrices. *Appl. Comput. Harmon. Anal.* **2012**, *32*, 242–254. [CrossRef]
21. Qaisar, S.; Bilal, R.M.; Iqbal, W.; Naureen, M.; Lee, S. Compressive Sensing: From Theory to Applications, a Survey. *J. Commun. Netw.* **2013**, *15*, 443–456. [CrossRef]
22. Mohimani, H.; Babaie-Zadeh, M.; Jutten, C. A Fast Approach for Overcomplete Sparse Decomposition Based on Smoothed l^0 Norm. *IEEE Trans. Signal Process.* **2009**, *57*, 289–301. [CrossRef]
23. Sari, H.; Moridi, S. New Phase and Frequency Detectors for Carrier Recovery in PSK and QAM Systems. *IEEE Trans. Commun.* **1988**, *36*, 1035–1043. [CrossRef]
24. Belfiore, C.A.; Park, J.H. Decision Feedback Equalization. *Proc. IEEE* **1979**, *67*, 1143–1156. [CrossRef]
25. Takiguchi, K. 40 Gsymbol/s Channel-Based Nyquist Wavelength Division Multiplexing Communication in a Terahertz-Band Using Optical-Domain Reception Signal Processing. *OSA Contin.* **2020**, *3*, 2308. [CrossRef]
26. Azarnia, G.; Sharifi, A.A.; Emami, H. Compressive Sensing Based PAPR Reduction in OFDM Systems: Modified Orthogonal Matching Pursuit Approach. *ICT Express* **2020**, *6*, 368–371. [CrossRef]

micromachines MDPI

Review

Research on Tbps and Kilometer-Range Transmission of Terahertz Signals

Jianjun Yu * and Jiali Chen

Key Laboratory for Information Science of Electromagnetic Waves (MoE), Fudan University, Shanghai 200433, China; 23210720141@m.fudan.edu.cn
* Correspondence: jianjun@fudan.edu.cn

Abstract

THz communication stands as a pivotal technology for 6G networks, designed to address the critical challenge of data demands surpassing current microwave and millimeter-wave (mmWave) capabilities. However, realizing Tbps and kilometer-range transmission confronts the "dual attenuation dilemma" comprising severe free-space path loss (FSPL) (>120 dB/km) and atmospheric absorption. This review comprehensively summarizes our group′s advancements in overcoming fundamental challenges of long-distance THz communication. Through systematic photonic–electronic co-optimization, we report key enabling technologies including photonically assisted THz signal generation, polarization-multiplexed multiple-input multiple-output (MIMO) systems with maximal ratio combining (MRC), high-gain antenna–lens configurations, and InP amplifier systems for complex weather resilience. Critical experimental milestones encompass record-breaking 1.0488 Tbps throughput using probabilistically shaped 64QAM (PS-64QAM) in the 330–500 GHz band; 30.2 km D-band transmission (18 Gbps with 543.6 Gbps·km capacity–distance product); a 3 km fog-penetrating link at 312 GHz; and high-sensitivity SIMO-validated 100 Gbps satellite-terrestrial communication beyond 36,000 km. These findings demonstrate THz communication′s viability for 6G networks requiring extreme-capacity backhaul and ultra-long-haul connectivity.

Keywords: terahertz communication; long-distance transmission; atmospheric absorption; photonic-assisted; polarization multiplexing; maximal ratio combining; high-gain antennas; power amplifiers; coherent detection; satellite–ground communication

1. Introduction

THz communication stands as a core candidate technology for next-generation wireless communication. Its necessity stems from the fundamental contradiction between the exponential growth in wireless data rates and the limitations of existing spectrum resources [1–5]. As 5G mmWave bands gradually become saturated, traditional microwave bands (<6 GHz) face inherent drawbacks including excessively wide beams, high power consumption, and insufficient bandwidth, pushing their spectral efficiency near the theoretical limit [6–10]. Furthermore, their practical deployment is hindered by large device size and low integration density. While laser communication offers high bandwidth potential, its practical implementation is severely constrained by difficulties in aligning its extremely narrow beams and its high sensitivity to adverse weather conditions [11,12]. Occupying the frequency gap between microwaves and optical bands, the THz band bridges this technological void with its unique physical properties, becoming crucial for future 6G developments and beyond [13,14]. The exponential increase in wireless data rates, coupled with the inability of mega-bit-per-second microwave speeds to meet the demands of massive data transmission, has driven exploration and research into the THz band with its rich spectrum resources [15–17]. Its enormous contiguous bandwidth lays the physical layer foundation for building 6G networks with high data rates, low latency, and high-density ubiquitous connectivity, enabling potential speeds reaching gigabits per second (Gbps) or even terabits per second (Tbps) [18–20].

The sub-terahertz band (typically 0.1–0.3 THz), as standardized in ITU-R V.431 [21] and IEEE 802.15.3d [22], utilizes solid-state electronics such as Complementary Metal Oxide Semiconductors (CMOSs) and SiGe Heterojunction Bipolar Transistors (HBTs) for integrated circuit implementation to support moderate-bandwidth communications. In contrast, the multi-terahertz band (3–10 THz) falls within the 'optically-oriented spectrum' per IRMMW-THz Conference definitions, where quantum cascade lasers (QCLs) and photoconductive antennas dominate, enabling high-resolution spectroscopy despite severe molecular absorption losses. Bridging these extremes, the 0.3–3 THz 'gap' predominantly employs hybrid approaches combining Schottky diode multipliers with photo-mixers [23–30].

THz communication emerges as a technology capable of breaking the performance limits of next-generation wireless networks due to its distinct advantages. Spectrum resources: the THz band offers hundreds of GHz of contiguous bandwidth, enabling single-link Tbps-level transmission without complex spectral efficiency enhancement techniques, bridging the technological gap between mmWave and Free-Space Optical (FSO) communication. Physical properties: its short wavelength not only suppresses multi-path interference and eavesdropping risks but also supports ultra-highly directional beams, exhibiting superior stability compared to FSO in complex environments [31–37].

However, achieving long-distance THz signal transmission faces significant challenges. As indicated by Figure 1, FSPL increases exponentially with frequency; THz signals experience path loss exceeding 120 dB per kilometer, far surpassing attenuation levels in mmWave bands. Simultaneously, atmospheric attenuation effects are particularly severe in specific frequency regions: molecular resonance absorption by water vapor (H_2O) and oxygen (O_2) creates multiple attenuation peaks, where attenuation at sea level can exceed 10 dB/km. Under rain or fog conditions, this loss can surge to tens of dB/km. The combined effect of these two physical limitations creates a "dual attenuation dilemma" for long-distance THz transmission: signals must overcome both energy dissipation due to free-space spreading and avoid strong absorption bands outside atmospheric windows [33,38–42]. The engineering exploration of THz communication technology began in the 1970s. First, in 1970, Bell Labs developed the first continuous-wave THz laser (methanol laser, 1–5 THz) using gas laser pumping technology. Although constrained by cryogenic operation and low efficiency, this breakthrough marked the inception of artificially controllable THz signal generation. Subsequently, Nippon Telegraph and Telephone Corporation (NTT) achieved the world's first long-distance THz communication experiment in 2006, demonstrating 10 Gbps over 15 km line-of-sight (LoS) using the 120 GHz band, validating THz technology feasibility for satellite backhaul and wide-area coverage scenarios. Further, during the 2008 Beijing Olympics, NTT deployed a 120 GHz THz communication system to successfully achieve real-time broadcast of 8K ultra-high-definition video signals, proving its commercial potential for applications with extreme bandwidth demands for the first time. Finally, in 2017, the IEEE published the first international THz communication standard (802.15.3d), specifying the physical layer protocol and frame structure for the 252–325 GHz band, laying the groundwork for chip design, device manufacturing, and the 6G standardization process [22,43–45].

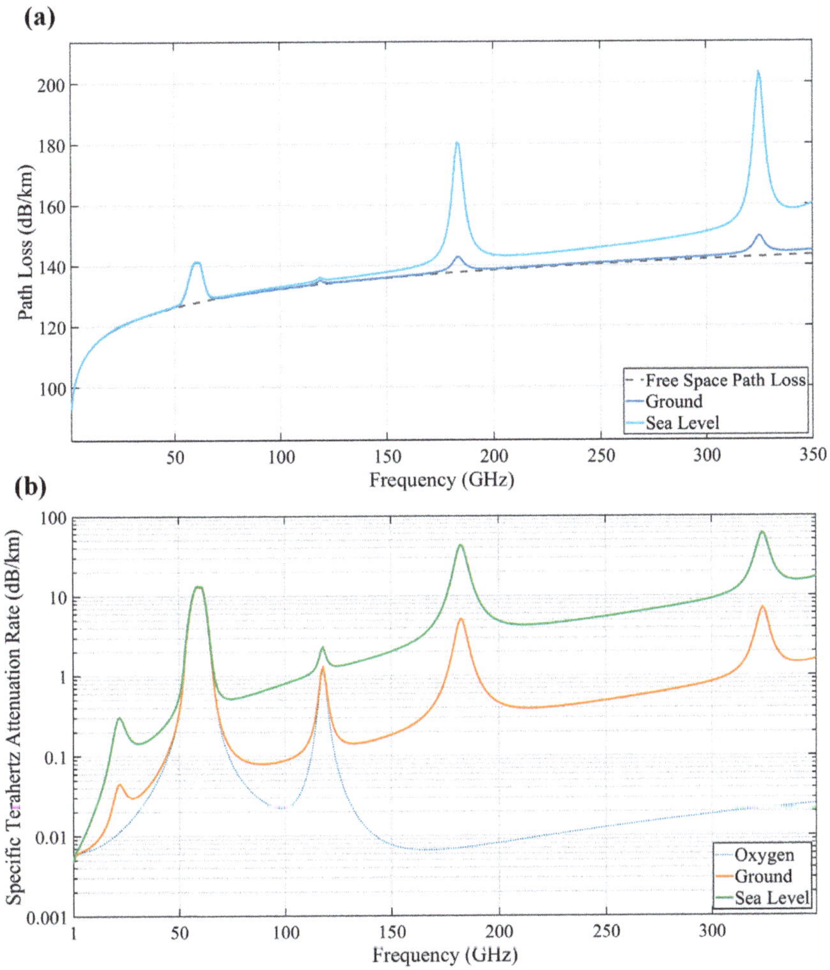

Figure 1. Path loss and atmospheric loss: (**a**) free space path loss; (**b**) atmospheric loss.

This paper summarizes a series of advanced methods and architectures that achieved historic, record-breaking ultra-broadband THz signal transmissions, providing critical technological support and spectrum expansion for the transition from 5G-Advanced to 6G. In Section 2, we discuss the generation of broadband THz signals using simple and cost-effective schemes. Section 3 summarizes key enabling technologies comprehensively considering power consumption, equalization, algorithms, and structure to increase the transmission distance of THz communication systems. To explore the role of THz systems within the integrated Space–Air–Ground (SAG) strategy, Section 4 presents simulations of satellite-to-ground communications. The final section provides a summary of the paper and outlines predictions and plans for the next phase of long-distance THz communication system development.

2. Terahertz Signal

Realizing an efficient, stable, and high-power THz signal source is key to the development of related applications. Especially in scenarios pursuing long-distance, high-capacity wireless communications, the performance of the signal source directly determines the system's link budget and transmission capabilities [46,47]. Currently, as shown in Figures 2 and 3, the generation of THz signals primarily relies on two main technological pathways: electronic approaches and photonic approaches.

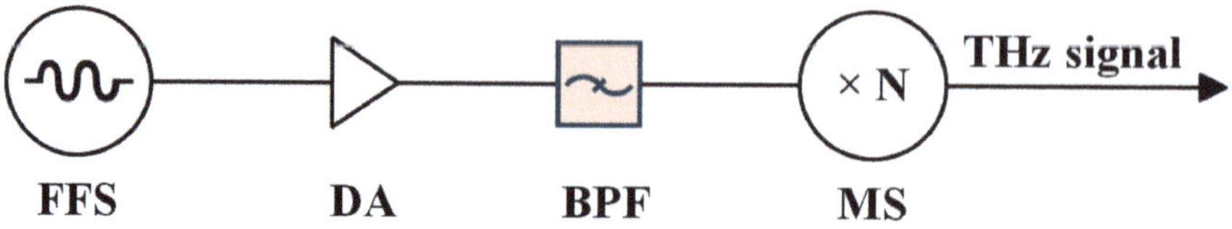

Figure 2. Multiplier-based terahertz signal; FFS: Fundamental Frequency Source; DA: Drive Amplifier; BPF: Band Pass Filter; MS: Multiplier Stage.

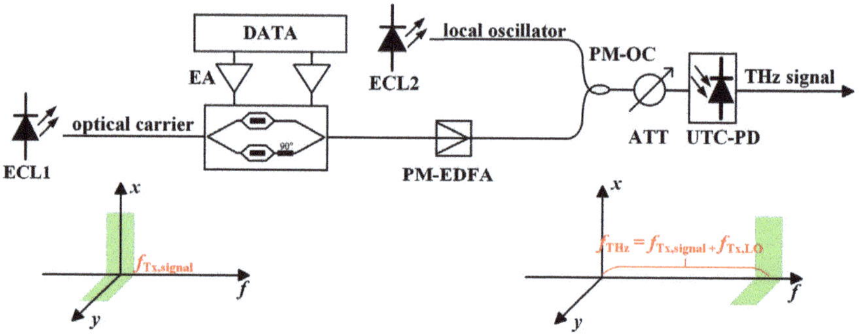

Figure 3. Photonic-assisted terahertz signal; ECL: external cavity laser; EA: electrical amplifier; UTC-PD: Uni-Traveling-Carrier Photodiode; PM-EDFA: polarization-maintaining erbium-doped fiber amplifier; PM-OC: polarization-maintaining optical coupler; ATT: attenuator.

2.1. Electronic Approaches

Electronic approaches primarily rely on silicon-based devices [48]. In recent years, significant progress has been made in technologies based on advanced semiconductor materials such as III-V compound semiconductors with Indium Phosphide (InP) and Gallium Nitride (GaN), including High Electron Mobility Transistors (HEMTs), Monolithic Millimeter-Wave Integrated Circuits (MMICs), and HBTs. For instance, HEMTs exhibit ultra-high electron saturation velocity and cut-off frequencies reaching 1.5 THz, laying the hardware foundation for THz communication systems. The core implementation path for all-solid-state-electronic solutions is the cascaded frequency multiplication technique [49–51]. An oscillator first generates a microwave/millimeter-wave fundamental frequency signal. This signal is then progressively up-converted into the THz band through frequency multiplier chains based on Schottky diodes or HBTs. Finally, an HEMT power amplifier (PA) outputs the effective radiated power [52–56].

In paper [57], researchers successfully demonstrated real-time transmission of 34-GBd polarization division multiplexing (PDM) quadrature phase shift keying (QPSK) signals using a purely electronic THz system integrated with a commercial Digital Coherent Optical (DCO) real-time modem. The experimental setup is shown in Figure 4. THz signal generation was achieved via electronic frequency multiplication and direct conversion mixing techniques: First, a signal generator produced an 8.311 GHz local oscillator (LO) signal, which was multiplied to ~300 GHz using a 36-fold frequency multiplier. Subsequently, the four analog baseband signals (corresponding to the $X_I/X_Q/Y_I/Y_Q$ components of the 34-GBd PDM-QPSK optical signal) demodulated from the first fiber span were directly fed into two MMIC-based direct-conversion in-phase and quadrature (I/Q) mixer transmitter frontends. Here, the baseband signals were up-converted to the THz band by mixing with the 300 GHz LO. Spatial multiplexing technology (using two independent antenna pairs) transmitted the dual-polarization data synchronously over a 50 cm wireless link. This method achieved the first real-time hybrid transmission with a net data rate of 100 Gb/s: Within a system comprising two fiber spans (total length 103 km) and the 300 GHz THz link, the Bit Error Rate (BER) remained stably below the forward error correction (FEC) threshold (3.4×10^{-2}) by employing single-carrier real-time Digital Signal Processing (DSP) to jointly compensate for accumulated dispersion, polarization rotation, and phase noise. This validates the transparency of the THz wireless extender to linear impairments inherent in optical fibers and provides a multi-hundred-gigabit wireless fiber extension solution for 5G/6G backhaul.

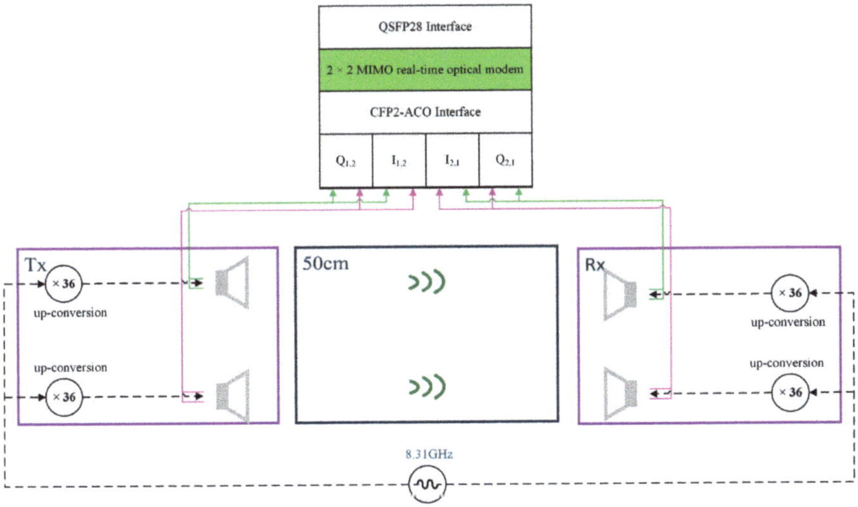

Figure 4. Pure-electronic terahertz real-time transmission system; Tx: transmission; Rx: reception.

2.2. Photonic Approaches

Photonic approaches fundamentally employ the dual-laser optical heterodyne beating technique for high-speed signal generation [58–61]. This technology leverages an opto-electro-optical seamless bridging architecture for signal generation and transmission: Two independent continuous-wave (CW) optical waves—one serving as the LO and the other as the modulated optical carrier—are generated by lasers. They are combined in an optical coupler for heterodyne beating, producing a THz signal whose frequency equals the difference between the two optical wavelengths. A high-speed photodetector then performs the optical-to-electrical (O/E) conversion, and the resulting THz signal is radiated by an antenna. At the receiver, a reverse conversion process reconstructs the optical carrier from the THz signal, which is finally received by a coherent receiver. This Fiber–Wireless (FiWi) integrated access system effectively combines the bandwidth and long-distance transmission advantages of optical fiber communications with the mobility and seamless coverage benefits of wireless communications. By utilizing the relatively simple and cost-effective photonically assisted THz technology, it overcomes the bandwidth bottleneck inherent in electronic devices to generate ultra-high-speed wireless THz signals.

Prior to reference [62], real-time fiber-THz-fiber link transmissions exceeding 100 Gbit/s at frequencies above 350 GHz remained unreported. This gap stemmed from the immense challenges associated with achieving high-speed, stable, end-to-end real-time transmission at such high frequencies, including the generation, modulation, transmission loss, and real-time signal processing of high-frequency signals. Addressing this challenge, through the experimental setup in Figure 5, we successfully constructed and demonstrated, for the first time, a photonically assisted THz wireless transmission system operating in the 330–500 GHz band achieving an unprecedented Aggregated Interface User Rate (AIUR) of 1.0488 Tbps, alongside seamless integration of 10 km of fiber and a 10 m wireless link.

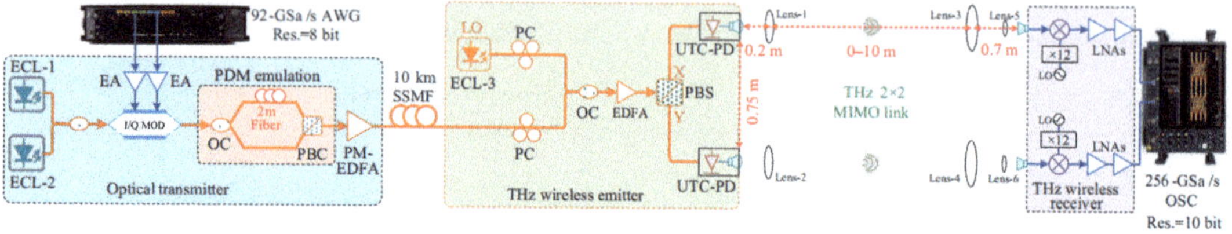

Figure 5. Photonic-assisted terahertz 1 Tbps transmission system; LNA: low-noise amplifier; OC: optical coupler; PC: polarization controller [62].

The core innovation lies in fully leveraging the unique advantages of photonics: The system employs free-running tunable ECLs as high-purity, low-phase-noise optical carriers (ECL-1 @ 193.47 THz, ECL-2 @ 193.565 THz). A high-speed optical I/Q modulator (Fujitsu FTM7961EX/301, 3 dB bandwidth > 35 GHz) efficiently modulates the 46 GBaud PS-64QAM baseband signal directly in the optical domain. PDM enables parallel transmission of dual-polarization signals within a single fiber. After 10 km of Standard Single-Mode Fiber (SSMF) transmission, a third ECL (ECL-3 @ 193.1 THz) serves as the LO at the receiver for optical heterodyne beating with the incoming signal. The critical O/E conversion is performed by a high-performance UTC-PD (NTT IOD-PMAN-13001). This device features an ultra-wide response bandwidth (300–2500 GHz typical), high saturation power (up to 15 dBm optical input), and good responsivity (0.15 A/W @ 1550 nm), enabling efficient, linear conversion of the heterodyne-generated difference frequency signal into powerful THz radiation. This photonically assisted THz generation method overcomes the bandwidth constraints faced by purely electronic approaches (>300 GHz), offering superior frequency flexibility, tunability, and potential bandwidth.

Finally, employing a meticulously designed 2 × 2 MIMO wireless link and an innovative real-time signal processing algorithm based on a Vector Quantized Variational Autoencoder (VQ-VAE)—specifically addressing the limitations of conventional equalizers for high-order PS-QAM signals—we successfully achieved a landmark AIUR of 1.0488 Tbps with a net rate of 828 Gbps. This groundbreaking result powerfully demonstrates the core value and immense potential of photonic technology in constructing future 6G networks demanding Tbps-level extreme connectivity.

3. Enabling Technologies for THz Communication Systems

In this section, to enhance the effective transmission distance of photonic-assisted THz communication systems, we introduce five key enabling technologies, specifically, improving receiver sensitivity; adopting low-order modulation formats characterized by large constellation point spacing and strong noise immunity; leveraging MIMO architectures and MIMO diversity gain for polarization-multiplexed optical THz signals; utilizing high-gain antennas and high-gain lenses; and employing PAs with high gain and high saturated output power coupled with a heterodyne coherent detection configuration at the receiver. Through this optoelectronic co-design optimization strategy, which simultaneously reduces the signal baud rate and the bandwidth requirements of the devices, we provide a feasible path for the long-distance deployment of photonic-assisted THz communications [63].

3.1. Selecting Appropriate Modulation Formats

In THz communication systems, achieving both long wireless transmission distance and high spectral efficiency presents a fundamental trade-off. While utilizing higher-order QAM signals rationally when extending transmission distance can yield extremely high spectral efficiency, the inherent burdens on the THz system cannot be ignored. For the same transmit power, higher-order QAM signals exhibit a smaller Euclidean distance between adjacent constellation points. Figure 6 shows the constellation diagrams of QPSK, 16QAM, and 64QAM under normalized power. Consequently, they require either higher receiver sensitivity, increased input power, or a higher signal-to-noise ratio (SNR) at the receiver side [64–71]. Furthermore, these signals are significantly constrained by the THz channel's path loss and nonlinear effects, necessitating the integration of advanced DSP techniques and system optimization strategies [19,72–74].

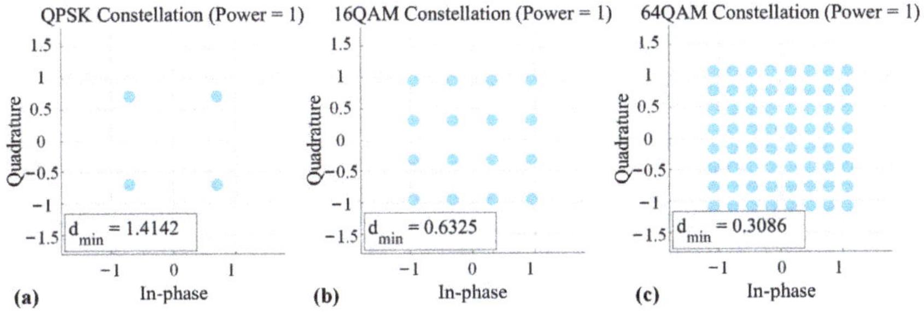

Figure 6. QPSK, 16QAM, and 64QAM constellation diagrams under normalized power: (**a**) QPSK; (**b**) 16QAM; (**c**) 64QAM.

Given our primary focus on achieving longer wireless transmission distances, we predominantly employ lower-order QAM formats, such as QPSK and 16QAM. Using 16QAM modulation, we achieved 920 Gbps·km, and using QPSK, we demonstrated transmissions spanning 30 km in the sub-THz band [75,76]. While higher-order modulation significantly boosts spectral efficiency by carrying more bits per symbol, its application is typically restricted to short-range scenarios for achieving ultra-high data rates due to the limiting characteristics of the THz channel.

3.2. Adopting MIMO Architecture Combining Polarization Multiplexing and MRC

In long-distance THz wireless communication systems, single-antenna architectures face severe challenges. Limitations in optoelectronic conversion efficiency cause significant energy loss during signal transmission [77]. Concurrently, the inherently poor penetration characteristics of high-frequency electromagnetic waves and severe atmospheric attenuation collectively result in extremely weak signal strength at the receiver, critically degraded SNR, and fundamental constraints on communication quality and transmission capacity [78–80].

To overcome these bottlenecks and dramatically enhance system performance, MIMO technology emerges as a critical solution. As shown in Figure 7, by deploying antenna arrays at both the transmitter and receiver ends, MIMO technology enables deep spatial exploitation, effectively addressing insufficient received power and low SNR issues [81,82]. Our primary enabling pathways include: MIMO combined with PDM (MIMO-PDM) and MIMO with MRC (MIMO-MRC) [83–87].

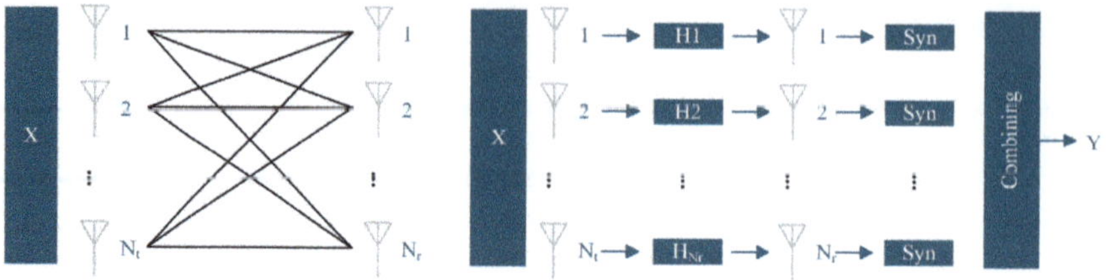

Figure 7. MIMO architecture diagram and MIMO diversity reception architecture.

For MIMO-PDM technology, THz signals are first generated using the photonic-assisted approach introduced in Section 2, employing two optical THz signals sharing the same optical frequency but possessing orthogonal polarization states. These two signals can be generated from either (a) a common CW laser source, subsequently combined via an OC, or (b) a PDM I/Q modulator, which directly produces orthogonally polarized optical THz signals.

Prior to the photodetection (optical-to-electrical conversion) of the polarization-multiplexed optical signal, its orthogonal polarization components must be separated using optical components. Separation options include schemes based on optical couplers combined with polarization beam splitters (PBSs) or, alternatively, using a polarization-diverse phase-diversity 90-degree optical hybrid. So first, the data is modulated onto one optical carrier wave. After passing through the PBS, it beats with another LO light source (also passed through a PBS) to generate a terahertz signal.

Subsequently, in the detection stage, the dual-diode balanced structure within balanced photodetectors (BPDs) effectively suppresses common-mode noise while enhancing tolerance to input signal power and improving SNR. After passing through two high-speed balanced photodetectors, two electrical THz signals are produced, collectively forming the polarization-division-multiplexed electrical THz signal. Wireless transmission of this signal is realized via a 2×2 MIMO link.

At the receiver, the received THz signal must first undergo analog down-conversion. This is achieved using balanced mixers driven by an electrical LO operating in the THz frequency band, converting the THz carrier signal down to a lower intermediate frequency (IF) signal. These IF signals are then captured by a real-time oscilloscope equipped with two input ports.

Finally, the data recorded by the real-time oscilloscope is fed into an advanced DSP module for analysis and processing to recover the original information. The optical baseband transmitter generates the original optical baseband signal, whose X- and Y-polarization components can be represented as the following column vector:

$$\mathbf{E}_{in} = \begin{pmatrix} E_{in,x} \\ E_{in,y} \end{pmatrix}. \quad (1)$$

The received signal after fiber transmission can be represented as follows:

$$\mathbf{E}_{out1} = \mathbf{J}\mathbf{E}_{in} = \begin{pmatrix} J_{xx} & J_{xy} \\ J_{yx} & J_{yy} \end{pmatrix} \begin{pmatrix} E_{in,x} \\ E_{in,y} \end{pmatrix}. \quad (2)$$

where $\mathbf{J}$ is the 2×2 Jones matrix; the diagonal elements (J_{xx}, J_{yy}) are principal polarization channels; and off-diagonal elements (J_{xy}, J_{yx}) are the polarization crosstalk from fiber impairments.

At the Rx, the received THz signal is

$$\mathbf{E}_{out2} = \mathbf{G}\mathbf{E}_{out1}\cos(\omega t) = \begin{pmatrix} G_{xx} & G_{xy} \\ G_{yx} & G_{yy} \end{pmatrix} \mathbf{E}_{out1}\cos(\omega t). \quad (3)$$

where **G** is the 2×2 MIMO gain matrix; ω is THz carrier frequency; and off-diagonal elements (W_{xy}, W_{yx}) are wireless crosstalk.

Combining fiber and wireless stages,

$$\mathbf{E}_{out2} = \mathbf{G}\mathbf{J}\mathbf{E}_{in}\cos(\omega t) = \mathbf{H}\mathbf{E}_{in}\cos(\omega t). \quad (4)$$

where the composite channel matrix **H** is

$$\begin{aligned} \mathbf{H} = \mathbf{G}\mathbf{J} &= \begin{pmatrix} H_{xx} & H_{xy} \\ H_{yx} & H_{yy} \end{pmatrix}, \\ H_{ij} &= \sum_{k=1}^{2} G_{ik}J_{kj} \quad (i,j \in \{x,y\}). \end{aligned} \quad (5)$$

The received signals are demultiplexed via a butterfly equalizer.

In reference [88], we successfully achieved a major breakthrough in high-speed long-distance wireless transmission in the D-band using MIMO-PDM, innovatively constructing a photonics-assisted 2×2 MIMO wireless transmission system. The experimental setup is shown in Figure 8. At the transmitter, dual external cavity lasers (ECL-1 @1550 nm, ECL-2 @1551 nm) provide the optical carrier and local oscillator, respectively; a 25-Gbaud 16-QAM baseband signal (generated by an arbitrary waveform generator (AWG) and amplified by an EA) is loaded via an I/Q modulator; and PDM is used to generate two independent signals. Critical opto-electronic conversion is performed by a high-performance UTC-PD, up-converting the optical signal to a 125 GHz D-band millimeter wave. Subsequently, the signal is amplified by cascaded LNAs and PAs (total gain ~31 dB, output power 13 dBm), transmitted through a pair of orthogonally polarized (horizontal/vertical) horn antennas (gain 25 dBi), and innovatively collimated and focused using large-aperture (60 cm) and small-aperture (10 cm) plano-convex lens assemblies to overcome high free-space transmission losses in the D-band. The receiver similarly employs orthogonally polarized horn antennas and lens assemblies to capture the signal, which is amplified by LNAs, down-converted to a 13 GHz intermediate frequency, and acquired by a high-speed oscilloscope. The receiver-side DSP employs a sophisticated algorithm chain including MIMO–constant modulus algorithm (CMA) equalization, frequency offset estimation, carrier phase recovery, advanced 51-tap MIMO-CMA equalization, a 183-core second-order MIMO Volterra nonlinear equalizer (MIMO-VNLE), and 155-tap decision-directed least mean square (DD-LMS) equalization, effectively mitigating impairments induced by long-distance transmission. Ultimately, over a 4.6 km real-world urban wireless link (between campuses of Fudan University), a total transmission rate of 200 Gbps (100 Gbps per channel) was successfully achieved, with a measured minimum BER of 2.05×10^{-2} meeting the 20% soft-decision forward error correction (SD-FEC) threshold requirement. This accomplishment established a record-breaking capacity–distance product of 920 Gbps·km.

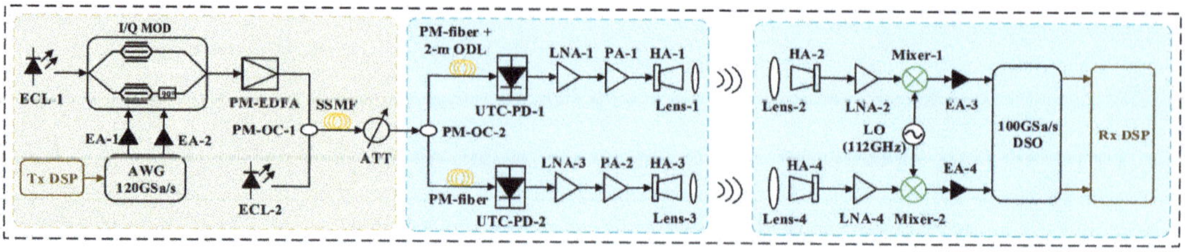

Figure 8. MIMO-PDM-based long-distance radio-over-fiber (RoF) terahertz transmission system; HA: horn antenna [88].

In single-polarization MIMO systems, multiple transmitter antennas simultaneously transmit identical signals. The predominant challenge in this configuration is maximizing gain while simultaneously improving receiver accuracy (i.e., signal fidelity). MIMO-MRC addresses this by applying beamforming (BF) principles. It determines the weights at the receiver side based on the received power level of each distinct channel. Subsequently, it combines and sums the signals from all channels using these weighted coefficients to achieve diversity gain. For a MIMO system with t transmit antennas and r receive antennas, the received signal Y can be expressed as

$$\mathbf{Y} = \mathbf{HX} + \mathbf{n}. \tag{6}$$

where $\mathbf{X}$ is the $N_t \times 1$ transmit vector, $\mathbf{H}$ is the $N_r \times N_t$ THz channel matrix, and $\mathbf{n}$ is the $N_r \times 1$ additive white Gaussian noise (AWGN) vector satisfying E{$\mathbf{nn}^{\mathrm{H}}$} = N_0I. At the receiver, the MRC algorithm is employed, and the receiver applies a complex weight vector $\mathbf{W} = [w_1, w_2, \ldots, w_{N_r}]^T (N_r \times 1)$ to the received signals and combines them:

$$\mathbf{W}^H\mathbf{Y} = \mathbf{W}^H\mathbf{HX} + \mathbf{W}^H n. \tag{7}$$

The SNR after MRC combining is derived as

$$SNR = \frac{\left|\mathbf{W}^H\mathbf{H}\right|^2 E_s}{\|\mathbf{W}\| N_0}. \tag{8}$$

where E_S is symbol energy, and N_0 is noise power spectral density. Under this constraint, maximizing SNR becomes equivalent to minimizing the effective noise power $\|\mathbf{W}\|N_0$. The SNR maximization under unit-norm constraint $\|\mathbf{W}\| = 1$ is formulated as

$$\max_{\mathbf{W}} \gamma = \frac{\left|\mathbf{W}^H\mathbf{H}\right|^2 E_s}{N_0}, \text{ s.t. } \|\mathbf{W}\|^2 = 1. \tag{9}$$

Using the Lagrange multiplier λ,

$$\mathbf{L} = \mathbf{W}^H\mathbf{H}\mathbf{H}\mathbf{W}^H\mathbf{W} - \lambda(\mathbf{W}^H\mathbf{W} - 1). \tag{10}$$

Taking gradient w.r.t. $\mathbf{W}$,

$$\nabla \mathbf{wL} = 2\mathbf{H}\mathbf{H}^H\mathbf{W} - 2\lambda\mathbf{W} = 0. \tag{11}$$

Thus, the maximum received SNR is denoted as

$$SNR = \frac{\|\mathbf{H}\| E_s}{N_0}. \tag{12}$$

where the channel gain is $g = \|\mathbf{H}\|$.

We compared systems transmitting identical symbols across all paths to exploit diversity gain through signal combining at the receiver. As shown in Figure 9, in a single-transmitter antenna system, MRC provides superior gain compared to Selection Combining (SC) and Equal Gain Combining (EGC). In our commonly deployed 2 × 2 MIMO systems, employing MRC technology can achieve up to 9 dB of diversity gain [89,90].

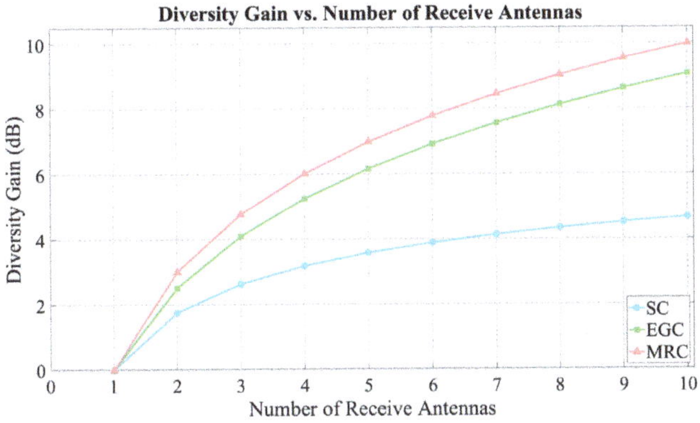

Figure 9. Diversity reception gain vs. number of receiver antennas diagram.

By combining PDM with a MIMO architecture, signals transmitted on two orthogonal polarizations preserve inter-channel incoherence. When identical signals are transmitted over both channels, MRC can be applied at the receiver to achieve diversity gain. This approach not only prevents the disruption of coherence caused by phase differences induced by wavelength discrepancies during transmission but also simultaneously obtains diversity gain.

In reference [91], we effectively resolved coherence disruption and enhanced performance in a 300 GHz terahertz wireless link over 200 m using a MIMO-PDM transmitter architecture combined with a MIMO-MRC receiver architecture. The system configuration is depicted in Figure 10. At the transmitter, PDM generates two orthogonally polarized 300 GHz terahertz signals (via a PDM module and dual UTC-PD opto-terahertz conversion), emitted by a high-gain cylindrical lens antenna array. At the receiver, dual antennas capture these independent polarization signals, with the MRC algorithm applied for optimal signal combining. This architecture innovatively leverages orthogonal polarization to ensure signal incoherence between paths, fundamentally preventing coherence disruption caused by phase differences in same-frequency/same-polarization signals, while enabling effective MRC implementation. Experimental results demonstrate successful achievement of a total data rate of 100 Gbit/s (50 Gbaud PDM-QPSK) over a 200 m wireless distance. The key performance improvement lies in the following: compared to single-channel transmission, this 2 × 2 MIMO-PDM-with-MRC solution achieves up to 1.2 dB SNR gain while successfully overcoming coherence disruption phenomena prevalent in long-distance MIMO transmission. This represents the first breakthrough solution simultaneously delivering performance gains and effective coherence disruption suppression in photonics-assisted terahertz MIMO systems.

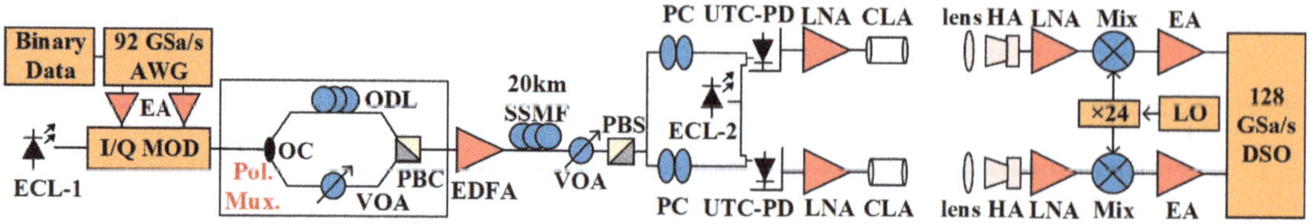

Figure 10. MIMO terahertz transmission system using PDM and MRC; VOA: variable optical attenuator; PBC: polarization beam combiner; Pol.Mux.: polarization multiplexer; CLA: cylindrical lens antenna [90].

3.3. High-Gain Antennas and Lenses

THz waves combine the advantages of millimeter waves and optical waves. Compared to millimeter waves, they offer broader usable frequency bands and superior beam directivity; they also provide enhanced security and superior anti-interference performance. Relative to optical waves, they are more efficient, exhibit stronger penetration capabilities, and demonstrate greater adaptability to harsh environments [92,93].

The main types of THz waveguides include conventional metallic rectangular waveguides, dielectric fiber waveguides, novel THz photonic crystal waveguides, surface plasmon polariton waveguides, and emerging two-dimensional material-based waveguides. The metallic rectangular waveguide, with its advantages of low dispersion and high power capacity, is employed in our experimental system for connecting with Cassegrain antennas [94–96]. THz antennas are indispensable devices in terahertz wireless communication systems for radiating and detecting terahertz waves. Driven by the unique properties of terahertz waves, key performance metrics of these antennas—particularly their broad operating bandwidth and gain—critically impact the overall system quality [97,98].

Antenna gain typically refers to the gain in the direction of maximum radiation. It can be expressed as

$$G = \frac{4\pi A_e}{\lambda^2}. \tag{13}$$

where A_e is the antenna effective aperture area and $\lambda = c/f$ is the signal wavelength. As the frequency f increases to 0.1–10 THz, λ becomes significantly shorter than millimeter wavelengths. For a fixed physical aperture, THz antenna gain improves by >10× over mmWave systems, making aperture efficiency critical. Enhancing effective radiation area requires metasurfaces or diffractive optics at THz frequencies. Path loss in long-distance THz communication systems is a dominant attenuation factor modeled as follows [38,99]:

$$P_L = 20\log_{10}\left(\frac{4\pi d f}{c}\right) + \alpha_{gas} d. \tag{14}$$

where d is the wireless transmission distance and α_{gas} is the molecular absorption loss. In the THz frequency band, absorption peaks of water vapor and oxygen cause additional attenuation, increasing α_{gas}. Thus, THz base station antennas require extremely high gain to overcome this enhanced path loss and improve system performance. While HAs serve as a low-cost, high-performance reference prototype—achieving up to 25 dBi gain with a simple structure—they prove insufficient for long-range THz wireless systems operating in complex environments [100]. Consequently, dual-reflector Cassegrain antennas (CAs) are widely adopted in THz systems.

Figure 11 illustrates schematic diagrams of a rectangular HA, CA and lens antenna. The horn antenna features an intuitive and classic design—essentially a flared waveguide opening that gradually expands to form a trumpet-shaped radiating aperture. Signals feed into the narrow end of the waveguide; as electromagnetic waves propagate through the expanding horn and reach the broad aperture, the beam widens and shapes into a directional pattern. Here, a and b denote the width and height of the rectangular waveguide, while a_E and b_H represent the aperture side lengths in the E-plane and H-plane, respectively; R is the distance between the upper and lower horn apertures [101].

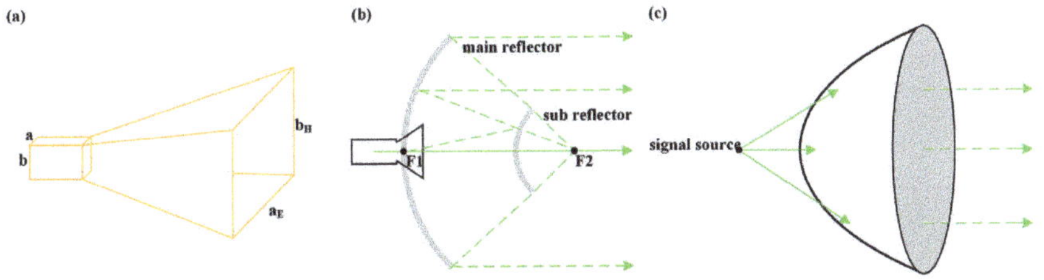

Figure 11. Terahertz antenna structure diagram: (**a**) HA; (**b**) CA; (c) lens antenna.

In contrast, the Cassegrain antenna employs a sophisticated dual-reflector system. Signals radiate from the feed source, first illuminating the sub-reflector. The sub-reflector redirects the signals to the primary reflector, which then collimates the beam for long-distance transmission. This structure "folds" the electromagnetic path via the sub-reflector, positioning the feed near the vertex of the primary reflector (or behind it) rather than at the distant focal point. This shortens the antenna's physical length (effective focal length $\geq 2\times$ physical length), reduces transmission loss, and maintains high gain and superior directivity. The sub-reflector effectively "relocates" the feed's phase center to the focus of the parabolic surface [102].

The most prevalent antenna in our long-range THz transmission system is the THz lens antenna. Leveraging its focusing and imaging capabilities, this antenna enhances performance by achieving excellent directivity and high gain. Our lens adopts a plano-convex structure—simple in design and mature in manufacturing—optimized via geometric optics principles. Constructed from polytetrafluoroethylene (PTFE), it utilizes THz-wave refraction to collimate divergent radiation from the feed into directional narrow beams. PTFE is the ideal material for high-performance THz lens antennas, offering unmatched ultra-low loss and a low dielectric constant in THz bands; its refractive index is 1.4 in a wide wavelength range [103]. In long-range THz communication experiments, the lens gain at both the transmitter and receiver reaches 70 dB.

In the experimental setup diagram shown in Figure 12, to address the impact of Earth's curvature on the transmission system, our transmitter is located atop the Guanghua Tower in Fudan University's Handan Campus, while the receiver is situated at an open coastal area. by integrating a Cassegrain antenna (gain: 58 dBi, diameter: 60 cm), a dual-cascaded lens antenna (gain: 40 dB + 6 dB, diameter: 60 cm and 10 cm), and a horn antenna (gain: 25 dBi), we constructed a high-performance D-band wireless link. We use the Friis transmission equation for link power budget analysis. The transmit power is 23 dBm, the transmit antenna gain is 58 dBi, the signal frequencies are 125 GHz and 128 GHz, the transmission distance is 30.2 km, and the atmospheric losses are 28 dB at 125 GHz and 27.6 dB at 128 GHz, respectively. The received signal powers measure −40 dBm at 125 GHz and −39.8 dBm at 128 GHz. This system successfully achieved stable transmission of an 18 Gbps signal over 30.2 km (comprising 10.8 km terrestrial and 19.4 km sea-surface paths) under complex rain/fog conditions, establishing a new record rate-distance product of 543.6 Gbps·km. The integrated antenna system significantly enhanced signal transceiving capability, with the innovative dual-lens design at the receiver contributing an additional 6 dB gain compared to single-lens solutions, effectively overcoming the substantial path loss induced by ultra-long distances and severe weather.

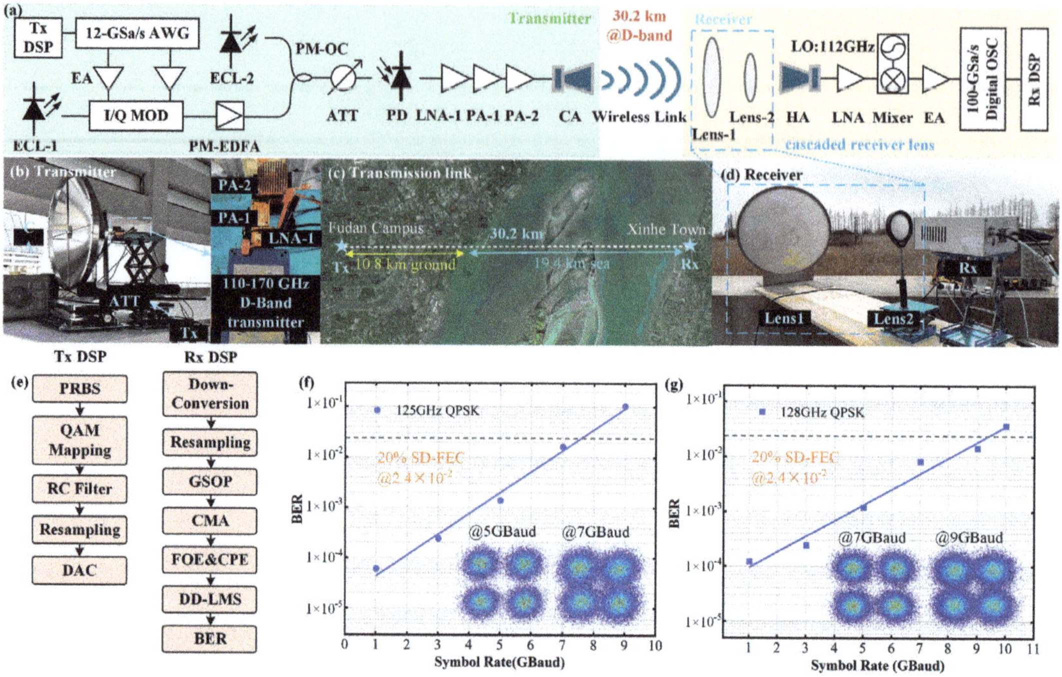

Figure 12. Ultra-long-distance terahertz wireless transmission system enabled by fusion of multiple antennas: (**a**) experimental setup of the ultra-long-distance D-band transmission system over 30.2 km; (**b**) photo of the transmitter; (**c**) photo of the transmission link; (**d**) photo of the receiver; (**e**) DSP structure at the transmitter side and receiver side; (**f**) BER curve of 125 GHz QPSK signal; and (**g**) BER curve of 128 GHz QPSK signal.

3.4. Power Amplifiers

In THz systems, PAs serve as the core engine for overcoming path loss in ultra-high-frequency bands. According to the Friis transmission equation, the PA's output power (P_t) and antenna gain jointly determine the available link budget. Notably, the path loss term in the Friis equation scales inversely with the square of the distance d and the square of the wavelength λ. Given the extremely short wavelength of THz waves, path loss over the same distance is drastically higher than in microwave bands. A 1 mW PA can support a 10 m link in the 1 THz band [104]. Without a PA providing sufficient transmit power, the received signal would be entirely buried in noise, rendering the system inoperable. Thus, PAs are essential for counteracting extreme free-space loss and atmospheric attenuation during THz signal propagation and are critical for compensating the gain limitations of passive components like antennas.

The PA's output power, efficiency, and linearity directly dictate the feasible transmission distance, communication quality, and system power consumption, often representing the primary bottleneck in THz implementations. Common amplifiers in our experiments include PAs, LNAs, and InP semiconductor amplifiers [105–107]. Their core function is to deliver significant power gain, requiring both voltage amplification and robust current-driving capability to achieve high output power. However, when input power exceeds a threshold, gain compression occurs, eventually leading to output power saturation. In reference [88], the gain and noise figure of the PA and LNA are illustrated; in reference [108], a total amplifier gain of 99 dB was achieved for a 30 km ultra-long-range THz transmission system. Furthermore, we recently demonstrated a 3000 m wireless link at 312 GHz using an InP semiconductor amplifier—marking the first kilometer-range transmission reported above 300 GHz. As shown in Figure 13 of our experimental results, leveraging an InP semiconductor amplifier enabled kilometer-range transmission of terahertz waves under foggy conditions.

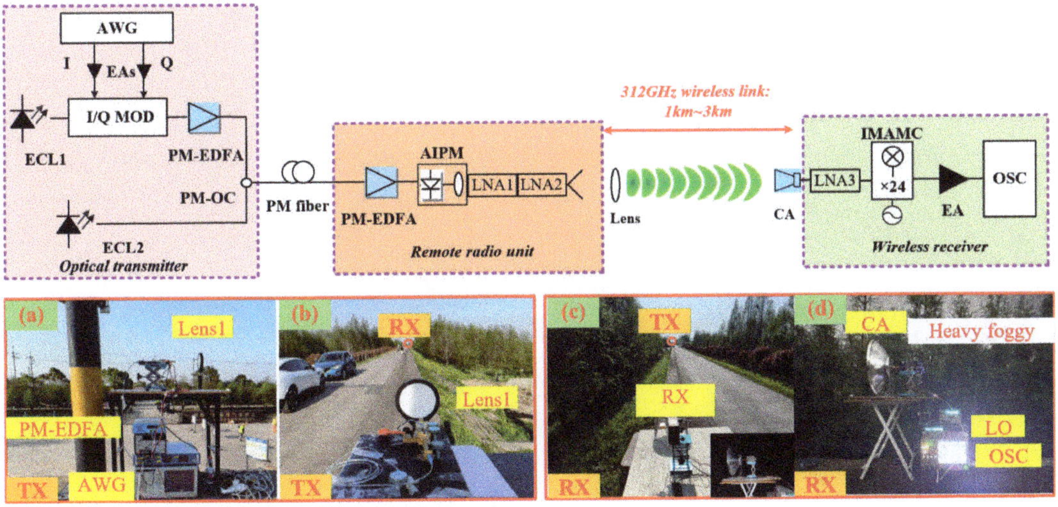

Figure 13. Complex environment transmission system based on InP semiconductor amplifier: (**a**) transmitter setup; (**b**) photo of transmitter; (**c**) photo of receiver; (**d**) receiver setup.

3.5. Heterodyne Coherent Detection Architecture

Signal detection in terahertz communication systems is divided into two main schemes: direct detection and heterodyne detection. The direct detection scheme amplifies THz signals received by antennas through EAs and then directly down-converts the high-frequency signal to baseband using an envelope detector. It has a simple structure and low cost, and it is insensitive to phase noise, but it only supports demodulation of intensity-modulated signals and suffers from nonlinear distortion and power wastage, leading to limited signal quality, making it suitable for low-speed, low-cost scenarios [109,110]. Therefore, in long-range wireless terahertz communication systems, we prefer the heterodyne detection approach. The structure of the heterodyne detection scheme in THz communication is shown in Figure 14. This scheme primarily mixes an LO signal with the received signal to generate an IF signal. After digital down-conversion and capture by a digital oscilloscope, phase correction via DSP fully recovers the I/Q components of the vector-modulated signal [111].

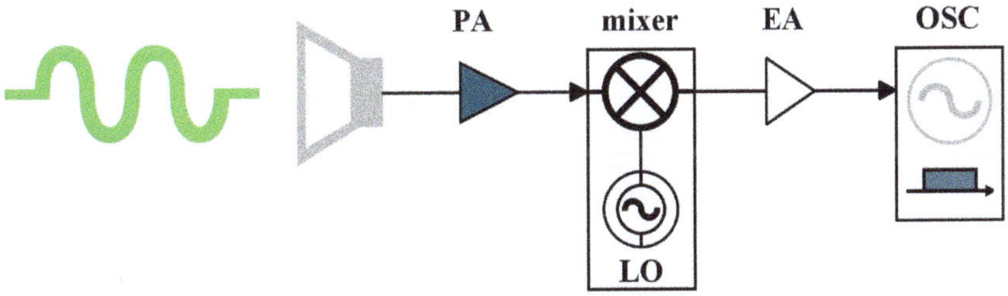

Figure 14. Heterodyne coherent detection schematic diagram.

The vector signal received by the THz receiver can be expressed as

$$S(t) = A[I(t)\sin(2\pi f_c t + \theta_c(t)) + Q(t)\cos(2\pi f_c t + \theta_c(t))]. \quad (15)$$

Here, A, f_c, and θ_c(t) denote the amplitude, frequency, and phase of the THz signal, respectively. I(t) and Q(t) represent the in-phase (I) and quadrature (Q) components of the vector signal carried by the millimeter wave. The RF LO signal can be expressed as

$$S_{LO}(t) = A_{LO}\cos(2\pi f_{LO}t + \theta_{LO}(t)). \quad (16)$$

where A_{LO}, f_{LO}, and θ_{LO} represent the amplitude, frequency and phase of the RF LO signal, respectively. The THz signal is down-converted to a lower IF bandwidth via mixing with the LO signal, facilitating subsequent signal conditioning. The resulting IF signal after mixing is

$$S_{IF}(t) = -jA_{LO}A[I(t) + jQ(t)]\exp(j2\pi f_{IF}t + j\theta_{IF}(t)). \quad (17)$$

The final output is obtained after digital down-conversion processing:

$$\text{Output} = j\cdot(-j)A_{LO}A[I(t) + jQ(t)] = A_{LO}A[I(t) + jQ(t)]. \quad (18)$$

Although this scheme requires an LO and mixer—leading to increased system complexity and cost—it supports high-order modulation formats and enables distortion cancellation through DSP algorithms, significantly enhancing receiver sensitivity and spectral efficiency. Consequently, it serves as a core technology for long-distance terahertz communications.

4. High-Sensitivity SIMO Space–Ground Communication Architecture

With technological advancements and the integration of satellite communications, the internet, and mobile wireless networks, the Space–Terrestrial Integrated Network (STIN) has emerged as a strategic framework [112,113]. Satellite–ground communication constitutes the core of STIN, driven by three fundamental requirements: global coverage, capability enhancement, and service evolution.

While microwave bands remain the most widely used spectrum for satellite communications, they suffer from inherent limitations—namely, broad beamwidth and high power consumption—making them unsuitable for high-rate, large-bandwidth transmissions over long distances. FSO communication employs near-infrared optical carriers to establish terrestrial links within Earth's atmosphere, but its long-range performance is severely constrained by atmospheric absorption, scattering, and turbulence, rendering it more suitable for inter-satellite links than satellite-to-ground scenarios demanding high data rates [114,115].

THz wireless communication emerges as a promising alternative to FSO for satellite-to-ground applications. Operating in the 0.1–10 THz range, it offers superior environmental robustness while preserving advantages of FSO-like bandwidth capacity [116–118].

4.1. Communication Models and System Designs for Terahertz Satellite–Ground Links

For the ultra-long-distance communication scenario between a geostationary orbit satellite (Fengyun-4) and a ground station (Shanghai), we demonstrated 100 Gbps high-speed simulated transmission over 36,000 km in the 300 GHz frequency band with 16QAM modulation through system-level simulation. System parameters are detailed in Table 1. Based on the Friis equation, the received signal power at the ground station is approximately −52.1 dBm. The receiver employs a 25 dB gain THz LNA followed by a 30 dB gain PA. After down-conversion (with 10 dB insertion loss), the signal passes through an additional

30 dB gain LNA. Accounting for the receiver's 10 dB noise figure, the final signal power before demodulation reaches ~13 dBm.

Table 1. High-sensitivity SIMO communication architecture system parameters.

Parameter	Value
Frequency	300 GHz
Distance	36,000 km
Symbol Rate	25 GBaud
Symbol Length	32,768
Sampling Points per Symbol	30
Modulation Order	16
TX Antenna Diameter	4.5 m
RX Antenna Diameter	2.06 m
Aperture Efficiency	0.6
Number of RX Antennas	4
Transmit Power	1 W

To enable higher data rates, we implement the MIMO-MRC technique described in Section 3.2, replacing a single 4.5 m large-aperture antenna with four smaller-aperture antennas. With four receiver antennas, MRC combining theoretically quadruples the SNR (6 dB gain). Equivalent aperture calculations show each smaller antenna requires a diameter of 2.06 m. After diversity reception, advanced DSP techniques—including carrier demodulation, matched filtering, carrier recovery, and MRC—are applied. The system's workflow and schematic are illustrated in Figure 15.

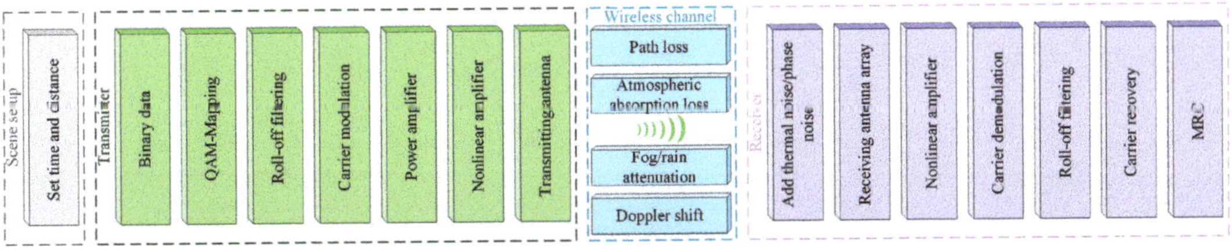

Figure 15. SIMO satellite–ground communication flowchart.

4.2. Simulation Results and Discussion

This Earth–satellite communication simulation under the SIMO architecture successfully modeled the dynamic satellite–ground station distance variation over 40 min caused by Earth's revolution and rotation, along with its impact on communication performance, as shown in Figure 16a,b. The distance increased from 3.79999×10^7 m to 3.80007×10^7 m, exhibiting a significant decline in P_r with increasing range.

As shown in Figure 16, the simulation results robustly validate the effectiveness of the core implemented technologies: DSP and MRC diversity reception techniques substantially enhanced communication quality. At the receiver, the constellation diagram effectively restored phase alignment after DSP processing and demonstrated sharper, more distinct clustering characteristics following MRC combining. The BER prior to diversity reception (Figure 16c) fluctuated around the 10^{-2} level, while post MRC combining (Figure 16d), it significantly reduced to the 10^{-3} order of magnitude. Concurrently, a system gain of approximately 4 dB in SNR was achieved.

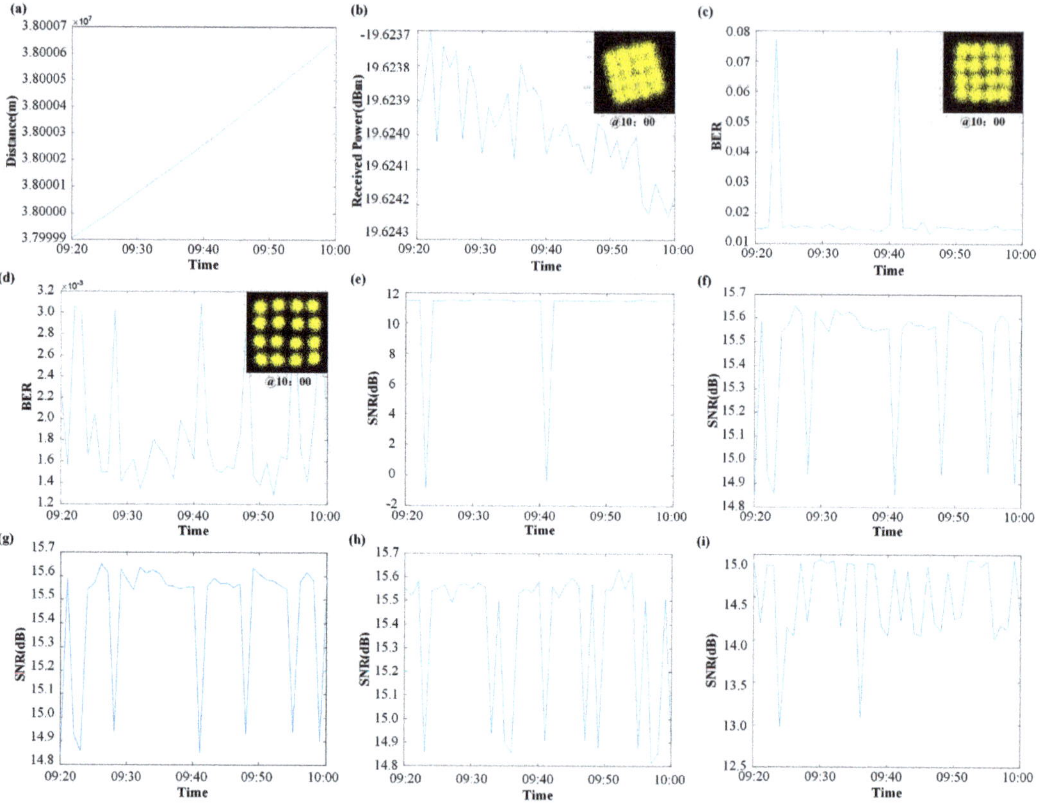

Figure 16. Simulation results of high-sensitivity SIMO satellite–ground communication architecture: (**a**) time vs. communication distance; (**b**) time vs. received power; (**c**) BER before diversity combining; (**d**) BER after diversity combining; (**e**) SNR before diversity combining; (**f**) SNR after diversity combining; (**g**) SNR under clear sky conditions; (**h**) SNR under fog attenuation; (**i**) SNR under rain attenuation.

Addressing weather-induced attenuation in the satellite-to-Earth link, the simulation contrasted performance under clear sky, fog attenuation, and rain attenuation conditions. Under fog attenuation (Figure 16h), SNR fluctuations remained below 0.1 dB compared to the clear sky baseline (Figure 16g). Rain attenuation caused more noticeable SNR degradation, though the maximum attenuation was controlled at approximately 1 dB.

This simulation not only demonstrates the technical feasibility of utilizing the terahertz band for Earth–satellite communications under dynamic link conditions and typical weather scenarios but also quantitatively proves the critical role of DSP and diversity reception techniques in enhancing link robustness and reliability.

5. Conclusions

This paper summarizes our advances in long-distance terahertz communication, systematically tackling the 'dual attenuation dilemma' of path loss and atmospheric attenuation through co-designed photonic–electronic architectures and adaptive signal processing. By developing hybrid THz generation schemes—including photonically assisted probabilistic 64QAM achieving a record 1.0488 Tbps throughput in the 330–500 GHz band—we established seamless FiWi integration capable of bridging fiber and Tbps wireless links. To overcome severe path loss exceeding 120 dB/km, our multi-pronged approach synergizes high-gain antenna–lens systems, InP semiconductor amplifiers enabling the first 3 km fog-penetrating link at 312 GHz, and modulation-robust DSP algorithms like MIMO-VNE. These innovations culminated in a landmark 30.2 km D-band transmission with 543.6 Gbps·km capacity–distance product, representing an 8-fold improvement over prior records. For dynamic channel adaptability, the proposed MIMO-MRC architecture demonstrated 9 dB diversity gain through polarization-multiplexed transmission and coherent combining, while real-time VQ-VAE-based processing maintained a BER below 3.4×10^{-2} under orbital dynamics and weather impairments. Critically, our satellite–ground simulation validated THz viability for space–terrestrial integration, achieving 100 Gbps over 36,000 km at 300 GHz with MRC diversity reducing the BER from the order of 10^{-2} to 10^{-3}. Collectively, these breakthroughs provide a scalable framework for 6G integrated SAG networks, delivering three foundational capabilities: terabit-per second-capable ultra-high-capacity backhaul via fiber–THz convergence; robust long-haul connectivity across distances reaching 30 km over land and extending to 36,000 km for space links, enabled through photonic–electronic co-optimization; and dynamic channel resilience enabled by intelligent MIMO-DSP adaptation. Future work will extend to non-line-of-sight scenarios using reconfigurable intelligent surfaces and standardization pathways for THz-enabled 6G deployments.

Author Contributions: Conceptualization, J.Y.; Methodology, J.Y.; Writing original draft, J.C.; Data curation, J.C.; Formal analysis, J.C.; Writing review and editing, J.Y. and J.C.; Supervision, J.Y.; Project administration, J.Y. All authors have read and agreed to the published version of the manuscript.

Funding: This work was supported by the National Key R&D Program of China (2024YEE0200202), National Natural Science Foundation of China (No. 62127802, 62331004, 62305067, U24B20142, U24B20168, and 62427815), and Key Project of Jiangsu Province of China (No. BE2023001-4).

Data Availability Statement: The data presented in this study is available on request from the corresponding author. The data is not publicly available due to privacy restrictions.

Acknowledgments: The authors sincerely thank colleagues from the Key Laboratory for Information Science of Electromagnetic Waves (MoE), Fudan University, and the Purple Mountain Laboratory for their invaluable contributions to this research.

Conflicts of Interest: The authors declare no conflicts of interest.

References

1. Elayan, H.; Amin, O.; Shihada, B.; Shubair, R.M.; Alouini, M.-S. Terahertz band: The last piece of RF spectrum puzzle for communication systems. *IEEE Open J. Commun. Soc.* **2019**, *1*, 1–32. [CrossRef]
2. O'Hara, J.F.; Ekin, S.; Choi, W.; Song, I. A perspective on terahertz next-generation wireless communications. *Technologies* **2019**, *7*, 43. [CrossRef]
3. Rao, A.S.; SS, S.M.; Yelasani, S.; Kumar, B.S. Terahertz communications: Applications, challenges and open research issues for next generation 6G wireless networks. *Int. J. Innov. Eng. Manag. Res.* **2023**, *12*, 295–315.
4. Dehos, C.; González, J.L.; De Domenico, A.; Ktenas, D.; Dussopt, L. Millimeter-wave access and backhauling: The solution to the exponential data traffic increase in 5G mobile communications systems? *IEEE Commun. Mag.* **2014**, *52*, 88–95. [CrossRef]
5. Zander, J.; Mähönen, P. Riding the data tsunami in the cloud: Myths and challenges in future wireless access. *IEEE Commun. Mag.* **2013**, *51*, 145–151. [CrossRef]
6. Rappaport, T.S.; Xing, Y.; Kanhere, O.; Ju, S.; Madanayake, A.; Mandal, S.; Alkhateeb, A.; Trichopoulos, G.C. Wireless communications and applications above 100 GHz: Opportunities and challenges for 6G and beyond. *IEEE Access.* **2019**, *7*, 78729–78757. [CrossRef]
7. Rodenbeck, C.T.; Jaffe, P.I.; Strassner, B.H., II; Hausgen, P.E.; McSpadden, J.O.; Kazemi, H.; Shinohara, N.; Tierney, B.B.; DePuma, C.B.; Self, A.P. Microwave and millimeter wave power beaming. *IEEE J. Microw.* **2021**, *1*, 229–259. [CrossRef]
8. Akyildiz, I.F.; Kak, A.; Nie, S. 6G and beyond: The future of wireless communications systems. *IEEE Access* **2020**, *8*, 133995–134030. [CrossRef]

9. Tripathi, S.; Sabu, N.V.; Gupta, A.K.; Dhillon, H.S. Millimeter-wave and terahertz spectrum for 6G wireless. In *6G Mobile Wireless Networks*; Springer: Berlin, Germany, 2021; pp. 83–121.
10. Rizzo, L.; Federici, J.F.; Gatley, S.; Gatley, I.; Zunino, J.L.; Duncan, K.J. Comparison of terahertz, microwave, and laser power beaming under clear and adverse weather conditions. *J. Infrared Millim. Terahertz Waves* **2020**, *41*, 979–996. [CrossRef]
11. Song, H.-J.; Lee, N. Terahertz communications: Challenges in the next decade. *IEEE Trans. Terahertz Sci. Technol.* **2021**, *12*, 105–117. [CrossRef]
12. Federici, J.F.; Ma, J.; Moeller, L. Review of weather impact on outdoor terahertz wireless communication links. *Nano Commun. Netw.* **2016**, *10*, 13–26. [CrossRef]
13. Salahdine, F.; Han, T.; Zhang, N. 5G, 6G, and Beyond: Recent advances and future challenges. *Ann. Telecommun.* **2023**, *78*, 525–549. [CrossRef]
14. Wang, M.; Yu, J.; Li, W.; Wei, Y.; Yang, X.; Shi, J.; Bian, C.; Zhou, W.; Wang, K. SNR improved digital-analog radio-over-fiber scheme for a millimeter wireless fronthaul. *J. Light. Technol.* **2023**, *42*, 3531–3539. [CrossRef]
15. Cherry, S. Edholm's law of bandwidth. *IEEE Spectr.* **2004**, *41*, 58–60. [CrossRef]
16. Jain, P.; Gupta, A.; Kumar, N.; Guizani, M. Dynamic and efficient spectrum utilization for 6G with THz, mmWave, and RF band. *IEEE Trans. Veh. Technol.* **2023**, *72*, 3264–3273. [CrossRef]
17. Wang, M.; Yu, J.; Zhao, X.; Li, W.; Wei, Y.; Yang, X.; Shi, J.; Bian, C.; Xie, T.; Zhao, F.; et al. SNR improved digital-delta-sigma-modulation radio-over-fiber scheme for D-band 4.6-km photonics-aided wireless fronthaul. *Opt. Lett.* **2023**, *48*, 3997–4000. [CrossRef] [PubMed]
18. Liu, J.; Yu, J.; Zhao, X.; Bian, C.; Yang, X.; Zhang, L.; He, W.; Long, J.; Zhang, Y.; Zhang, Y.; et al. W-band photonics-aided ISAC wireless system sharing OFDM signal as communication and sensing. In Proceedings of the Optical Fiber Communication Conference, San Diego, CA, USA, 24–28 March 2024. Tu3K. 4.
19. Wang, M.; Yu, J. *104-m Wireless Transmission of 4096-QAM for Mobile Backhaul Based on Delta-Sigma Modulation in 145 GHz*; Springer: Berlin, Germany, 2023.
20. Wang, K.; Yu, J.; Bian, C.; Ding, J.; Zhang, J.; Zhu, M.; Yang, X.; Li, W.; Wei, Y.; Wang, M.; et al. Terahertz transmitter demonstration of the simultaneous transmission of optical and electrical terahertz band signals at 1 Tb/s bit rate and 50 m wireless delivery. *Sci. China Technol. Sci.* **2025**, *68*, 1380901. [CrossRef]
21. *ITU-R V.431*; Nomenclature of the frequency and wavelength bands used in telecommunications. ITU: Geneva, Switzerland, 2015.
22. *IEEE 802.15.3d*; IEEE Standard for High Data Rate Wireless Multi-Media Networks-Amendment 2: 100 Gb/s Wireless Switched Point-to-Point Physical Layer. IEEE: New York, NY, USA, 2017.
23. Petrov, V.; Kurner, T.; Hosako, I. IEEE 802.15.3d: First Standardization Efforts for Sub-Terahertz Band Communications toward 6G. *IEEE Commun. Mag.* **2020**, *58*, 28–33. [CrossRef]
24. Banerjee, A.; Zhang, L.; Wang, H.; Wambacq, P. Sub-THz and THz Signal Generation Using Photonic and Electronic Techniques (Invited Paper). In Proceedings of the IEEE MTT-S International Microwave Conference on Hardware and Systems for 5G and Beyond (IMC-5G), Atlanta, GA, USA, 15–16 August 2019; pp. 1–3.
25. Freeman, J.R.; Brewer, A.; Madéo, J.; Cavalié, P.; Dhillon, S.S.; Tignon, J.; Beere, H.E.; Ritchie, D.A. Heterogeneous THz Quantum Cascade Lasers: Broadband Operation. In Proceedings of the International Conference on Infrared, Millimeter, and Terahertz Waves, Houston, TX, USA, 2–7 October 2011; pp. 1–2.
26. Guo, H.; Guo, K.; Lin, Z.; Shum, K.M.; Chan, C.H. A 190-217-GHz Frequency Multiplier Chain With 13.2 dB Conversion Gain in 65-nm CMOS. *IEEE Trans. Circuits Syst. II Express Briefs* **2024**, *71*, 4859–4863. [CrossRef]
27. Freudenstein, J.; Borsch, M.; Meierhofer, M.; Afanasiev, D.; Schmid, C.P.; Sandner, F.; Liebich, M.; Girnghuber, A.; Knorr, M.; Kira, M.; et al. Attoclocking Delocalized Bloch Electrons with Multi-Terahertz Fields. In Proceedings of the International Conference on Infrared, Millimeter, and Terahertz Waves (IRMMW-THz), Montreal, QC, Canada, 17–22 September 2023; pp. 1–2.
28. Razeghi, M.; Lu, Q.Y.; Bandyopadhyay, N.; Slivken, S.; Bai, Y. Room Temperature Compact THz Sources Based on Quantum Cascade Laser Technology. In Proceedings of the SPIE Conference on Terahertz Emitters, Receivers, and Applications IV, San Diego, CA, USA, 25–26 August 2013; Volume 8846, p. 884602.
29. Ren, Y.; Wallis, R.; Jessop, D.S.; Degl'Innocenti, R.; Klimont, A.; Beere, H.E.; Ritchie, D.A. Fast Terahertz Imaging Using a Quantum Cascade Amplifier. *Appl. Phys. Lett.* **2015**, *107*, 011107. [CrossRef]
30. Halbauer, H.; Wild, T. Towards Power Efficient 6G Sub-THz Transmission. In Proceedings of the Joint European Conference on Networks and Communications&6G Summit (EuCNC/6G Summit), Porto, Portugal, 8–11 June 2021; pp. 25–30.
31. Song, H.-J.; Nagatsuma, T. Present and future of terahertz communications. *IEEE Trans. Terahertz Sci. Technol.* **2011**, *1*, 256–263. [CrossRef]
32. Wu, Y.; Singh, S.; Taleb, T.; Roy, A.; Dhillon, H.S.; Kanagarathinam, M.R.; De, A. *6G Mobile Wireless Networks*; Springer: Berlin, Germany, 2021.
33. Sarieddeen, H.; Alouini, M.-S.; Al-Naffouri, T.Y. An overview of signal processing techniques for terahertz communications. *Proc. IEEE* **2021**, *109*, 1628–1665. [CrossRef]

34. Tan, J.; Long, J.; Yu, J.; Yang, X.; Liu, J.; Tian, P.; Xu, S.; Han, Y.; Zhang, B.; Zhang, Y. 4096-QAM Signal Transmission by An IM/DD System at THz Band Using Delta-Sigma Modulation. *IEEE Photonics Technol. Lett.* 2024, *in press*. [CrossRef]
35. Zhao, X.; Yu, J.; Yang, X.; Wei, Y.; Li, W.; Tan, J.; Wang, M.; Zhang, Q.; Zhang, B.; Bian, C.; et al. D-band 4600 m wireless transmission with rates exceeding 100 Gbit/s based on photonics aided technologies. *Sci. China* **2024**, *67*, 2968–2970. [CrossRef]
36. Wang, M.; Yu, J.; Zhao, X.; Yang, X.; Bian, C.; Zhang, Q.; Zhou, W.; Wang, K.; Li, W. First Demonstration of OTFS in a D-band Indoor Wireless Communication System Based on Photonics-aided Scheme. In Proceedings of the Optical Fiber Communication Conference, San Diego, CA, USA, 23–27 March 2025. M1I. 2.
37. Wang, M.; Yu, J.; Zhao, X.; Zhou, W.; Chen, J.; Yang, X.; Bian, C.; Wei, Y.; Zhang, Q.; Han, Y.; et al. Research on Orthogonal Time Frequency Space in a 125-GHz mmWave Indoor Wireless Communication System. *J. Light. Technol.* **2025**, *43*, 5762–5772. [CrossRef]
38. Erdem, M.; Saleem, A.; Gurbuz, O.; Ecemis, C.; Saeed, A. A simple analytical model for THz band path loss. *IEEE Commun. Lett.* **2023**, *27*, 996–1000. [CrossRef]
39. Abbasi, N.A.; Hariharan, A.; Nair, A.M.; Molisch, A.F. Channel measurements and path loss modeling for indoor THz communication. In Proceedings of the 14th European Conference on Antennas and Propagation, Copenhagen, Denmark, 15–20 March 2020; pp. 1–5.
40. Han, C.; Gao, W.; Yang, N.; Jornet, J.M. Molecular absorption effect: A double-edged sword of terahertz communications. *IEEE Wirel. Commun.* **2022**, *30*, 140–146. [CrossRef]
41. Shi, J.; Wang, Y.; Zhang, J.; Zhao, X.; Zhu, M.; Zhou, W.; Yu, J. Beyond 500 GHz THz wireless links based on heterodyne photomixing and absolute operation pruned two-stage MIMO–volterra. *IEEE Trans. Terahertz Sci. Technol.* **2024**, *14*, 364–376. [CrossRef]
42. Kokkoniemi, J.; Lehtomäki, J.; Juntti, M. A discussion on molecular absorption noise in the terahertz band. *Nano Commun. Netw.* **2016**, *8*, 35–45. [CrossRef]
43. Hecht, J. History of gas lasers, Part 1-Continuous wave gas lasers. *Opt. Photonics News.* **2010**, *21*, 16–23. [CrossRef]
44. Nagatsuma, T.; Hirata, A.; Kukutsu, N.; Kado, Y. Multiplexed transmission of uncompressed HDTV signals using 120-GHz-band millimeter-wave wireless link. In Proceedings of the International Topical Meeting on Microwave Photonics, Victoria, BC, Canada, 3–5 October 2007; pp. 237–240.
45. Hirata, A.; Kosugi, T.; Takahashi, H.; Takeuchi, J.; Togo, H.; Yaita, M.; Kukutsu, N.; Aihara, K.; Murata, K.; Sato, Y.; et al. 120-GHz-band wireless link technologies for outdoor 10-Gbit/s data transmission. *IEEE Trans. Microw. Theory Tech.* **2012**, *60*, 881–895. [CrossRef]
46. Mukherjee, P.; Gupta, B. Terahertz (THz) frequency sources and antennas-A brief review. *Int. J. Infrared Millim. Waves* **2008**, *29*, 1091–1102. [CrossRef]
47. Montanaro, A.; Piccinini, G.; Mišeikis, V.; Sorianello, V.; Giambra, M.A.; Soresi, S.; Giorgi, L.; D'Errico, A.; Watanabe, K.; Taniguchi, T.; et al. Sub-THz Wireless Transmission Based on Graphene-Integrated Optoelectronic Mixer. *Nat. Commun.* **2023**, *14*, 6471. [CrossRef] [PubMed]
48. Han, R.; Hu, Z.; Wang, C.; Holloway, J.; Yi, X.; Kim, M.; Mawdsley, J. Filling the gap: Silicon terahertz integrated circuits offer our best bet. *IEEE Microw. Mag.* **2019**, *20*, 80–93. [CrossRef]
49. Mei, X.; Yoshida, W.; Lange, M.; Lee, J.; Zhou, J.; Liu, P.-H.; Leong, K.; Zamora, A.; Padilla, J.; Sarkozy, S.; et al. First demonstration of amplification at 1 THz using 25-nm InP high electron mobility transistor process. *IEEE Electron Device Lett.* **2015**, *36*, 327–329. [CrossRef]
50. Hacker, J.; Seo, M.; Young, A.; Griffith, Z.; Urteaga, M.; Reed, T.; Rodwell, M. THz MMICs based on InP HBT technology. In Proceedings of the IEEE MTT-S International Microwave Symposium, Anaheim, CA, USA, 23–28 May 2010; pp. 1126–1129.
51. Leuther, A.; Tessmann, A.; Doria, P.; Ohlrogge, M.; Seelmann-Eggebert, M.; Maßler, H.; Schlechtweg, M.; Ambacher, O. 20 nm metamorphic HEMT technology for terahertz monolithic integrated circuits. In Proceedings of the 9th European Microwave Integrated Circuit Conference, Rome, Italy, 6–7 October 2014; pp. 84–87.
52. Hamza, K.H.; Nirmal, D. A review of GaN HEMT broadband power amplifiers. *AEU Int. J. Electron. Commun.* **2020**, *116*, 153040. [CrossRef]
53. Crupi, G.; Vadalà, V.; Colantonio, P.; Cipriani, E.; Caddemi, A.; Vannini, G.; Schreurs, D.M. Empowering GaN HEMT models: The gateway for power amplifier design. *Int. J. Numer. Model. Electron. Netw. Devices Fields.* **2017**, *30*, e2125. [CrossRef]
54. Komiak, J.J. GaN HEMT: Dominant force in high-frequency solid-state power amplifiers. *IEEE Microw. Mag.* **2015**, *16*, 97–105. [CrossRef]
55. Samoska, L.A. An Overview of Solid-State Integrated Circuit Amplifiers in the Submillimeter-Wave and THz Regime. *IEEE Trans. Terahertz Sci. Technol.* **2011**, *1*, 9–24. [CrossRef]
56. Hacker, J.; Urteaga, M.; Seo, M.; Skalare, A.; Lin, R. InP HBT Amplifier MMICs Operating to 0.67 THz. In Proceedings of the IEEE MTT-S International Microwave Symposium Digest (MTT), Seattle, WA, USA, 2–7 June 2013; pp. 1–3.

57. Castro, C.; Elschner, R.; Merkle, T.; Schubert, C.; Freund, R. 100 Gb/s real-time transmission over a THz wireless fiber extender using a digital-coherent optical modem. In Proceedings of the Optical Fiber Communication Conference, San Diego, CA, USA, 8–12 March 2020. M4I. 2.
58. Yu, J.; Li, X.; Zhou, W. Tutorial: Broadband fiber-wireless integration for 5G+ communication. *APL Photonics* **2018**, *3*, 111101. [CrossRef]
59. Yu, J. Photonics-assisted millimeter-wave wireless communication. *IEEE J. Quantum Electron.* **2017**, *53*, 1–17. [CrossRef]
60. Wang, M.; Yu, J.; Zhao, X.; Yang, X.; Wei, Y.; Bian, C.; Han, Y.; Tian, P.; Xu, S.; Zhou, W.; et al. Demonstration of 1.2-km D-band Wireless Fronthaul using Digital-Differential-PCM Scheme. *IEEE Photonics Technol. Lett.* **2025**, *37*, 183–186. [CrossRef]
61. Li, W.; Yu, J.; Wang, Y.; Wang, F.; Zhu, B.; Zhou, W.; Zhao, F.; Yu, J. Photonics-based high-speed long-distance fiber-wireless-integration communication at the W-band. *Sci. China Inf. Sci.* **2023**, *66*, 127301. [CrossRef]
62. Zhu, M.; Zhang, J.; Liu, X.; Hua, B.; Cai, Y.; Ding, J.; Lei, M.; Zou, Y.; Tian, L.; Wang, Y.; et al. Photonics-assisted THz wireless transmission with air interface user rate of 1-Tbps at 330–500 GHz band. *Sci. China Inf. Sci.* **2023**, *66*, 199302. [CrossRef]
63. Li, X.; Yu, J.; Chang, G.-K. Photonics-assisted technologies for extreme broadband 5G wireless communications. *J. Light. Technol.* **2019**, *37*, 2851–2865. [CrossRef]
64. Wang, M.; Yu, J.; Zhao, X.; Wang, C.; Long, J.; Zhao, F.; Zhou, W.; Wang, K. Non-uniform-digital-analog radio-over-fiber scheme for future fronthaul based on a low-complexity 4-fold symmetric non-uniform quantization method. *J. Light. Technol.* **2024**, *42*, 6798–6807. [CrossRef]
65. Singya, P.K.; Shaik, P.; Kumar, N.; Bhatia, V.; Alouini, M.-S. A survey on higher-order QAM constellations: Technical challenges, recent advances, and future trends. *IEEE Open J. Commun. Soc.* **2021**, *2*, 617–655. [CrossRef]
66. Jia, Z.; Chien, H.-C.; Zhang, J.; Cai, Y.; Yu, J. Performance comparison of dual-carrier 400G with 8/16/32-QAM modulation formats. *IEEE Photonics Technol. Lett.* **2015**, *27*, 1414–1417. [CrossRef]
67. Zhou, W.; Wang, S.; Xu, S.; Wang, Q.; Ou, Z.; Ma, Y.; Lin, J.; Ge, J.; Yu, J. Frequency-Domain Complex-Valued NN Equalizer Based on Deep Reinforcement Learning for Photonics-Assisted W-Band 16QAM-DMT Signals System Over 50 m. *J. Light. Technol.* **2024**, *43*, 2651–2663. [CrossRef]
68. Yang, X.; Yu, J.; Zhao, X.; Wei, Y.; Li, W.; Shi, J.; Bian, C.; Zheng, T.; Zhao, F.; Zhou, W.; et al. 41.7-Gb/s D-band signals wireless delivery over 4.6 km distance based on photonics-aided technology. *Opt. Laser Technol.* **2024**, *174*, 110660. [CrossRef]
69. Jia, S.; Zhang, L.; Wang, S.; Li, W.; Qiao, M.; Lu, Z.; Idrees, N.M.; Pang, X.; Hu, H.; Zhang, X.; et al. 2 × 300 Gbit/s line rate PS-64QAM-OFDM THz photonic-wireless transmission. *J. Light. Technol.* **2020**, *38*, 4715–4721. [CrossRef]
70. Tong, W.; Ding, J.; Meng, J.; Xu, X.; Zhang, J.; Hua, B.; Cai, Y.; Lei, M.; Xin, Z.; Yu, J.; et al. 400-Gbit/s Net Rate DP-OFDM Signals Fiber-THz-Fiber Seamless Transmission Employing Hybrid Time-Frequency-Polarization Domain Equalization. *J. Light. Technol.* **2024**, *42*, 5148–5155. [CrossRef]
71. Wang, M.; Yu, J.; Zhao, X.; Zhu, B.; Wang, C.; Zhou, W.; Wang, K. SNR improved truncated-digital-DSM radio-over-fiber scheme for future mobile fronthaul. *J. Light. Technol.* **2023**, *42*, 1934–1940. [CrossRef]
72. Tong, W.; Ding, J.; Zhu, M.; Xin, Z.; Yang, X.; Xie, Z.; Meng, J.; Cai, Y.; Hua, B.; Zhang, J.; et al. 200-m Photonics-Aided Terahertz Wireless Transmission of 253-Gbit/s DP-OFDM Signals Utilizing Multidimensional Nonlinear Equalization. *J. Light. Technol.* **2024**, *43*, 214–221. [CrossRef]
73. Wang, M.; Yu, J.; Zhao, X.; Bian, C.; Zhou, W.; Wang, K. K-means non-uniform-quantization digital–analog radio-over-fiber scheme for THz-band photonics-aided wireless fronthaul. *Opt. Lett.* **2024**, *49*, 2801–2804. [CrossRef] [PubMed]
74. Wang, K.; Zhu, B.; Wei, Y.; Wang, C.; Zhou, W.; Zhao, L.; Yu, J. *Demonstration of DSM-1024QAM Signals Transmission in A 96 Channel WDM-FSO System for Mobile Fronthaul*; Springer: Berlin, Germany, 2023.
75. Wei, Y.; Yu, J.; Zhao, X.; Yang, X.; Wang, M.; Li, W.; Tian, P.; Han, Y.; Zhang, Q.; Tan, J.; et al. Demonstration of a photonics-aided 4600-m wireless transmission system in the sub-THz band. *J. Light. Technol.* **2024**, *42*, 8564–8576. [CrossRef]
76. Wei, Y.; Yang, X.; Yu, J.; Zhao, X.; Wang, M.; Tan, J.; Li, W.; Zhao, F.; Bian, C.; Ma, H.; et al. Experimental demonstration of D-band signals wireless transmission over 30 km. *Sci. China Technol. Sci.* **2025**, *68*, 1280901. [CrossRef]
77. Nagatsuma, T.; Horiguchi, S.; Minamikata, Y.; Yoshimizu, Y.; Hisatake, S.; Kuwano, S.; Yoshimoto, N.; Terada, J.; Takahashi, H. Terahertz wireless communications based on photonics technologies. *Opt. Express* **2013**, *21*, 23736–23747. [CrossRef] [PubMed]
78. Jornet, J.M.; Akyildiz, I.F. Channel modeling and capacity analysis for electromagnetic wireless nanonetworks in the terahertz band. *IEEE Trans. Wirel. Commun.* **2011**, *10*, 3211–3221. [CrossRef]
79. Li, W.; Wang, K.; Sang, B.; Yu, J. Dual-channel WDM fiber-wireless integrated terahertz transmission system. *Acta Electron. Sin.* **2022**, *50*, 2311–2317.
80. Long, J.; Yu, J.; Wang, C.; Zhang, L.; Sang, B.; Wu, Y.; Yang, X.; Wei, Y.; Wang, K.; Zhou, W.; et al. 510-Gbps 322-GHz photonics-aided terahertz wireless transmission system with MIMO embedded adaptive phase recovery equalizer. *IEEE Trans. Microw. Theory Tech.* **2024**, *73*, 3597–3607. [CrossRef]
81. Kürner, T.; Priebe, S. Towards THz communications-status in research, standardization and regulation. *J. Infrared Millim. Terahertz Waves* **2014**, *35*, 53–62. [CrossRef]

82. Paulraj, A.J.; Gore, D.A.; Nabar, R.U.; Bolcskei, H. An overview of MIMO communications-a key to gigabit wireless. *Proc. IEEE* **2004**, *92*, 198–218. [CrossRef]
83. Li, W.; Yu, J.; Zhu, B.; Zhang, J.; Zhu, M.; Wang, C.; Zhou, W.; Xie, T.; Yu, J.; Zhao, F. Photonics 60 GBaud PDM-16QAM fiber-wireless 2 × 2 MIMO delivery at THz-band. *Chin. Opt. Lett.* **2023**, *21*, 073901. [CrossRef]
84. Wang, C.; Yu, J.; Zhang, J.; Zhu, M.; Tan, Y.; Wang, K.; Ding, J.; Zhou, W.; Xie, T.; Yu, J.; et al. Beyond 300-Gbps/λ photonics-aided THz-over-fiber transmission employing MIMO single-carrier frequency-domain equalizer. *Opt. Lett.* **2023**, *48*, 1363–1366. [CrossRef] [PubMed]
85. Cai, Y.; Yang, X.; Zhu, M.; Hua, B.; Xie, Z.; Tong, W.; Xin, Z.; Li, J.; Zhang, J.; Ding, J.; et al. Photonics-aided exceeding 200-Gb/s wireless data transmission over outdoor long-range 2 × 2 MIMO THz links at 300 GHz. *Opt. Express.* **2024**, *32*, 33587–33602. [CrossRef] [PubMed]
86. Li, W.; Yu, J.; Ji, X.; Liu, J.; Wang, F.; Zhu, B.; Zhou, W.; Zhao, F.; Yu, J. Over 100 Gb/s mm-wave delivery with 4600 m wireless distance based on dual polarization multiplexing. *Sci. China Inf. Sci.* **2023**, *66*, 224301. [CrossRef]
87. Zhang, Q.; Yu, J.; Zhao, X.; Li, W.; Wei, Y.; Yang, X.; Tan, J.; Zhang, B.; Wang, K.; Zhang, Y.; et al. 4.6-km D-band photonic-assisted terahertz wireless communication employing SIMO and MRC technology. *IEEE Trans. Microw. Theory Tech.* **2024**, *72*, 6657–6668. [CrossRef]
88. Wei, Y.; Yu, J.; Wang, M.; Zhao, X.; Yang, X.; Li, W.; Tian, P.; Han, Y.; Zhang, Q.; Tan, J.; et al. Demonstration of 200 Gbps D-band Wireless Delivery in a 4.6 km 2 × 2 MIMO system. In Proceedings of the Optical Fiber Communication Conference, San Diego, CA, USA, 24–28 March 2024. M2F. 4.
89. Simon, M.K.; Alouini, M.-S. *Digital Communication over Fading Channels*; John Wiley&Sons: Hoboken, NJ, USA, 2004.
90. Zhang, Q.; Yu, J.; Zhao, X.; Li, W.; Zhu, M.; Zhang, J.; Wei, Y.; Wang, K.; Zhou, W.; Liu, B.; et al. 2 × 2 MIMO W-Band Millimeter Wireless Signal Delivery Employing CMA and MRC Technology. *J. Light. Technol.* **2024**, *42*, 4485–4492. [CrossRef]
91. Zhang, Q.; Yu, J.; Li, W.; Zhang, B.; Zhou, W.; Wang, K.; Wei, Y.; Yang, X.; Wang, M.; Liu, B.; et al. Demonstration of a 200-m 300 GHz photon-assisted MIMO THz system using PDM and MRC technologies with SNR gain and coherence destruction mitigated. *Sci. China Technol. Sci.* **2025**, *68*, 1680901. [CrossRef]
92. Abadal, S.; Han, C.; Jornet, J.M. Wave propagation and channel modeling in chip-scale wireless communications: A survey from millimeter-wave to terahertz and optics. *IEEE Access* **2019**, *8*, 278–293. [CrossRef]
93. Sakhnini, I.; Dominic, O. Performance Comparison Between Optical Wireless and mmWave/THz RF Communications Systems. *IEEE J. Sel. Areas Commun.* **2025**, *43*, 1635–1645. [CrossRef]
94. Nallappan, K.; Guerboukha, H.; Cao, Y.; Xu, G.; Nerguizian, C.; Mittleman, D.M.; Skorobogatiy, M. Terahertz Waveguides for Next Generation Communication Network: Needs, Challenges and Perspectives. In *Next Generation Wireless Terahertz Communication Networks*, 1st ed.; Ghafoor, S., Rehmani, M.H., Davy, A., Eds.; CRC Press: Boca Raton, FL, USA, 2021; pp. 379–410.
95. Atakaramians, S.; Afshar, S.V.; Monro, T.M.; Abbott, D. Terahertz Dielectric Waveguides. *Adv. Opt. Photonics* **2013**, *5*, 169–215. [CrossRef]
96. Tan, Z.; Zhang, Q.-Y.; Lei, Y.-L.; Zhao, Y.; Ding, J.-Q. Terahertz Waveguide Multiplexers: A Review. *Microw. Opt. Technol. Lett.* **2023**, *65*, 1925–1935. [CrossRef]
97. Xu, R.; Gao, S.; Izquierdo, B.S.; Gu, C.; Reynaert, P.; Standaert, A.; Gibbons, G.J.; Bösch, W.; Gadringer, M.E.; Li, D. A review of broadband low-cost and high-gain low-terahertz antennas for wireless communications applications. *IEEE Access.* **2020**, *8*, 57615–57629. [CrossRef]
98. He, Y.; Chen, Y.; Zhang, L.; Wong, S.-W.; Chen, Z.N. An overview of terahertz antennas. *China Commun.* **2020**, *17*, 124–165. [CrossRef]
99. Maddalena, R.J. *Theoretical Ratio of Beam Efficiency to Aperture Efficiency*; GBT Memo: Green Bank, WV, USA, 2010.
100. Lu, X.; Yu, J.; Li, W.; Zhao, X.; Zhu, M.; Zhang, J.; Yang, W.; Wang, M.; Wei, Y.; Zhang, Q.; et al. Realizing 10 Gbps Frequency Continuously Tunable D-Band Signal Transmission With 20 Kilometers Wireless Distance Based on Photon Assistance. *J. Light. Technol.* **2025**, *43*, 5709–5717. [CrossRef]
101. Balanis, C.A. *Antenna Theory: Analysis and Design*; John Wiley&Sons: Hoboken, NJ, USA, 2016.
102. Olver, A.D. *Microwave Horns and Feeds*; IET: London, UK, 1994; Volume 39.
103. Tian, P.; Han, Y.; Li, W.; Yang, X.; Wang, M.; Yu, J. New Design Scheme for and Application of Fresnel Lens for Broadband Photonics Terahertz Communication. *Sensors* **2024**, *24*, 7592. [CrossRef] [PubMed]
104. Akyildiz, I.F.; Jornet, J.M.; Han, C. Terahertz band: Next frontier for wireless communications. *Phys. Commun.* **2014**, *12*, 16–32. [CrossRef]
105. Urteaga, M.; Griffith, Z.; Seo, M.; Hacker, J.; Rodwell, M.J. InP HBT technologies for THz integrated circuits. *Proc. IEEE* **2017**, *105*, 1051–1067. [CrossRef]
106. Weimann, N. InP HBT technology for THz applications. In Proceedings of the IEEE International Symposium on Radio-Frequency Integration Technology, Hiroshima, Japan, 2–4 September 2020; pp. 190–192.

107. Wei, Y.; Yu, J.; Wang, M.; Yang, X.; Zhang, Q.; Tan, J.; Zhang, B.; Wu, Y.; Han, Y.; Tian, P.; et al. Experimental demonstration of 220-GHz terahertz signals wireless transmission over 4.6 km. *Sci. China Inf. Sci.* **2025**, *68*, 1–3. [CrossRef]
108. Wang, M.; Yu, J.; Yang, X.; Wei, Y.; Bian, C.; Zhao, X.; Tian, P.; Han, Y.; Zhang, Q.; Tan, J.; et al. Demonstration of D-Band Wireless Transmission Over 30.2 km Distance Based on Photonic-assisted Scheme. In Proceedings of the Optical Fiber Communication Conference, San Diego, CA, USA, 23–27 March 2025. W4F. 5.
109. Brown, E.R.; Segovia-Vargas, D. Principles of THz direct detection. In *Semiconductor Terahertz Technology: Devices and Systems at Room Temperature Operation*; Wiley Online Library: Hoboken, NJ, USA, 2015; pp. 212–253.
110. Zhang, L.; Yu, J.; Zhu, M.; Zhu, Z.; Yang, X.; Tan, J.; Yu, J.; Ding, J.; Zhang, J. Real-time demonstration of 27.6-Gbit/s 0.3 THz signal wireless transmission over 54 meters based on IM/envelop detection. *Sci. China Technol. Sci.* **2025**, *68*, 1480901. [CrossRef]
111. Wang, M.; Yu, J.; Zhao, X.; Zhou, W.; Wang, K. SNR improved digital-cascaded-pulse-code-modulation radio-over-fiber scheme supporting 16,777,216 QAM for mobile fronthaul. *J. Opt. Commun. Netw.* **2023**, *15*, 948–957. [CrossRef]
112. Yao, H.; Wang, L.; Wang, X.; Lu, Z.; Liu, Y. The space-terrestrial integrated network: An overview. *IEEE Commun. Mag.* **2018**, *56*, 178–185. [CrossRef]
113. Chao, H.-C.; Comer, D.E.; Kao, O. Space and terrestrial integrated networks: Emerging research advances, prospects, and challenges. *IEEE Netw.* **2019**, *33*, 6–7. [CrossRef]
114. Chen, S.; Sun, S.; Kang, S. System Integration of Terrestrial Mobile Communication and Satellite Communication—The Trends, Challenges and Key Technologies in B5G and 6G. *China Commun.* **2021**, *17*, 156–171. [CrossRef]
115. Samy, R.; Yang, H.-C.; Rakia, T.; Alouini, M.-S. Space-Air-Ground FSO Networks for High-Throughput Satellite Communications. *IEEE Commun. Mag.* **2022**, *60*, 82–87. [CrossRef]
116. Han, C.; Gao, W.; Yin, Z.; Yang, C.; Peng, M.; Zhang, W. Joint Communication and Radar Sensing for Terahertz Space-Air-Ground Integrated Networks (SAGIN). *arXiv* **2025**, arXiv:2502.17811.
117. Zhang, Q.; Yu, J.; Long, J.; Wang, C.; Chen, J.; Lu, X. A Hybrid RF/FSO Transmission System Based on a Shared Transmitter. *Sensors* **2025**, *25*, 2021. [CrossRef] [PubMed]
118. Zhang, Q.; Yu, J.; Long, J.; Wang, C.; Chen, J.; Lu, X.; Liu, B.; Yang, X.; Wei, Y.; Zhao, F.; et al. A THz/FSO Fusion Transmission System with Shared Transmitter and Communication Link. In Proceedings of the Optical Fiber Communication Conference, San Diego, CA, USA, 23–27 March 2025. W4F. 8.

micromachines

MDPI

Review

Integrated THz/FSO Communications: A Review of Practical Constraints, Applications and Challenges

Jingtian Liu *, Xiongwei Yang, Yi Wei and Feng Zhao *

School of Electronic Engineering, Xi'an University of Posts and Telecommunications, Xi'an 710121, China; 22110720137@m.fudan.edu.cn (X.Y.); weiyi_email@163.com (Y.W.)
* Correspondence: jingtian.liu@xupt.edu.cn (J.L.); hfengzhao@xupt.edu.cn (F.Z.)

Abstract

This paper presents a comprehensive review of integrated terahertz (THz) and free-space optical (FSO) communication systems, focusing on their potential to address the escalating demands for high-capacity, long-distance, and ultra-reliable transmission in future six-generation (6G) and space–air–ground integrated networks (SAGIN). The study systematically examines recent advancements in three critical areas: channel modeling, transmission performance, and integrated system architectures. Specifically, it analyzes THz and FSO channel characteristics, including attenuation mechanisms, turbulence effects, pointing errors, and noise sources, and compares their complementary strengths under diverse atmospheric conditions. Key findings reveal that THz communication achieves transmission rates up to several Tbps over distances of several kilometers but is constrained by molecular absorption and weather-induced attenuation, while FSO offers superior bandwidth-distance products yet suffers from turbulence-induced fading, posing significant reliability challenges. The integration of THz and FSO through adaptive switching strategies (e.g., hard and soft switching) demonstrates enhanced reliability and spectral efficiency, with experimental results showing seamless data rates exceeding Tbps in hybrid systems. However, challenges persist in transceiver hardware integration, algorithmic optimization, and dynamic resource allocation. The review concludes by identifying future research directions, including the development of unified channel models, shared architectures, and intelligent switching algorithms to achieve robust integrated communication infrastructures.

Keywords: terahertz communication; free space optical communication; integrated convergence; reliability enhancement; switching strategies; atmospheric turbulence; long-distance transmission; atmospheric absorption; photonic-assisted

1. Introduction

Next-generation 6G networks targeting 2030+ applications necessitate Tbps-level throughput while establishing ultra-massive connectivity and ultra-reliable low-latency communication infrastructures. Satellite-assisted SAGIN addresses these demands through coordinated multi-layer architectures [1–4], enabling enhanced transmission capabilities via spaceborne, aerial, and terrestrial multi-access convergence. This integration paradigm ensures continuous service availability with exceptional reliability and quality-of-service (QOS) parameters, ultimately delivering seamless global communication infrastructure.

Contemporary inter-satellite and satellite-ground communications predominantly employ electromagnetic wave transmission, with microwave- and laser-based systems

 https://doi.org/10.3390/mi16111297

constituting the primary technical approaches. While microwave systems demonstrate proven reliability and extensive coverage, their spectrum constraints within the 3–40 GHz range prove insufficient for emerging high-speed, large-capacity transmission requirements. FSO communications, benefiting from abundant spectrum resources as illustrated in Figure 1, offers distinct advantages [5], including broad bandwidth and long-range transmission capabilities [6]. Operating in unregulated spectrum regions, this technology proves particularly effective for intersatellite and satellite-ground link establishment. However, FSO system performance degrades under atmospheric disturbances such as fog, smoke, and turbulence, with additional susceptibility to background radiation-induced link instability [7–10].

The THz band presents complementary strengths through enhanced atmospheric penetration in adverse conditions compared to FSO systems, while maintaining immunity to background radiation and turbulence effects [11,12]. Nevertheless, current technological limitations in high-power THz signal generation and amplification restrict the practical implementation of high-capacity THz communication to kilometer-scale transmission distances.

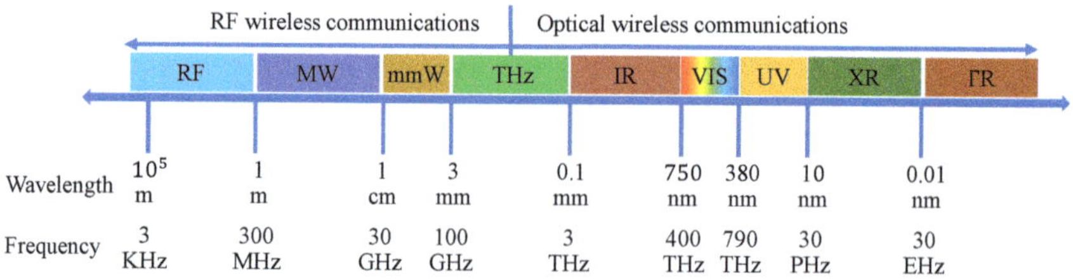

Figure 1. Illustration of wireless communication frequency bands. The commonly used THz band in free-space communication is between 100–300 GHz, and FSO is between 780–1550 nm; RF: radio frequency; MW: microwave; mmW: millimeter wave; IR: infrared light; VIS: visible light; UV: ultraviolet; XR: X-ray; ΓR: Γ-ray.

The strategic integration of THz and FSO communication systems presents a transformative approach to address evolving demands in satellite communications, 6G networks, and mission-critical applications [13–15] as conceptually illustrated in Figure 2. This hybrid architecture capitalizes on complementary spectral and propagation characteristics to achieve four fundamental scientific advancements:

- Spectral resource optimization: Integrated THz/FSO systems transcend single-band limitations through multi-dimensional multiplexing schemes, enabling simultaneous seamless ultra high speed, long range, and high capacity transmission. The integrated spectrum utilization paradigm effectively resolves frequency congestion in conventional electromagnetic bands.
- Channel robustness enhancement: By implementing adaptive switching mechanisms between THz's fog/smoke penetration capabilities and FSO's directional precision, the system dynamically optimizes transmission efficiency across diverse atmospheric conditions, significantly improving link stability compared to standalone implementations.
- Device architecture innovation: The convergence necessitates breakthroughs in three core technological domains: Coherent source–receiver alignment for heterogeneous frequency operation, advanced signal processing algorithms for hybrid waveform decoding, photonic–electronic integration for efficient beamforming architectures.
- Interdisciplinary synergy: System integration drives theoretical advancements across electromagnetic theory, atmospheric photonic, and quantum-limited detection, establishing novel frameworks for next-generation communication physical layer design.

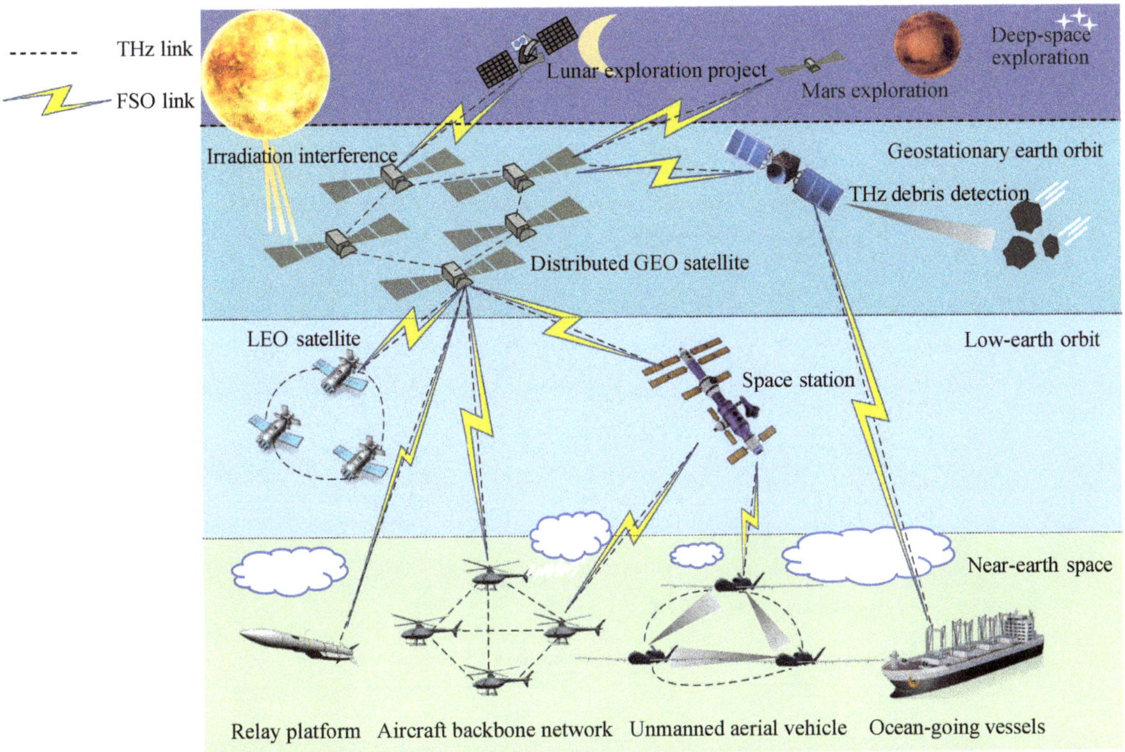

Figure 2. Illustration of integrated THz/FSO SAGIN applications. GEO: geostationary Earth orbit; LEO: low-Earth orbit.

This study comprehensively explores THz and FSO communication systems for high-capacity, long-distance transmission, systematically assesses their integration potential, and outlines future research directions. In Section 2, we rigorously characterize wave propagation behavior under different channel conditions, quantifying key impairment mechanisms such as molecular absorption, aerosol-induced scattering, meteorological interference, and turbulence-induced signal distortion. Then, Section 3, we critically examine the latest advancements and fundamental limitations of both transmission technologies in achieving ultra-high throughput over extended operating ranges. In Section 4, we first systematically review the development history and current status of hybrid transmission systems. Based on these analyses, we then systematically discuss integrated THz/FSO system architectures, with particular focus on adaptive switching techniques. These methods optimize spectral efficiency and connectivity robustness through intelligent transmission switching algorithms. In Section 5, we summarize the research challenges and future directions for advancing next-generation space-terrestrial integrated communication infrastructure. Finally, in Section 6, we provide a detailed summary.

2. Current Progress in THz and FSO Channel Model

Many effects influence the transmission performance of THz and FSO transmission. To help readers better understand the impact of these effects on transmission, this chapter details the channel models for both. Figures 3 and 4 list the attenuation and noise effects primarily considered in previous studies, and underline effects that are negligible in the SAGIN system.

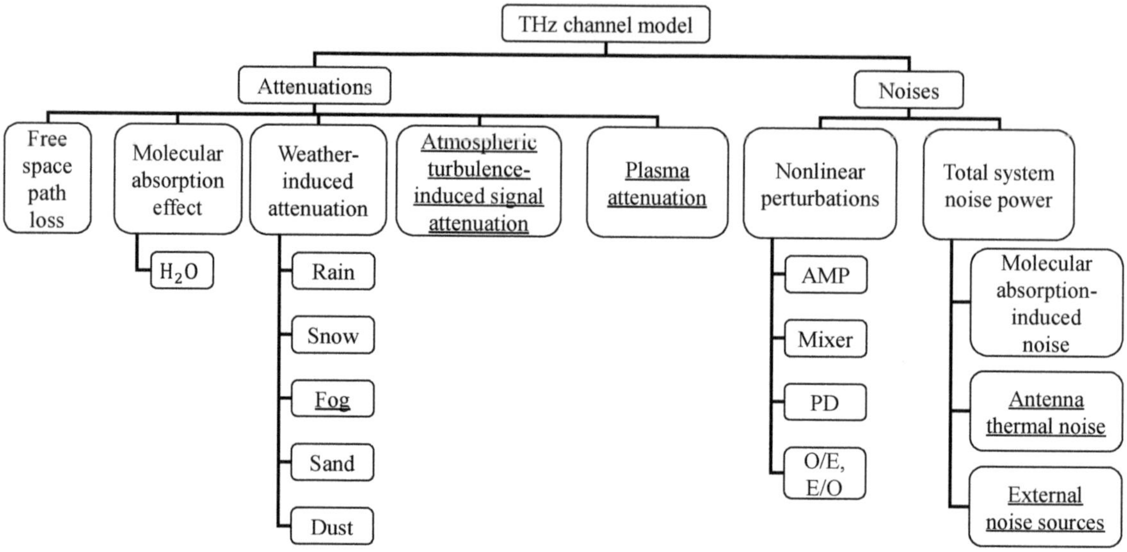

Figure 3. Classification of THz channel impact based on attenuation and noise. Underline: Effects that can be ignored in this paper's model.

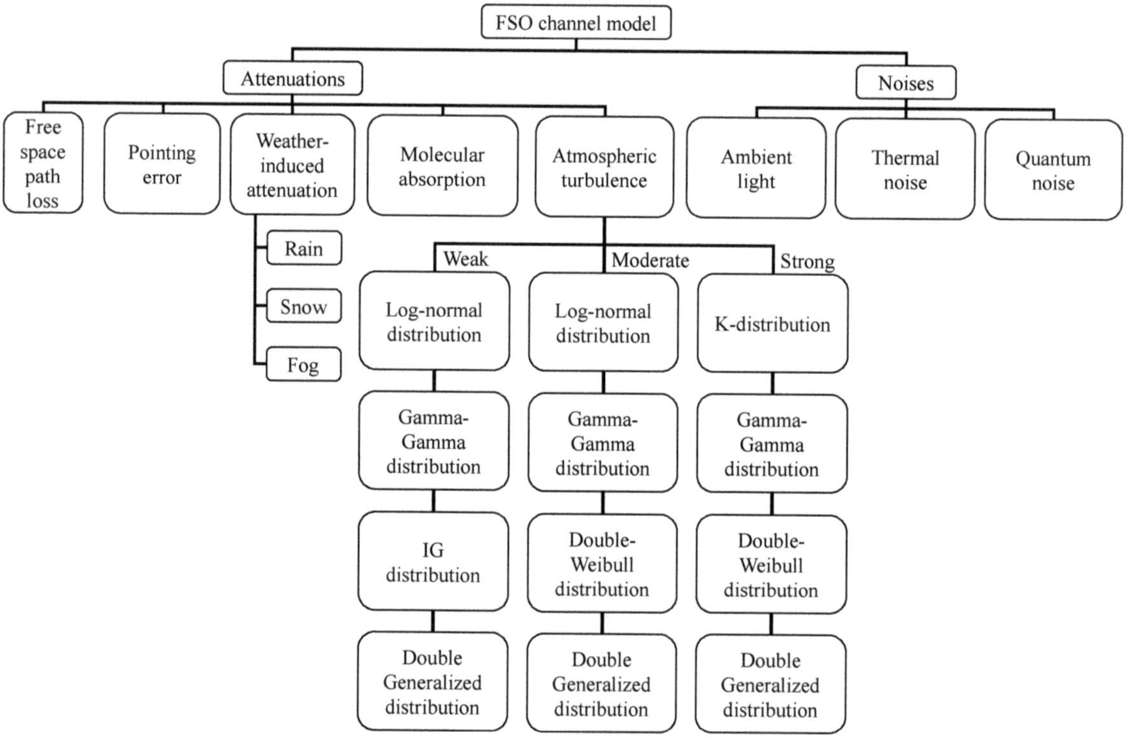

Figure 4. Classification of FSO channel impact based on attenuation and noise.

2.1. THz Channel Model

THz wave propagation is predominantly governed by four attenuation mechanisms: free space path loss (FSPL), molecular absorption, weather-induced attenuation, and nonlinear distortions from devices/channels as illustrated in Figure 3. In terrestrial scenarios, dense obstacles and reflectors generate significant multipath propagation. Conversely, in non-terrestrial environments (e.g., aerial/space links), multipath effects are minimal, often reducing to a single dominant line-of-sight (LOS) path between transmitter and receiver [16]. Consequently, we adopt the LOS path loss model for channel gain calculation.

Further, the extended beam coverage area at THz frequencies substantially exceeds the physical aperture of receiving antennas. This configuration renders pointing-error-induced fading negligible under typical alignment conditions. If required, such effects can be incorporated via the FSO alignment error model detailed in Section 2.2.3.

2.1.1. Free Space Path Loss

In the THz band, the FSPL is calculated according to the Friis transmission equation [17] to account for reflection or diffraction as follows:

$$L_{\mathrm{p}} = \left(\frac{4\pi f d}{c}\right)^{2}, \tag{1}$$

where d is the distance between transmitter and receiver (m), f is the frequency (Hz), c is the velocity of light (m/s).

2.1.2. Molecular Absorption Effect

Atmospheric absorption of THz waves arises from quantum-level energy transitions during photon-molecule collisions involving species such as O_2, N_2, and H_2O. Resonant absorption occurs when the incident photon energy precisely matches the energy gap between molecular quantum states, yielding a substantial absorption cross-section. Among these molecules, water vapor dominates atmospheric attenuation due to its broad resonant lines in the millimeter-wave and THz spectra. The aggregate molecular absorption gain across these frequency bands can be quantified using the Beer–Lambert law [17]:

$$G_{\mathrm{a}} = e^{-\kappa_{\mathrm{a}}(f)d}, \tag{2}$$

where $\kappa_a(f)$ denotes the molecular absorption coefficient, defined as the sum of absorption cross-sections per unit volume within the medium. This coefficient governs attenuation in the 100–450 propagation window, where distinct absorption peaks emerge at approximately 119, 183, 325 and 380 GHz, with the dominant peak occurring at 380 GHz [18,19].

2.1.3. Weather Effect

Tropospheric precipitation (including rain, fog, snow, and dust) induces wavelength-dependent signal attenuation. When particle radii are comparable to the THz wavelength, the Mie absorption-scattering model characterizes the cross section; otherwise, the Rayleigh model applies. Since raindrop radii (1 to 10 mm) are comparable to D-band THz wavelengths (∼3 mm), whereas fog droplets (20 μm to 2 mm) are significantly smaller, rain-induced Mie scattering causes severe signal degradation and heightened attenuation. The specific attenuation γ_{R} (dB/km) in the ITU-R recommended model is calculated from the rain rate R (mm/h), as shown in Table 1 via the power-law relationship [20]:

$$\gamma_{\mathrm{R}} = kR^{\alpha}, \tag{3}$$

where the coefficients k and α are frequency-dependent (see Table 5 in [20]), with typical values of 1.6286 and 0.6262 at 300 GHz. As an empirical formulation grounded in extensive experimental data, this equation demonstrates strong consistency with analytical droplet size distribution (DSD) models such as the MP, Weibull, and Gamma DSD models [21]. For example, at a rainfall rate of 20 mm/h, the rain attenuation reaches 10.6 dB/km.

Table 1. Intensity range of rain.

Intensity	**Range of R (mm/h)**
Very light	$R < 1$
Light	$1 \leq R < 2$
Moderate	$2 \leq R < 5$
Heavy	$5 \leq R < 10$
Very heavy	$10 \leq R < 20$
Extreme	$R \geq 20$

Rayleigh scattering from fog particles minimally impacts THz wave propagation [4] unless the fog is extremely dense, characterized by very low visibility, where attenuation can exceed 100 dB/km [22]. Snow and sleet represent additional attenuation sources, primarily due to the water vapor they contain. However, compared to rain and fog, accurately predicting the attenuation caused by snow or sleet presents greater challenges due to the complex and variable size distribution and geometries of snowflakes. Measurements indicate that within the 300 GHz band and at equivalent precipitation rates (e.g., 20 mm/h), sleet induces higher attenuation than snow, which, in turn, exceeds that caused by rain [23,24]. Under these conditions, dry snow attenuation is approximately 15 dB/km. Furthermore, dust and sand also contribute to THz signal loss, with the attenuation magnitude influenced by the particle density within aerosols and the size of the sand grains [25,26].

2.1.4. Nonlinear Perturbations

In high-speed communication systems, nonlinear distortions manifest due to hardware imperfections in components such as amplifiers (AMP), mixers, photodetectors (PD), and optoelectronic converters (O/E, E/O) [27–29]. For THz transceivers specifically, these imperfections primarily consist of in-phase/quadrature imbalance (IQI), phase noise (PHN), and amplifier nonlinearities, which collectively degrade performance in high-data-rate operation [30,31]. The aggregate distortion η can be statistically modeled as a circularly symmetric complex Gaussian process ($\mathcal{CN}$) [32]:

$$\eta \sim \mathcal{CN}(0, \kappa^2 P), \tag{4}$$

where κ denotes the aggregate impairment coefficient (typically 0 to 0.3), this coefficient corresponds directly to the error vector magnitude (EVM)—a metric that quantifies waveform distortion and is experimentally measurable [33]. P represents the average transmitted signal power.

2.1.5. Other Discussions

In addition to the effects mentioned above, there are numerous other physical effects that can attenuate or add noise to THz transmissions. As established in [34], atmospheric turbulence-induced signal attenuation exhibits a frequency dependence proportional to $f^{\frac{7}{6}}$. However, for THz links under typical non-extreme conditions, this attenuation remains negligible. Plasma attenuation is equally insignificant, measuring below 10^{-3} dB/km and thus disregarded in THz systems.

Molecular absorption within the propagation medium introduces dual effects: it governs channel attenuation while simultaneously generating molecular absorption noise [35]. This noise originates from thermal re-emission of absorbed energy, where molecular vibrations excited by incident radiation produce electromagnetic waves at resonant frequencies. Beyond molecular noise, the antenna noise temperature incorporates contributions from thermal noise in antenna components, electromagnetic interference from integrated nanodevices, and system-level noise sources. Consequently, total receiver noise power comprises three components:

- Molecular absorption-induced noise;
- Antenna thermal noise;
- External noise sources.

Since nanomaterial-enabled receivers achieve ultralow intrinsic noise floors [36], our model focuses exclusively on channel-generated molecular absorption noise—consistent with foundational THz channel modeling frameworks [17]:

$$P_{\mathrm{n}}(f,d) = k_{\mathrm{B}} \int_{B} T_0(1 - G_{\mathrm{a}}(f,d))df, \tag{5}$$

where T_0 is the reference temperature, k_{B} is the Boltzmann constant, and B is the bandwidth.

2.1.6. Channel Model

According to the above effects, the received signal is conventionally modeled based on Equation (5) in [32]:

$$y = \sqrt{\frac{G_{\mathrm{a}}G_{\mathrm{t}}G_{\mathrm{r}}}{L_{\mathrm{p}}10^{\gamma_{\mathrm{R}}d/10}}}(x+\eta) + \nu, \tag{6}$$

where ν is modeled as a complex zero-mean additive white Gaussian process with variance equals P_{n}, G_{t} and G_{r} respectively represent the antenna orientation dependent transmission and reception gain.

2.2. FSO Channel Model

FSO communications experience six primary impairments, as illustrated in Figure 4: path loss, atmospheric turbulence, molecular absorption, weather-induced attenuation, pointing error, and noises from ambient light and thermal noise. For long-range links, path loss follows a deterministic model (Equation (1)) governed by beam divergence, for short-range links, path loss is incorporated within the pointing error component as geometric spread. These composite impairments cause signal attenuation that directly degrades the signal-to-noise ratio (SNR) at the receiver.

2.2.1. Atmospheric Turbulence

Atmospheric turbulence originates from temperature and pressure gradients that create refractive index inhomogeneities, these inhomogeneities form turbulent eddies that deflect optical beams through three distinct mechanisms based on eddy-to-beam size ratio [37,38]:

- Beam Wander (Eddies > Beam Width): Large eddies displace the entire beam centroid randomly, causing pointing errors that may misalign the beam from the receiver aperture.
- Scintillation (Eddies ≅ Beam Width): Eddy-induced focusing/defocusing creates random irradiance fluctuations at the receiver, degrading SNR through deep fading.
- Beam Spreading (Eddies < Beam Width): Small-scale diffraction and scattering reduce received power density while distorting the wavefront.

Scintillation, manifested as intensity fluctuations, dominates fading effects in typical FSO channels [39]. Statistical models for scintillation-induced fading include [40,41]:

- Log-normal distribution: Weak-to-moderate turbulence [41].

- Inverse Gaussian (IG) distribution: Weak turbulence [42].
- Double Weibull distribution: Moderate-to-strong turbulence [43].
- Double Generalized distribution: Weak-to-strong turbulence [44].
- K-distribution: Strong turbulence [45].
- Gamma-Gamma (GG) distribution: Weak-to-strong turbulence [46].

The GG distribution, which is in close agreement with measurements under a variety of turbulence conditions, is the product of two random variables of large scale and small scale eddies. The probability density function (PDF) of the GG distribution with Rytov variance σ_{R}^2 [47] for the intensity h_{a} is given as [46]:

$$f(h_{\mathrm{a}}) = \frac{2(\alpha\beta)^{\frac{\alpha+\beta}{2}} h_{\mathrm{a}}^{\frac{\alpha+\beta}{2}-1}}{\Gamma(\alpha)\Gamma(\beta)} \mathrm{K}_{\alpha-\beta}\left(2\sqrt{\alpha\beta h_{\mathrm{a}}}\right), h_{\mathrm{a}} > 0 \tag{7}$$

and the parameters α and β are given as:

$$\alpha = \left[\exp\left(\frac{0.49\sigma_{\mathrm{R}}^2}{\left(1+1.11\sigma_{\mathrm{R}}^{\frac{12}{5}}\right)^{\frac{7}{6}}}\right) - 1\right]^{-1}, \tag{8}$$

$$\beta = \left[\exp\left(\frac{0.51\sigma_{\mathrm{R}}^2}{\left(1+0.69\sigma_{\mathrm{R}}^{\frac{12}{5}}\right)^{\frac{5}{6}}}\right) - 1\right]^{-1}, \tag{9}$$

where α and β are the effective number of the small scale and large scale eddies of the scattering process, and $K_a(.)$ is the modified Bessel function of order a.

The fading process in atmospheric turbulence channels exhibits a coherence time of 10^{-3} to 10^{-2} s, orders of magnitude longer than typical symbol durations (10^{-9} s) [48,49]. Consequently, the channel gain h_{a} remains effectively constant during individual transmissions. This slow variation renders interleaving techniques impractical for averaging over fading states. Such channels exhibit block-fading characteristics (slow-fading/non-ergodic), this fading can be efficiently modeled via exponentially correlated processes using first-order stochastic differential equations [50].

2.2.2. Weather Effects

The atmospheric attenuation coefficient represents the aggregate loss resulting from the absorption and scattering coefficients associated with both atmospheric molecular constituents and aerosols. At the wavelength of 1550 nm, molecular absorption is approximately 0.01 dB/km and can therefore be neglected. The primary contributor to atmospheric attenuation is particle scattering, which occurs across various hydrometeors such as haze particles, fog droplets, snow, rain, and hail. Among these, Mie scattering by fog particles induces significant attenuation, as characterized by the following empirical relationship [51]:

$$\alpha_{\mathrm{fog}}(V) = \frac{13}{V}\left(\frac{\lambda}{0.55}\right)^{-q(V)} [\mathrm{dB/km}], \tag{10}$$

$$q(V) = \begin{cases} 1.6 & \text{for } V > 50 \text{ km} \\ 1.3 & \text{for } 6 < V < 50 \text{ km} \\ 0.16V + 0.34 & \text{for } 1 < V < 6 \text{ km} \\ V - 0.5 & \text{for } 0.5 < V < 1 \text{ km} \\ 0 & \text{for } V < 0.5 \text{ km.} \end{cases} \tag{11}$$

where α_{fog} denotes the attenuation coefficient due to fog particle scattering in dB/km; λ is the wavelength in micrometers (μm); $q(V)$ represents a fog particle size distribution parameter dependent on meteorological visibility; and V signifies the meteorological visibility range, which is defined as the distance over which atmospheric transmittance at 550 nm decreases to a reference threshold value of $\epsilon = 0.05$ [51]. (Note: Increased fog concentration is positively correlated with decreased turbulence intensity [52].)

Beyond fog, rain constitutes another significant source of signal attenuation in FSO communication systems. Because raindrop radii typically exceed the wavelengths used in FSO, Mie scattering governs the attenuation process. Compared to fog, however, attenuation due to rain is generally less severe. The rain-induced specific attenuation, dependent on the rainfall rate R (as shown in Table 1) is given by [53]:

$$\alpha_{\text{rain}} = kR^{\alpha} \text{ [dB/km]}, \tag{12}$$

where the coefficients k and α are characteristic of the rain type and climate; typical values, for example, are $k = 1.076$ and $\alpha = 0.67$ [53].

Attenuation caused by snow exhibits considerable variability, influenced by characteristics such as snowflake size morphology and snowfall intensity. Notably, large snowflakes (exceeding 20 mm) can potentially induce severe attenuation or even completely block an optical path, contingent upon the beam width [54].

2.2.3. Pointing Error

In line-of-sight FSO communication links, random building vibrations induced by wind loads or thermal expansion can introduce pointing errors, leading to signal fading at the receiver [55]. For a Gaussian beam, the pointing error loss h_{p} due to misalignment can be modeled by the following distribution under the assumption that elevation and horizontal sway displacements follow independent and identical Gaussian distributions [39]:

$$f_{h_{\text{p}}}(h_{\text{p}}) = \frac{\gamma^2}{A_0^{\gamma^2}} h_{\text{p}}^{\gamma^2 - 1}, \quad 0 \leq h_{\text{p}} \leq A_0 \tag{13}$$

where γ denotes the ratio of the equivalent beam radius at the receiver to the standard deviation of the pointing error displacement (jitter) at the receiver. The parameter A_0 represents the fraction of collected power at zero distance ($d = 0$), accounting for inherent beam spreading effects within the model.

Beyond the macroscopic building sway discussed above, pointing errors in FSO systems are critically influenced by two additional factors: the intrinsic properties of the laser beam and the limitations of the beam steering apparatus itself.

Firstly, beam divergence is a fundamental source of geometric loss. Even with perfect alignment, the natural spreading of the Gaussian beam over distance means that only a fraction of the transmitted power is collected by the finite-sized receiver aperture, which is the parameter A_0 in Equation (13). This effect is compounded by any lateral displacement, leading to a rapid decrease in received power. Therefore, the pointing error model must be considered in conjunction with the geometric path loss inherent to divergent beam propagation.

Secondly, the precision of the steering devices—such as fast steering mirrors, galvanometers, and spatial light modulators (SLMs)—introduces micro-scale errors. The limited resolution, step accuracy, and hysteresis of these mechanisms can cause deviations that are often more deterministic and persistent than random jitter. For instance, the finite pixel size of an SLM used for wavefront correction or beam steering imposes a quantization limit on the achievable pointing angle. These device-level inaccuracies can indeed be a dominant error source in high-precision, long-distance FSO links, as they directly limit the system's ability to achieve and maintain optimal alignment. Addressing these combined challenges necessitates advanced acquisition, tracking, and pointing (ATP) systems, sophisticated control algorithms are required to mitigate the compound effects of random platform vibrations, inherent beam divergence, and the finite precision of steering hardware.

2.2.4. Noise Sources

The dominant noise sources in FSO receivers include ambient light noise, quantum noise, and thermal noise. Ambient light noise manifests as shot noise induced by solar and atmospheric radiation, with its variance expressed as:

$$\sigma_{\mathrm{Am}}^2 = 2qB\Re(I_{\mathrm{sky}} + I_{\mathrm{sun}}) \tag{14}$$

where B denotes the electrical bandwidth; $\Re$ is the photodetector responsivity; q represents the electron charge; and I_{sky} and I_{sun} denote sky and solar irradiance (power per unit area), respectively.

Quantum (shot) noise arises from the stochastic nature of photon detection and yields a variance of:

$$\sigma_{\mathrm{Qtm}}^2 = 2qB\Re I \tag{15}$$

Thermal noise originates from electron fluctuations in the receiver circuitry, characterized by:

$$\sigma_{\mathrm{Th}}^2 = \frac{4k_B T_e B}{R_{\mathrm{L}}} \tag{16}$$

where R_{L} is the equivalent load resistance, T_{e} is the effective temperature, and k_{B} is Boltzmann's constant. When ambient light noise and thermal noise components dominate over the received signal, the aggregate noise can be accurately modeled as additive white Gaussian noise (AWGN) that is statistically independent of the transmitted signal [56,57]. The total noise variance at the receiver is then:

$$\sigma_{\mathrm{n}}^2 = \sigma_{\mathrm{Qtm}}^2 + \sigma_{\mathrm{Am}}^2 + \sigma_{\mathrm{Th}}^2 \tag{17}$$

2.2.5. Channel Model

According to the above effects, the received signal is conventionally modeled as:

$$y = h_{\mathrm{l}} h_{\mathrm{p}} h_{\mathrm{a}} \Re x + n, \tag{18}$$

where x is the transmitted intensity; $h_{\mathrm{l}} = 10^{\frac{-\alpha d}{10}}$ is the atmospheric loss based on actual weather conditions by Equations (10) and (12); h_{p} and h_{a} are random variables (RVs) following the distribution in Equations (7) and (13), respectively; and n is a signal-independent AWGN with variance σ_{n}^2.

2.3. *Discussions and Conclusions*

This chapter has systematically examined the channel models for THz and FSO communications, elucidating their distinct attenuation mechanisms, noise sources, and performance under varied atmospheric conditions. As summarized in Table 2, a comparative analysis from a modeling perspective reveals their complementary characteristics: THz communication offers extremely high spectral resources enabling Tbps-level rates but is constrained by deterministic attenuation from molecular absorption and weather effects, posing challenges for transmitter amplifiers over long distances. In contrast, FSO channels provide abundant spectrum and high capacity with virtual immunity to molecular absorption, yet they suffer from random fading induced by atmospheric turbulence and pointing errors, which severely disrupts link reliability.

Table 2. Performance comparison of different communication systems.

	FSO	RF	THz	RF/FSO	THz/FSO
Spectrum resources	High	Low	High	Low/High	High/High
Transmission rate	High	Low	High	Low/High	High/High
Atmospheric turbulence impact	High	Low	Low	Low	Low
Rain impact	Medium	Low	High	Low	Medium
Fog impact	High	Low	Low	Low	Low
Perception capability	Low	High	High	High	High
Confidentiality	High	Low	Relatively high	Relatively low	Relatively high
Reliability	Low	High	Relatively high	High	High

Importantly, the analysis demonstrates that FSO and THz systems exhibit complementary attenuation profiles under different weather conditions. For instance, THz waves offer better penetration in foggy conditions where FSO links degrade, while FSO excels in clear weather with minimal turbulence. This inherent complementarity, as conceptually illustrated in Table 2, forms the foundational rationale for integrated THz/FSO systems. Through dynamic cross-layer resource allocation, such integrated architectures can achieve robust resilience across diverse propagation environments by fully leveraging the ultra-wideband potential of both technologies.

However, these distinct channel characteristics also unveil significant hurdles. The deterministic but high path loss of THz and the stochastic, turbulence-induced fading of FSO present complex challenges for achieving seamless, high-capacity, and ultra-reliable transmission in integrated systems. These fundamental limitations, rooted in the physical layer properties analyzed in this chapter, necessitate innovative solutions that transcend simple channel modeling. Consequently, the following chapter will dedicate a comprehensive analysis to the practical constraints, applications, and future research directions for THz/FSO communications, as well as integrated systems, building upon the foundational insights established in this review of channel characteristics.

3. Current Progress in THz and FSO High-Capacity Long-Distance Communications

3.1. High-Capacity Communications

Both THz and FSO communications have the potential for ultra-large-capacity data transmission on the order of Tbps, as evidenced by recent representative research progress summarized in Table 3. Consequently, THz and FSO are recognized as key enabling technologies for future 6G large-capacity, ultra-high-speed communications [58].

Table 3. Representative progress in THz and FSO high-capacity communication research.

Wavelength	Techniques	Capacity	Distance	Reference
THz	Deep neural network equalization, high-order modulation, MIMO and link optimization	107.52 Gbps	36 m	[59]
	High-order QAM combined with PCS	109 Gbps	1.1 m	[60]
	Kramers–Kronig receiver and optical injection locking technique	120 Gbps	1.2 m	[61]
	Optimizing Links with Cassegrai Antennas	220 Gbps	214 m	[62]
	Channel equalization using likelihood-aware vector quantization variational autoencoder	366.4 Gbps	6.5 m	[63]
	Airy beam, multi-stream low interference design	400 Gbps	0.7 m	[64]
	High order QAM, multi-band	938 Gbps	0.12 m	[65]
	Frequency, polarization and space division multiplexing	1.034 Tbps	100 m	[66]
	80-channel WDM	6.4 Tbps	54 m	[67]
FSO	QPSK and 3-channel WDM	120 Gbps	1 km	[68]
	Link optimization using dual polarization and linear polarization combined with OCDMA	240 Gbps	1.41 km	[69]
	Using the time domain memory and channel estimation algorithm of the FSO link, combined with PCS	400 Gbps	55 m	[70]
	Link optimization using cylindrical vector beams, 8-channel WDM	640 Gbps	3 km	[71]
	PCS dynamic modulation, ATP active alignment, high-order modulation	1 Tbps	3 m	[72]
	High-order QAM and adaptive optics mitigate the effects of atmospheric turbulence	1 Tbps	53 km	[73]
	Link optimization and 102-channel WDM using micro-cavity soliton frequency comb	1.02 Tbps	1 km	[74]
	Delta–Sigma modulation 1024QAM combined with 96-channel WDM	7.68 Tbps	0.1 km	[75]
	9-hole transmit and single-hole receive link optimization, PCS, 35-channel WDM	14 Tbps	220 m	[76]

Extensive research has focused on optimizing THz communication links to achieve high capacity [59,61–64]. Critical enabling techniques include high-order modulation [60,65] and probabilistic constellation shaping (PCS) [77], which adaptively adjust transmission rates and enhance system capacity under limited SNR constraints, achieving net rates of hundreds of gigabits per second (Gbps). Furthermore, multiplexing schemes, implemented using sophisticated nonlinear compensation algorithms, enable the simultaneous transmission of multiple PCS signals within a single physical channel, thereby substantially increasing THz system capacity [66,67,78]. Significant achievements leveraging multiplexing include Zhang et al.'s demonstration of a 1.034 Tbps rate over 100 m using combined frequency, polarization, and spatial multiplexing [66], and Ding et al.'s use of wavelength division multiplexing (WDM) attaining 6.4 Tbps over 54 m [67].

Similarly, FSO systems employ analogous approaches—link optimization [69,71,74,79,80], high-order modulation [72,73], multiplexing [68,74–76,81], and PCS [70,73]—to enhance communication capacity. For instance, El-Mottaleb et al. achieved 240 Gbps over 1.41 using dual polarization and linear polarization modes [69]. Guiomar et al. demonstrated a 400 Gbps rate over 55 applying adaptive PCS to dynamically tune the symbol probability distribution, thereby improving FSO system throughput [70]. Substantial advances have also occurred in FSO multiplexing. Shao et al. replaced conventional semiconductor laser arrays with an emerging microcavity soliton optical frequency comb as a multi-carrier light source for WDM. This provided a more compact solution, enabling 1.02 Tbps over 1 km [74]. Additionally, Wang et al. reported transmitting 7.68 Tbps of polarization-division-multiplexed (PDM) orthogonal frequency-division multiplexed (OFDM) 1024-quadrature amplitude modulation (QAM) signals over 104 m using a 96-channel WDM-FSO system based on 1-bit Delta–Sigma modulation [75].

3.2. *Long-Distance Communications*

Table 4 summarizes recent representative long-distance transmission experiments conducted using THz and FSO communications. Although both technologies face challenges in achieving long-distance transmission, they exhibit distinct advantages. THz communication is significantly constrained by high attenuation. Consequently, given the current lack of high-power signal sources and amplifiers, considerable research effort has focused on transmission link optimization [82–92]. For example, Wei et al. successfully achieved a 4.6 km wireless transmission link in the D-band utilizing THz lens antennas, low-noise amplifiers, high-sensitivity receivers, and advanced digital signal processing (DSP) algorithms, achieving a transmission rate of 200 Gbps [87]. Similarly, Juan et al. developed a 140 GHz wireless communication system. By cascading solid-state power amplifiers with vacuum electron amplifiers, they attained a transmitter output power approaching 2 W, enabling a transmission rate of 500 Mbps over a 27 km sea surface link [88]. Yang et al. accomplished 30 km wireless transmission of 4 Gbps QPSK signals in the 125 GHz band using photon-assisted technology, advanced electronic devices, and DSP [89]. Notably, the vacuum environment characteristic of inter-satellite links drastically reduces THz wave propagation loss, making this technology a highly suitable candidate for such applications [93].

Conversely, long-distance FSO communication suffers from extreme sensitivity to adverse weather conditions and atmospheric turbulence, where dense fog and snow can induce severe attenuation or complete signal interruption. To extend the operational range of FSO systems, the development of sophisticated beam control [94–97] and turbulence compensation techniques [98,99] constitutes critical research areas for overcoming current limitations. For instance, Balasiano et al. employed an optical phased array to coherently synthesize 32 lasers. Real-time phase correction maintained beam drift within five times the diffraction-limited diameter on a 10 km link, achieving a data rate of 100 Gbps [94]. Technical optimization approaches for long-distance FSO communication also vary depending on the communication environment. In dynamic platform scenarios, Yu et al. established a robust 2.5 Gbps transmission link between airships 12 km apart. This system utilized GPS/INS laser orientation combined with an adaptive avalanche photodiode channel compensation mechanism designed to tolerate turbulence encountered between altitudes of 200 m and 1000 m [97]. For satellite communications, Horst et al. leveraged advanced adaptive optics to realize a highly reliable FSO link over 53.42 km, achieving a remarkably low link failure rate and a single-channel capacity of 1.008 Tbps [99].

Table 4. Representative progress in THz and FSO long-distance communication research.

Wavelength	Techniques	Capacity	Distance	Reference
THz	High-gain parabolic antenna, front-end hardware optimization, modulation error correction and environmental adaptability design	44 Gbps	1 km	[82]
	Coherent reception and combining of 8-channel FDM	160 Gbps	1.4 km	[83]
	8PSK and uses on-chip power synthesizer to optimize the link	8 Gbps	2.03 km	[84]
	Cassegrain antenna, efficient modulation, dynamic environmental adaptation	8.387 Gbps	2.92 km	[85]
	Dual-carrier superheterodyne wireless link optimization and FDMA	41.6 Gbps	3 km	[86]
	Leverage polarization multiplexing, MIMO and link optimization	200 Gbps	4.6 km	[87]
	Solid-state amplifiers and vacuum electronic amplifiers cascaded to optimize links and use variable transmission rates	0.5 Gbps	27 km	[88]
	Photon-assisted technology, cascaded electrical amplifiers, advanced DSP	2 Gbps	30.2 km	[89]
FSO	Coherent beam synthesis, combined with optical phased array	100 Gbps	10 km	[94]
	Coherent optical transmission and reception	111.8 Gbps	10.3 km	[98]
	Using fine steering mirrors to compensate for beam fluctuations combined with 40-channel DWDM	1.72 Tbps	10.45 m	[95]
	Closed-loop control of the beam combined with 54-channel DWDM	13.16 Tbps	10.45 km	[96]
	GPS/INS real-time precision pointing, coarse-precision combined tracking, adaptive turbulence compensation	2.5 Gbps	12 km	[97]
	Large aperture Cassegrain antenna, multimode fiber, FPGA real-time processing	2.5 Gbps	29 km	[100]
	High-order modulation and adaptive optical system to correct wavefront distortion caused by atmospheric turbulence	1.008 Tbps	53.42 km	[99]

4. Current Progress in Integrated THz/FSO Communications

4.1. Integrated THz/FSO Communications Capabilities

The signal fading induced by pointing errors, atmospheric turbulence, smoke, and fog significantly limits the performance of FSO links [25]. To address this challenge, high-reliability RF links are often deployed as backup solutions to maintain connectivity. However, RF links typically offer data rates limited to hundreds of Mbps, creating a bottleneck within hybrid RF/FSO systems and constraining their effectiveness and broader applicability. Integrated THz/FSO communication emerges as a superior alternative [13–15]. As summarized in Table 2, both THz and FSO technologies support Gbps-level data rates, enabling high-speed, seamless transmission. Furthermore, their complementary transmission characteristics enhance resilience under complex weather conditions. The high directivity inherent to both technologies also drastically reduces risks associated with signal interception and interference. The integration of THz and FSO systems combines the advantages of optical and wireless communication, significantly enhancing transmission capacity, reliability, and adaptability. Table 5 summarizes recent representative advancements in integrated THz/FSO fusion systems.

Earlier research investigated hybrid RF/FSO architectures. In 2015, Mai et al. introduced a parallel RF/FSO fusion concept [100]. Employing a Markov chain analytical framework, they theoretically demonstrated that a rate-adaptive RF/FSO system offers superior throughput and reliability compared to single-link FSO or RF systems, particularly when accounting for weather factors. Their work quantified the specific impact of weather conditions on performance and validated the reliability of the integrated transmission approach. Subsequently, in 2016, Zhang et al. experimentally validated the reliability gains of an integrated RF/FSO system under simulated fog, rain, and turbulence conditions using an adaptive diversity signal merging algorithm [101]. To further explore potential transmission gains, they proposed a novel signal coordinated signal mapping and combining (CMC) technique [102]. This method senses signal quality per data block to distribute the load accordingly, interleaving and repeating data in the frequency and time domains at the transmitter. At the receiver, adaptive weighting combines the repeated blocks, effectively mitigating system degradation caused by burst interference and power attenuation during severe weather, thereby enhancing robustness. However, these initial schemes primarily provided conceptual demonstrations of the parallel fusion architecture, without fully leveraging the substantial bandwidth resources and complementary characteristics inherent in FSO and THz technologies.

Table 5. Representative progress in integrated fusion communication research.

Wavelength	Techniques	Capacity	Distance	Reference
RF/FSO	Multi-dimensional multiplexing, Kramers–Kronig reception	1.196 Tbps	800 m	[103]
RF/FSO	Real-time neural network prediction and real-time adaptive diversity path selection based on FPGA	10 Gbps	500 m	[104]
RF/FSO	ARQ mechanism and adaptive signal combining algorithm	20 Gbps	4 m	[105]
RF/FSO	Adaptive diversity combination signal merging algorithm	4 Gbps	1.83 m	[101]
RF/FSO	CMC mechanism	2 Gbps	3 m	[102]
RF/FSO	High-order QAM combined with adaptive PCS rate control	200 Gbps	3 m	[106]
THz/FSO	Intelligent switching, DNN decision model	11.6 Gbps	20 m	[107]
THz/FSO	The integrated architecture of PCS, THz lens, low noise amplifier and shared transmitter	1 Tbps	54 m	[14]

Research advanced significantly in 2019 when Wang et al., building on the RF/FSO parallel architecture, integrated wavelength-division multiplexing, frequency-division multiplexing, polarization multiplexing, and Kramers–Kronig receiver techniques [103]. Their system achieved an aggregate transmission capacity of 1.196 Tbits/s over an 800 m link, offering the first experimental proof that flexible resource allocation within a fused architecture can yield substantial capacity gains and improved cost-effectiveness. Recognizing the complementary advantages under diverse channel conditions, Brandão et al. implemented adaptive rate control on a 200 Gbit/s RF/FSO link using probabilistic shaping control strategies [106]. This work demonstrated the complementary benefits of the fused link across different channel states.

Addressing real-time adaptability, Song et al. (2022) demonstrated adaptive real-time diversity path selection under varying weather conditions within an RF/FSO fusion system [104]. Leveraging a time attention-enhanced gated recurrent unit (GRU) neural network implemented on an FPGA platform, they developed a prediction-based mechanism to select the optimal path, confirming the feasibility of adaptive rate control and path selection for similar THz/FSO fusion systems. In 2023, Liu et al. combined an automatic repeat request (ARQ) mechanism with an adaptive signal merging algorithm [105]. Their experiments showed that an RF/FSO fusion architecture could significantly enhance link reliability under challenging channel conditions. In 2024, Chong et al. first deployed a deep neural network (DNN) decision model in a THz/FSO system [107]. The trained neural network model outputs switching decisions in real time, enabling early switching when the channel changes, thus achieving zero-outage communication without switching latency.

Seeking cost reduction, Wang et al. (2024) proposed an innovative THz/FSO fusion scheme based on a shared light source and shared optical path mechanism [14]. This integrated system featured a shared transmitter architecture enabling 1 Tbit/s signal transmission over a 50 m wireless link. By deeply integrating THz and FSO designs, the team maximized their complementary strengths in high-capacity transmission, significantly boosting communication capacity at a lower cost. This shared transmitter approach not only meets ultra-high data rate requirements but also offers novel design paradigms for future communication systems.

4.2. Integrated Communication Systems Switching Strategies

To address the constraints imposed by dynamic channel impairments on transmission performance in FSO communications, researchers have increasingly explored multi-band fusion communication in recent years. This has led to an evolution of technical approaches, from early RF/FSO hybrid systems toward the integration of THz/FSO communication systems. By leveraging frequency band complementarity and intelligent coordination mechanisms, these systems synergistically enhance transmission reliability and spectral efficiency—particularly under complex meteorological conditions (e.g., rain, fog, atmospheric turbulence) and in long-distance scenarios. Current research primarily focuses on two dominant technical paradigms: intelligent hard-switching mechanisms and adaptive soft-switching strategies, as illustrated in Figure 5.

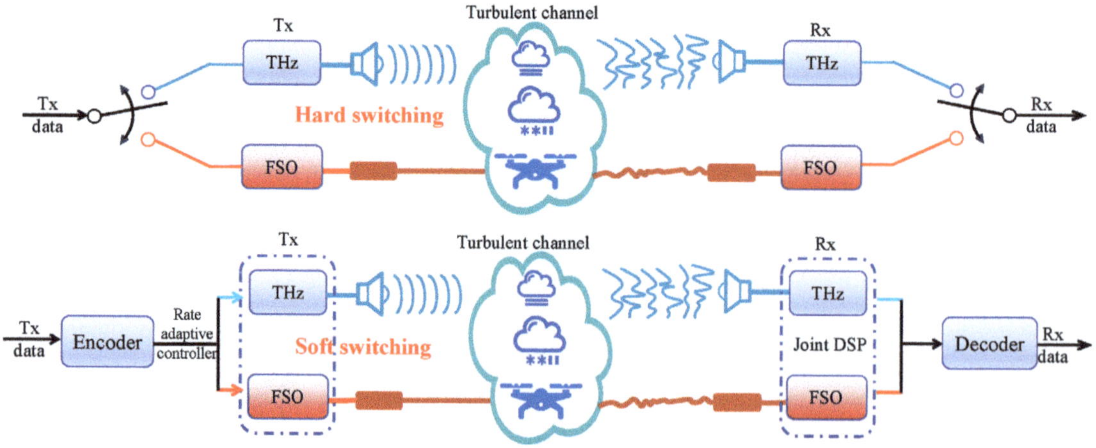

Figure 5. Schematic diagram of the hard/soft switching principle of the integrated fusion system.

4.2.1. Intelligent Hard Switching Mechanism

In a hard-switching fusion communication system, as shown in Figure 5, only one of the FSO or RF/THz links is used as a data transmission link, and the other is used as a backup link. The system determines the data transmission path based on the degree of impact of the two links under different weather conditions. To avoid frequent switching between the two links and ensure the maximum utilization of the FSO link, people have proposed a variety of switching algorithm designs. According to the threshold judgment principle, the hard switching mechanism can be divided into single threshold switching, multi-threshold switching, and intelligent threshold switching.

- **Single-threshold switching:** Single-threshold switching is one of the earlier research directions and has made a series of important progress. In 2009, Nadeem et al. conducted a study on an RF/FSO fusion system with single threshold switching based on bit error rate [108]. When the bit error rate of the FSO link is lower than the set threshold, RF is enabled as a backup link. Their research results show that the multi-band switching system can achieve reliability close to the operator level. In 2016, Touati et al. proposed a single threshold switching algorithm based on the received signal-to-noise ratio [109]. When the received signal-to-noise ratio of the FSO link is lower than the set threshold, RF is enabled as a backup link. The effectiveness of this method in overcoming the sensitivity of the FSO link to atmospheric environment changes and aiming errors was verified. However, when the channel quality of the FSO link hovers near the threshold, the single threshold switching scheme may cause frequent switching of the communication link, which will hurt the link performance.
- **Multi-threshold switching:** In 2019, Sharma et al. further proposed a multi-threshold RF/FSO dual-hop transmission system based on the received signal-to-noise ratio [110]. By setting a dual judgment threshold hysteresis interval for the FSO link, the problem of frequent link switching was effectively avoided. In 2021, Shah et al. introduced the maximum ratio merging algorithm into the hybrid link [111]. When the signal-to-noise ratio of the FSO link is lower than the threshold, the maximum ratio merging of the RF and FSO links will be performed to further obtain performance gains. In 2022, Singya et al. studied the theoretical performance of the multi-threshold switching system of THz/FSO [112]. The research results show that the THz/FSO system can better achieve seamless transmission and significantly reduce the average bit error rate of the hybrid system, providing important theoretical support for the subsequent research on THz/FSO fusion communication.
- **Intelligent threshold switching:** The aforementioned switching models generally adopt a passive switching mechanism, that is, the transmitting system needs to wait for feedback information before making corresponding threshold judgments and switching, which inevitably introduces system delays or link interruptions, neural networks (NNs), especially deep learning models, have become powerful tools for addressing the complex challenges of THz and FSO communication systems. In intelligent switching strategies, NNs provide an end-to-end learning framework, bringing intelligence and adaptability to THz/FSO systems. They play a significant role in channel modeling and channel state prediction, facilitating proactive, early switching. In 2016, Nock et al. introduced a Kalman filter to predict the change of received power [113], thereby switching in advance before the link deteriorates. This method effectively prevents frequent switching and significantly reduces the impact on link throughput. In 2017, Guoru et al. first introduced deep learning into the field of channel state prediction [114], showing performance superior to traditional prediction methods. In 2019, Amirabadi et al. verified the effectiveness of deep learning in FSO atmospheric turbulence channels [115]; in the same year, Yejun et al.

applied recurrent NNs to the prediction of FSO link received signal strength [116], and made threshold judgments and switching based on the predicted values, realizing a paradigm shift from "switching after failure" to "switching before degradation". In 2024, Chong et al. deployed a DNND decision model in a THz/FSO system for the first time [107]. They used fog sensor data and channel status as joint inputs and directly output switching decisions through the trained neural network model. In outdoor experimental verification, this method demonstrated extremely high reliability.

4.2.2. Adaptive Soft Switching Strategy

One problem with the hard switching method is the frequent switching between the FSO link and the RF/THz link, which affects the link performance and increases the implementation complexity. In the soft switching scheme, as shown in Figure 5, the FSO and RF/THz links transmit data simultaneously. The hybrid system judges the channel transmission environment based on the channel state information fed back by the receiving end, and then adjusts the transmission ratio of the data in the two links, the system data rate, or the coding rate, and other parameters to avoid switching delays. According to the coding principle, soft switching is mainly divided into hybrid channel coding, hybrid modulation, and adaptive coding.

- **Hybrid channel coding:** In 2009, Hranilovic et al. proposed a hybrid coding method that divides the original data into two parts and then performs short-length Raptor code encoding on each part [117]. This scheme does not require channel information and can adjust the code rate according to the FSO and RF channel conditions, thereby increasing the data transmission rate while reducing the decoding consumption. In 2012, Tang et al. seamlessly maximized the channel throughput by adaptively allocating the amount of data, adjusting the symbol rate, and the encoder rate [118]. In 2023, Amay et al. intelligently adjusted the code rates of the two channels in real time based on artificial intelligence for different channel states [119], maximizing mutual information while enhancing the reliability of the overall link.
- **Hybrid modulation:** In 2009, He et al. first applied joint bit-interleaved coded modulation to hybrid communication systems [120]. By using joint coding and interleaving to better combat the instantaneous impact of deep fading, they achieved seamless integration of RF and FSO channels and proved the excellent performance of hybrid modulation schemes under various weather conditions. In 2015, Tang et al. proposed a hybrid modulation scheme based on P-LDPC code [121]. The scheme maps two bits to the FSO link and one bit to the RF link to obtain three-dimensional dual-link hybrid modulation symbols, and further optimizes the symbol mapping to improve the performance of RF/FSO hybrid communication systems. In 2024, Merrouche et al. proposed adaptive modulation and combination technology [122], which improves the performance by 7 dB compared to hard switching and more than doubles the system service quality.
- **Adaptive coding:** In 2014, Mai et al. proposed a multi-modulation switching technology [123], which significantly outperformed the traditional switching scheme by combining hard switching between two links and adaptive rate soft switching on each link. Following this, in 2017, I.K.Son et al. proposed a dynamic rate allocation technology [124], which dynamically adjusted the dual-link rate through adaptive coding, which could maximize the system reliability and bandwidth utilization. On this basis, in 2019, Qian et al. first applied deep learning channel state prediction and adaptive coding to satellite-to-ground communication [125]. This study presets 9 different coding methods by changing the modulation format (from QPSK to 32 PSK) and adjusting the LDPC code rate, thereby obtaining an error-free threshold adjust-

ment range of nearly 16 dB. According to the channel prediction results, the system can adaptively adjust the coding method to prevent the bit error rate from being too high, significantly improving the channel spectrum utilization while reducing the communication interruption rate.

In recent years, adaptive coding technology based on PCS has shown broad prospects in this field. This technology can provide nearly continuous fine rate adjustment [126,127]. In 2020, Elzanaty et al. designed an asymmetric probability shaping for FSO backhaul links using intensity modulation [128] to counteract the impact of turbulence on the channel, significantly improving the robustness of the link. In 2020, Guiomarr et al. innovatively designed an adaptive coding algorithm based on probability shaping technology [70]. The algorithm predicts the state changes of the FSO channel in real time through the time window averaging algorithm, and adjusts the error-free communication threshold by adjusting the probability shaping parameters, realizing adaptive continuous rate control for the first time. In 2022, they further proposed an RF/FSO adaptive rate complementary system [106]. The system dynamically adjusts the RF link spectrum efficiency when the FSO link rate decreases through probability shaping technology, thereby keeping the total link rate constant, and successfully realizing a highly reliable constant-rate hybrid communication system. In 2025, Xin et al. used photodiode current as channel information feedback [129] and optimized probability distribution in real time based on a genetic algorithm, further demonstrating the excellent performance of adaptive probability shaping in combating atmospheric turbulence. At present, the research on soft switching technology mainly focuses on FSO and RF fusion systems. However, the transmission rate of FSO and THz is similar, which provides greater potential and broader development space for achieving seamless switching.

5. Challenges and Future Directions

While significant progress has been made in the integration of THz and FSO communication, several challenges and opportunities remain, summarized in Table 6, majorly in five parts.

Table 6. Challenges and future research directions.

Challenges	Key Issues	Future Research Directions
Channel Modeling & Attenuation	Lack of unified model; THz molecular absorption; FSO turbulence	Develop ML-based unified channel models for dynamic adaptation
Hardware Limitations	THz low power efficiency; FSO alignment precision; transceiver integration	Advance semiconductor tech (e.g., GaN/InP) and shared architectures
Resource Allocation & Optimization	Seamless switching challenges; resource utilization	AI-driven adaptive algorithms (e.g., reinforcement learning)
System Reliability & Resilience	Vulnerability to weather/mobility; link outages in SAGIN	Integrate sensing and AI for proactive resilience
Standardization & Deployment	Lack of standards; interoperability gaps	Collaborative standardization and real-world testbeds

- **Challenge 1: Channel Modeling and Attenuation Imperfections**
  Integrated THz/FSO systems face significant signal degradation due to distinct attenuation mechanisms. THz channels are primarily constrained by molecular absorption (e.g., water vapor resonance) and weather-induced attenuation (e.g., rain), which

cause high path loss and limit transmission distances to a few kilometers. Conversely, FSO channels suffer from atmospheric turbulence, fog, and pointing errors, leading to random fading and reliability issues. These impairments are exacerbated in dynamic environments like SAGIN, where factors such as ionospheric effects and mobility further complicate channel predictability. Current models often treat THz and FSO channels in isolation, lacking a unified framework to capture their complementary behaviors under varying conditions.

Future Direction: Development of Unified and Adaptive Channel Models
Future research should prioritize the creation of integrated channel models that combine THz and FSO propagation characteristics, incorporating machine learning and deep learning techniques for real-time adaptation. For instance, machine learning algorithms can dynamically predict attenuation patterns based on environmental sensors, enabling proactive compensation. Additionally, leveraging AI for turbulence tracking and absorption mitigation—such as using hybrid neural networks to model spatiotemporal variations—can enhance model accuracy. Standardized datasets from real-world deployments (e.g., SAGIN testbeds) will be crucial for benchmarking and validation, facilitating robust system design for 6G applications.

- **Challenge 2: Hardware Limitations and Transceiver Inefficiencies**
  Practical implementation of THz/FSO systems is hindered by hardware constraints. THz components, including high-power sources and amplifiers, struggle with low efficiency due to parasitic effects (e.g., capacitance losses), restricting output power and leading to limited link distances. FSO systems require precise alignment mechanisms (e.g., ATP), but conventional ATP methods are inadequate under turbulence or mobility. Both technologies face integration challenges, such as mismatched data rates between THz and FSO streams, and lack of compact, cost-effective transceivers for seamless coexistence.
  Future Direction: Advancements in Semiconductor Technologies and Shared Architectures
  Innovations in semiconductor devices, particularly III-V materials (e.g., GaN, InP), can yield high-gain antennas and efficient power amplifiers for THz bands, extending transmission ranges. For FSO, deep learning-based ATP systems—using convolutional neural networks for beam alignment—can improve robustness. Future work should also explore shared transmitter architectures (e.g., common light sources for THz and FSO) to reduce costs and enhance interoperability. Prototypes focusing on CubeSat-compatible hardware will address SAGIN needs, while photonic integration techniques can unify signal processing chains.
- **Challenge 3: Dynamic Resource Allocation and Switching Strategies**
  Integrated systems rely on adaptive switching (e.g., hard or soft switching) to maintain reliability, but current approaches suffer from latency, frequent unnecessary mode changes, and suboptimal resource utilization. Hard switching causes interruptions due to threshold-based triggers, while soft switching requires complex algorithms for simultaneous data flow, often leading to unstable BER in volatile environments. Moreover, resource allocation lacks intelligence in balancing throughput, spectral efficiency, and energy consumption under constraints like high mobility in SAGIN.
  Future Direction: AI-Driven Adaptive Algorithms and Cross-Layer Optimization
  Future efforts should emphasize AI-enabled strategies, such as reinforcement learning for real-time switching decisions and resource management. Deep neural networks can predict channel states to enable "switching before degradation," minimizing delays. Hybrid modulation and coding schemes (e.g., probabilistic shaping combined with orthogonal time frequency space waveforms) can optimize bandwidth utilization. Integration with edge computing will allow dynamic task scheduling, while federated

learning frameworks can ensure privacy in distributed networks. Standardization of switching protocols (e.g., via IEEE) will promote interoperability.

- **Challenge 4: Reliability and Resilience in Complex Environments**
  THz/FSO systems must achieve ultra-reliable transmission for critical applications (e.g., deep-space communication or urban 6G networks), but they remain vulnerable to environmental disruptions. FSO links are highly sensitive to fog and turbulence, while THz performance drops in humid conditions. In SAGIN, issues like satellite mobility, space debris, and interference amplify these risks, resulting in unpredictable link outages. Existing systems lack effective mitigation techniques for cross-domain threats, such as simultaneous weather and mobility impacts.
  Future Direction: Intelligent Sensing-Computing-Communication Integration and Debris Detection
  Research should focus on joint communication and radar sensing frameworks, where THz-based sensing provides environmental feedback (e.g., debris detection with centimeter resolution) to preemptively adjust communication parameters. AI-empowered networks can fuse data from multiple sensors (e.g., fog detectors) to enhance resilience. Additionally, waveform designs like affine frequency-division multiplexing can balance sensing accuracy and data rates. Future directions include developing THz-based debris monitoring systems and robust MAC protocols for SAGIN, leveraging digital twins for simulation-based validation.
- **Challenge 5: Standardization and Practical Deployment Gaps**
  Despite theoretical advances, integrated THz/FSO systems face hurdles in real-world deployment due to the absence of standardization, model architectures, and benchmarking. AI-based solutions often rely on simulated data, limiting applicability. Interoperability issues arise from heterogeneous hardware, and regulatory constraints (e.g., spectrum allocation) hinder scalability.
  Future Direction: Collaborative Standardization and Benchmarking Initiatives
  Accelerated efforts by bodies like IEEE, ITU, and ETSI are needed to define standards for parameters such as modulation classifications and channel estimation. Creating open-source datasets with real-world turbulence and weather effects will enable reproducible machine learning model training. Future work should also promote hardware-software co-design, ensuring algorithms are optimized for embedded systems in UAVs or satellites. Emphasis on green technologies (e.g., energy-efficient amplifiers) will align with sustainability goals.

6. Conclusions

This article provides a comprehensive review of channel transmission models, historical developments, and the current state of THz and FSO systems for high-capacity, long-haul converged transmission. It further explores reliability enhancement solutions for integrated transmission architectures. Through an in-depth analysis of representative advancements, we demonstrate that THz and FSO communications have achieved significant milestones, including Tbps transmission over kilometer-scale distances. Substantial progress has also been evidenced in hybrid RF/FSO system design and reliability-driven hard/soft switching technologies. Currently, research focus is shifting toward seamless THz/FSO systems, where initial breakthroughs have been validated. Our analysis reveals that integrated THz/FSO communications hold transformative potential for 6G networks and SAGIN, offering high capacity and reliability by leveraging complementary strengths. However, key challenges—spanning channel modeling, hardware limitations, resource allocation, resilience, and standardization—must be addressed to realize this potential. Critical findings indicate that AI and machine learning technologies are pivotal for dynamic adaptation, while advancements in semiconductors and shared architectures can overcome hardware bottlenecks. Future success hinges on interdisciplinary collaboration, emphasizing unified models, intelligent switching mechanisms, and practical deployments. By prioritizing these directions, integrated THz/FSO systems can evolve into robust, scalable infrastructures capable of meeting the demands of next-generation networks.

Author Contributions: Conceptualization, J.L.; methodology, J.L.; writing original draft, J.L.; data curation, X.Y.; formal analysis, Y.W.; writing—review and editing, X.Y. and Y.W.; supervision, F.Z.; project administration, F.Z.; funding acquisition, F.Z. All authors have read and agreed to the published version of the manuscript.

Funding: This research was funded by the Natural Science Basic Research Program of Shaanxi Department of Science and Technology (2023-JC-JQ-58).

Data Availability Statement: The data presented in this study is available on request from the corresponding author. The data is not publicly available due to privacy restrictions.

Conflicts of Interest: The authors declare no conflicts of interest.

References

1. Akyildiz, I.F.; Han, C.; Hu, Z.; Nie, S.; Jornet, J.M. Terahertz band communication: An old problem revisited and research directions for the next decade. *IEEE Trans. Commun.* **2022**, *70*, 4250–4285. [CrossRef]
2. Nie, S.; Akyildiz, I.F. Channel modeling and analysis of inter-small-satellite links in terahertz band space networks. *IEEE Trans. Commun.* **2021**, *69*, 8585–8599. [CrossRef]
3. Mao, T.; Zhang, L.; Xiao, Z.; Han, Z.; Xia, X.G. Terahertz-band near-space communications: From a physical-layer perspective. *IEEE Commun. Mag.* **2022**, *62*, 110–116. [CrossRef]
4. Han, C.; Gao, W.; Yin, Z.; Yang, C.; Peng, M.; Zhang, W. Joint Communication and Radar Sensing for Terahertz Space-Air-Ground Integrated Networks (SAGIN). *arXiv* **2025**, arXiv:2502.17811. [CrossRef]
5. Trichili, A.; Cox, M.A.; Ooi, B.S.; Alouini, M.S. Roadmap to free space optics. *J. Opt. Soc. Am. B* **2020**, *37*, A184–A201. [CrossRef]
6. Samy, R.; Yang, H.C.; Rakia, T.; Alouini, M.S. Reliable terabits feeder link for very high-throughput satellite systems with SAG-FSO transmission. *IEEE Wirel. Commun.* **2023**, *31*, 112–116. [CrossRef]
7. Khan, A.N.; Saeed, S.; Naeem, Y.; Zubair, M.; Massoud, Y.; Younis, U. Atmospheric turbulence and fog attenuation effects in controlled environment FSO communication links. *IEEE Photonics Technol. Lett.* **2022**, *34*, 1341–1344. [CrossRef]
8. Singh, M.; Elsayed, E.E.; Alayedi, M.; Aly, M.H.; Abd El-Mottaleb, S.A. Performance analysis in spectral-amplitude-coding-optical-code-division-multiple-access using identity column shift matrix code in free space optical transmission systems. *Opt. Quantum Electron.* **2024**, *56*, 795. [CrossRef]
9. Khalighi, M.A.; Uysal, M. Survey on free space optical communication: A communication theory perspective. *IEEE Commun. Surv. Tutor.* **2014**, *16*, 2231–2258. [CrossRef]
10. Kaushal, H.; Kaddoum, G. Optical communication in space: Challenges and mitigation techniques. *IEEE Commun. Surv. Tutor.* **2016**, *19*, 57–96. [CrossRef]
11. Boulogeorgos, A.A.A.; Riera, J.M.; Alexiou, A. On the joint effect of rain and beam misalignment in terahertz wireless systems. *IEEE Access* **2022**, *10*, 58997–59012. [CrossRef]
12. Liu, K.; Feng, Y.; Han, C.; Chang, B.; Chen, Z.; Xu, Z.; Li, L.; Zhang, B.; Wang, Y.; Xu, Q. High-speed 0.22 THz communication system with 84 Gbps for real-time uncompressed 8K video transmission of live events. *Nat. Commun.* **2024**, *15*, 8037. [CrossRef]
13. Sharma, S.; Madhukumar, A. On the performance of hybrid THz/FSO system. *IEEE Commun. Lett.* **2023**, *27*, 3394–3398. [CrossRef]
14. Wang, K.; Yu, J.; Bian, C.; Ding, J.; Zhang, J.; Zhu, M.; Yang, X.; Li, W.; Wei, Y.; Wang, M.; et al. Terahertz transmitter demonstration of the simultaneous transmission of optical and electrical terahertz band signals at 1 Tb/s bit rate and 50 m wireless delivery. *Sci. China Technol. Sci.* **2025**, *68*, 1–3. [CrossRef]
15. Zhang, Q.; Yu, J.; Long, J.; Wang, C.; Chen, J.; Lu, X.; Liu, B.; Yang, X.; Wei, Y.; Zhao, F.; et al. A THz/FSO Fusion Transmission System with Shared Transmitter and Communication Link. In Proceedings of the 2025 Optical Fiber Communications Conference and Exhibition (OFC), San Francisco, CA, USA, 30 March–3 April 2025 ; IEEE: Piscataway, NJ, USA , 2025; pp. 1–3.
16. Li, Y.; Li, N.; Han, C. Ray-tracing simulation and hybrid channel modeling for low-terahertz UAV communications. In Proceedings of the ICC 2021—IEEE International Conference on Communications, Montreal, QC, Canada, 14–23 June 2021; IEEE: Piscataway, NJ, USA, 2021; pp. 1–6.

17. Jornet, J.M.; Akyildiz, I.F. Channel modeling and capacity analysis for electromagnetic wireless nanonetworks in the terahertz band. *IEEE Trans. Wirel. Commun.* **2011**, *10*, 3211–3221. [CrossRef]
18. Papasotiriou, E.N.; Kokkoniemi, J.; Boulogeorgos, A.A.A.; Lehtomäki, J.; Alexiou, A.; Juntti, M. A new look to 275 to 400 GHz band: Channel model and performance evaluation. In Proceedings of the 2018 IEEE 29th Annual International Symposium on Personal, Indoor and Mobile Radio Communications (PIMRC), Bologna, Italy, 9–12 September 2018; IEEE: Piscataway, NJ, USA, 2018; pp. 1–5.
19. Da Nobrega, R.V.; Raddo, T.R.; Jurado-Navas, A.; Loiola, M.B.; Sanches, A.L.; Debbah, M. A channel loss model for THz networks from 100–600 GHz considering both molecular and water vapor continuum absorptions. *IEEE Open J. Veh. Technol.* **2023**, *4*, 475–489. [CrossRef]
20. ITU-R. Specific Attenuation Model for Rain for Use in Prediction Methods. Document Int. Telecommun. Union–Recommendation P.838-3. 2005. Available online: https://www.itu.int/dms_pubrec/itu-r/rec/p/r-rec-p.838-3-200503-i!!pdf-e.pdf (accessed on 16 November 2025).
21. Norouzian, F.; Marchetti, E.; Gashinova, M.; Hoare, E.; Constantinou, C.; Gardner, P.; Cherniakov, M. Rain attenuation at millimeter wave and low-THz frequencies. *IEEE Trans. Antennas Propag.* **2019**, *68*, 421–431. [CrossRef]
22. Golovachev, Y.; Etinger, A.; Pinhasi, G.A.; Pinhasi, Y. Propagation properties of sub-millimeter waves in foggy conditions. *J. Appl. Phys.* **2019**, *125*, 151612. [CrossRef]
23. Renaud, D.L.; Federici, J.F. Terahertz attenuation in snow and sleet. *J. Infrared Millim. Terahertz Waves* **2019**, *40*, 868–877. [CrossRef]
24. Amarasinghe, Y.; Zhang, W.; Zhang, R.; Mittleman, D.M.; Ma, J. Scattering of terahertz waves by snow. *J. Infrared Millim. Terahertz Waves* **2020**, *41*, 215–224. [CrossRef]
25. Su, K.; Moeller, L.; Barat, R.B.; Federici, J.F. Experimental comparison of performance degradation from terahertz and infrared wireless links in fog. *J. Opt. Soc. Am. A* **2012**, *29*, 179–184. [CrossRef] [PubMed]
26. Du, R.; Norouzian, F.; Marchetti, E.; Willetts, B.; Gashinova, M.; Cherniakov, M. Characterisation of attenuation by sand in low-THz band. In Proceedings of the 2017 IEEE Radar Conference (RadarConf), Seattle, WA, USA, 8–12 May 2017; IEEE: Piscataway, NJ, USA, 2017; pp. 0294–0297.
27. Wei, C.C.; Xie, Z.X.; Ting, P.H.; Huang, Z.W.; Lin, C.T. 3 × 3 MIMO 60-GHz OFDM-RoF system with long-distance wireless transmission enabled by MIMO volterra filtering. *J. Light. Technol.* **2022**, *40*, 6860–6866. [CrossRef]
28. Zhou, S.; Gao, R.; Zhang, Q.; Chang, H.; Xin, X.; Zhao, Y.; Liu, J.; Lin, Z. Data-defined naïve Bayes (DNB) based decision scheme for the nonlinear mitigation for OAM mode division multiplexed optical fiber communication. *Opt. Express* **2021**, *29*, 5901–5914. [CrossRef]
29. Liu, X.; Zhang, J.; Zhu, M.; Tong, W.; Xin, Z.; Wang, Y.; Lei, M.; Hua, B.; Cai, Y.; Zou, Y.; et al. Nonlinearity mitigation in a fiber-wireless integrated system based on low-complexity autoencoder and BiLSTM-ANN equalizer. *Opt. Express* **2023**, *31*, 20005–20018. [CrossRef]
30. Boulogeorgos, A.A.A. Interference Mitigation Techniques in Modern Wireless Communication Systems. Ph.D. Thesis, Aristotle University of Thessaloniki, Thessaloniki, Greece, 2016 ; p. 2016.
31. Schenk, T. *RF Imperfections in High-Rate Wireless Systems: Impact and Digital Compensation*; Springer: Berlin/Heidelberg, Germany, 2008.
32. Bjornson, E.; Matthaiou, M.; Debbah, M. A new look at dual-hop relaying: Performance limits with hardware impairments. *IEEE Trans. Commun.* **2013**, *61*, 4512–4525. [CrossRef]
33. Wang, A.; Ligmanowski, R.; Castro, J.; Mazzara, A. EVM simulation and analysis techniques. In Proceedings of the MILCOM 2006—2006 IEEE Military Communications Conference, Washington, DC, USA, 23–25 October 2006; IEEE: Piscataway, NJ, USA, 2006; pp. 1–7.
34. Gao, W.; Han, C.; Chen, Z. Attenuation and loss of spatial coherence modeling for atmospheric turbulence in terahertz UAV MIMO channels. *IEEE Trans. Wirel. Commun.* **2024**, *23*, 11636–11648. [CrossRef]
35. Box, F. Utilization of atmospheric transmission losses for interference-resistant communications. *IEEE Trans. Commun.* **2003**, *34*, 1009–1015. [CrossRef]
36. Pal, A.N.; Ghosh, A. Ultralow noise field-effect transistor from multilayer graphene. *Appl. Phys. Lett.* **2009**, *95*, 082105. [CrossRef]
37. Anbarasi, K.; Hemanth, C.; Sangeetha, R. A review on channel models in free space optical communication systems. *Opt. Laser Technol.* **2017**, *97*, 161–171. [CrossRef]
38. Ghassemlooy, Z.; Popoola, W.O. Terrestrial free-space optical communications. *Mob. Wirel. Commun. Netw. Layer Circuit Level Des.* **2010**, *17*, 355–391.
39. Farid, A.A.; Hranilovic, S. Outage capacity optimization for free-space optical links with pointing errors. *J. Light. Technol.* **2007**, *25*, 1702–1710. [CrossRef]
40. Andrews, L.C.; Phillips, R.L.; Hopen, C.Y.; Al-Habash, M. Theory of optical scintillation. *J. Opt. Soc. Am. A* **1999**, *16*, 1417–1429. [CrossRef]

41. Al-Habash, M.; Andrews, L.C.; Phillips, R.L. Mathematical model for the irradiance probability density function of a laser beam propagating through turbulent media. *Opt. Eng.* **2001**, *40*, 1554–1562. [CrossRef]
42. Baker, K.R. On the WMC density as an inverse Gaussian probability density. *IEEE Trans. Commun.* **1996**, *44*, 15–17. [CrossRef]
43. Chatzidiamantis, N.D.; Sandalidis, H.G.; Karagiannidis, G.K.; Kotsopoulos, S.A.; Matthaiou, M. New results on turbulence modeling for free-space optical systems. In Proceedings of the 2010 17th International Conference on Telecommunications, Doha, Qatar, 4–7 April 2010; IEEE: Piscataway, NJ, USA, 2010; pp. 487–492.
44. Kashani, M.A.; Uysal, M.; Kavehrad, M. A novel statistical model for turbulence-induced fading in free-space optical systems. In Proceedings of the 2013 15th International Conference on Transparent Optical Networks (ICTON), Cartagena, Spain, 23–27 June 2013; IEEE: Piscataway, NJ, USA, 2013; pp. 1–5.
45. Samimi, H. Distribution of the sum of K-distributed random variables and applications in free-space optical communications. *IET Optoelectron.* **2012**, *6*, 1–6. [CrossRef]
46. Uysal, M.; Li, J.; Yu, M. Error rate performance analysis of coded free-space optical links over gamma-gamma atmospheric turbulence channels. *IEEE Trans. Wirel. Commun.* **2006**, *5*, 1229–1233. [CrossRef]
47. Andrews, L.C.; Phillips, R.L.; Hopen, C.Y. *Laser Beam Scintillation with Applications*; SPIE Press: Bellingham, WA, USA, 2001; Volume 99.
48. Davidson, F.; Juan, C.J.; Hammons, A.R. Channel fade modeling for free-space optical links. In Proceedings of the 2010-MILCOM 2010 MILITARY COMMUNICATIONS CONFERENCE, San Jose, CA, USA, 31 October–3 November 2010; IEEE: Piscataway, NJ, USA, 2010; pp. 802–807.
49. Jurado-Navas, A.; Garrido-Balsellss, J.M.; Castillo-Vázquez, M.; Puerta-Notario, A. A computationally efficient numerical simulation for generating atmospheric optical scintillations. In *Numerical Simulations of Physical and Engineering Processes*; InTech: Tokyo, Japan, 2011; pp. 157–180.
50. Bykhovsky, D. Simple generation of gamma, gamma–gamma, and K distributions with exponential autocorrelation function. *J. Light. Technol.* **2016**, *34*, 2106–2110. [CrossRef]
51. Prokes, A. Atmospheric effects on availability of free space optics systems. *Opt. Eng.* **2009**, *48*, 066001. [CrossRef]
52. Kim, I.I.; McArthur, B.; Korevaar, E.J. Comparison of laser beam propagation at 785 nm and 1550 nm in fog and haze for optical wireless communications. In Proceedings of the Optical Wireless Communications III, Boston, MA, USA, 5–8 November 2000; Spie: Boston, MA, USA, 2001; Volume 4214, pp. 26–37.
53. ITU-R. Prediction Methods Required for the Design of Terrestrial Free-Space Optical Links. Document Int. Telecommun. Union–Recommendation 1814. 2007. Available online: https://www.itu.int/dms_pubrec/itu-r/rec/p/r-rec-p.1817-1-201202-i!!pdf-e.pdf (accessed on 16 November 2025).
54. Kaushal, H.; Jain, V.; Kar, S. Free-space optical channel models. In *Free Space Optical Communication*; Springer: Berlin/Heidelberg, Germany , 2017; pp. 41–89.
55. Arnon, S. Effects of atmospheric turbulence and building sway on optical wireless-communication systems. *Opt. Lett.* **2003**, *28*, 129–131. [CrossRef]
56. Zhu, X.; Kahn, J.M. Free-space optical communication through atmospheric turbulence channels. *IEEE Trans. Commun.* **2002**, *50*, 1293–1300. [CrossRef]
57. Leitgeb, E.; Plank, T.; Awan, M.S.; Brandl, P.; Popoola, W.; Ghassemlooy, Z.; Ozek, F.; Wittig, M. Analysis and evaluation of optimum wavelengths for free-space optical transceivers. In Proceedings of the 2010 12th International Conference on Transparent Optical Networks, Munich, Germany, 27 June–1 July 2010; IEEE: Piscataway, NJ, USA, 2010; pp. 1–7.
58. Zhang, J.; Lei, M.; Zhu, M.; Hua, B.; Cai, Y.; Zou, Y.; Tian, L.; Wang, Y.; Xiao, J.; Huang, Y.; et al. Optical-terahertz-optical seamless integration system for dual-λ 400 GbE real-time transmission at 290 GHz and 340 GHz. *Inf. Sci.* **2023**, *66*, 214301:1–214301:3. [CrossRef]
59. Zhou, L.; Tsutsui, M.; Nakasha, Y.; Ozaki, S.; Dateki, T.; Kumazaki, Y.; Ohki, T. Sub-THz SC-FDE MIMO Systems with Deep Neural Network Equalization for 100 Gbps of 6G. In Proceedings of the 2024 IEEE 100th Vehicular Technology Conference (VTC2024-Fall), Washington, DC, USA, 7–10 October 2024; IEEE: Piscataway, NJ, USA, 2024; pp. 1–5.
60. Baek, I.H.; Stiewe, O.; Nellen, S.; Schwanke, G.; Elschner, R.; Schubert, C.; Kohlhaas, R.B.; Schell, M.; Freund, R. 32 GBd, 109 Gbit/s probabilistically shaped THz wireless transmission using PIN-PD based photonic upconversion. In Proceedings of the 49th European Conference on Optical Communications (ECOC 2023), Hybrid Conference, Glasgow, UK, 1–5 October 2023; Volume 2023, pp. 1302–1305.
61. Wen, Z.Y.; Yang, F.; Yang, Z.; Wang, H.; Liu, Y.; Li, M.; Zhang, J. Spectrally Efficient Photonic Terahertz Transmitter for Kramers-Kronig Reception Without Guard-Band Based on Optical Injection Locking. *J. Light. Technol.* **2024**, *42*, 5466–5475. [CrossRef]
62. Maekawa, K.; Yoshioka, T.; Nakashita, T.; Ohara, T.; Nagatsuma, T. Single-carrier 220-Gbit/s sub-THz wireless transmission over 214 m using a photonics-based system. *Opt. Lett.* **2024**, *49*, 4666–4668. [CrossRef]

63. Liu, X.; Zhang, J.; Zhu, M.; Tong, W.; Wang, Y.; Xin, Z.; Hua, B.; Lei, M.; Cai, Y.; Yu, J. 366.4-gbit/s pcs-64qam thz transmission enhanced by likelihood-aware vector-quantized variational autoencoders. In Proceedings of the 49th European Conference on Optical Communications (ECOC 2023), Hybrid Conference, Glasgow, UK, 1–5 October 2023; Volume 2023, pp. 1154–1157.
64. Lee, D.; Yagi, Y.; Suzuoki, K.; Kudo, R. Experimental Demonstration of Wireless Transmission Using Airy Beams in Sub-THz Band. *IEEE Open J. Commun. Soc.* **2025**, *6*, 1091–1102. [CrossRef]
65. Zhou, Z.; Kassem, A.; Seddon, J.; Sillekens, E.; Darwazeh, I.; Bayvel, P.; Liu, Z. 938 Gb/s, 5–150 GHz Ultra-Wideband Transmission Over the Air Using Combined Electronic and Photonic-Assisted Signal Generation. *J. Light. Technol.* **2024**, *42*, 7247–7252.
66. Zhang, H.; Yang, Z.; Lyu, Z.; Yang, H.; Zhang, L.; Ozolins, O.; Pang, X.; Zhang, X.; Yu, X. 300 GHz photonic-wireless transmission with aggregated 1.034 Tbit/s data rate over 100 m wireless distance. In Proceedings of the Optical Fiber Communication Conference, San Diego, CA, USA, 24–28 March 2024; Optica Publishing Group: Washington, DC, USA, 2024; p. M2F.1.
67. Ding, J.; Li, W.; Zhang, L.; Wang, Y.; Liu, J.; Wang, K.; Zhao, L.; Zhou, W.; Zhang, J.; Zhu, M.; et al. Demonstration of 6.4-Tbit/s THz-wave signal transmission over 20-km wired and 54-m wireless distance. In Proceedings of the 2022 European Conference on Optical Communication (ECOC), Basel, Switzerland, 18–22 September 2022; IEEE: Piscataway, NJ, USA, 2022; pp. 1–4.
68. Feng, X.; Wu, Z.; Wang, T.; Jiang, H.; Su, Y.; Jiang, H.; Gao, S. 120 Gbit/s high-speed WDM-QPSK free-space optical transmission through 1 km atmospheric channel. *Electron. Lett.* **2018**, *54*, 1082–1084. [CrossRef]
69. Abd El-Mottaleb, S.A.; Bencheikh, A.; Singh, M. 240 Gbps SI-FMF/FSO communication system based on LP modes and PV code with the impact of foggy weather in Alexandria and Setif. *Appl. Opt.* **2025**, *64*, C79–C87. [CrossRef]
70. Guiomar, F.P.; Lorences-Riesgo, A.; Ranzal, D.; Rocco, F.; Sousa, A.N.; Fernandes, M.A.; Brandão, B.T.; Carena, A.; Teixeira, A.L.; Medeiros, M.C.; et al. Adaptive probabilistic shaped modulation for high-capacity free-space optical links. *J. Light. Technol.* **2020**, *38*, 6529–6541. [CrossRef]
71. Yu, Y.; Xu, M.; Pu, M.; Ding, J.; Chen, S.; Zhang, Y.; Shi, S.; Guo, Y.; Li, X.; Ma, X.; et al. 640 Gbit/s FSO turbulence-resilient field trial utilizing the cylindrical vector beam. *Opt. Lett.* **2025**, *50*, 237–240. [CrossRef] [PubMed]
72. Bitachon, B.I.; Horst, Y.; Kulmer, L.; Blatter, T.; Keller, K.; Bonnefois, A.M.; Conan, J.M.; Lim, C.; Montri, J.; Perrault, P.; et al. Tbit/s single channel 53 km free-space optical transmission-assessing the feasibility of optical GEO-satellite feeder links. In Proceedings of the European Conference and Exhibition on Optical Communication, Basel, Switzerland, 18–22 September 2022; Optica Publishing Group: Washington, DC, USA, 2022; p. Th3A.6.
73. Fernandes, M.A.; Monteiro, P.P.; Guiomar, F.P. Free-space terabit optical interconnects. *J. Light. Technol.* **2022**, *40*, 1519–1526. [CrossRef]
74. Shao, W.; Wang, Y.; Jia, S.; Xie, Z.; Gao, D.; Wang, W.; Zhang, D.; Liao, P.; Little, B.E.; Chu, S.T.; et al. Terabit FSO communication based on a soliton microcomb. *Photonics Res.* **2022**, *10*, 2802–2808. [CrossRef]
75. Wang, K.; Zhu, B.; Wei, Y.; Wang, C.; Zhou, W.; Zhao, L.; Chen, Y.; Zhao, X.; Yu, J. Experimental demonstration of a 96 channel WDM-FSO system based mobile fronthaul with 1024QAM delta-sigma-modulation. In Proceedings of the 2023 Optical Fiber Communications Conference and Exhibition (OFC), San Diego, CA, USA, 5–9 March 2023; IEEE: Piscataway, NJ, USA, 2023; pp. 1–3.
76. Matsuda, K.; Binkai, M.; Koshikawa, S.; Yoshida, T.; Sano, H.; Konishi, Y.; Suzuki, N. Demonstration of a real-time 14 Tb/s multi-aperture transmit single-aperture receive FSO system with class 1 eye-safe transmit intensity. *J. Light. Technol.* **2022**, *40*, 1494–1501. [CrossRef]
77. Tong, W.; Ding, J.; Meng, J.; Xu, X.; Zhang, J.; Hua, B.; Cai, Y.; Lei, M.; Xin, Z.; Yu, J.; et al. 400-Gbit/s Net Rate DP-OFDM signals fiber-THz-fiber seamless transmission employing hybrid time-frequency-polarization domain equalization. *J. Light. Technol.* **2024**, *42*, 5148–5155. [CrossRef]
78. Ding, J.; Zhang, L.; Liu, J.; Li, W.; Wang, Y.; Wang, K.; Zhao, L.; Zhou, W.; Zhang, J.; Zhu, M.; et al. THz-over-fiber transmission with a net rate of 5.12 Tbps in an 80 channel WDM system. *Opt. Lett.* **2022**, *47*, 3103–3106. [CrossRef]
79. Guiomar, F.P.; Fernandes, M.A.; Nascimento, J.L.; Rodrigues, V.; Monteiro, P.P. Coherent free-space optical communications: Opportunities and challenges. *J. Light. Technol.* **2022**, *40*, 3173–3186. [CrossRef]
80. Jia, S.; Xie, Z.; Shao, W.; Wang, Y.; He, Y.; Zhang, D.; Liao, P.; Wang, W.; Gao, D.; Wang, W.; et al. 150 Gbit/s 1 km high-sensitivity FSO communication outfield demonstration based on a soliton microcomb. *Opt. Express* **2022**, *30*, 35300–35310. [CrossRef] [PubMed]
81. Fernandes, M.A.; Fernandes, G.M.; Brandao, B.T.; Freitas, M.M.; Kaai, N.; Tomeeva, A.; van Der Wielen, B.; Reid, J.; Raiteri, D.; Monteiro, P.P.; et al. Achieving multi-terabit FSO capacity with coherent WDM transmission over a 1.8 km field trial. In Proceedings of the IET Conference Proceedings CP839, Yunlin, Taiwan, 21–23 October 2023; Volume 2023, pp. 1238–1241.
82. Castro, C.; Elschner, R.; Merkle, T.; Schubert, C.; Freund, R. Long-range high-speed THz-wireless transmission in the 300 GHz band. In Proceedings of the 2020 Third International Workshop on Mobile Terahertz Systems (IWMTS), Essen, Germany, 1–2 July 2020; IEEE: Piscataway, NJ, USA, 2020; pp. 1–4.

83. Blatter, T.; Kulmer, L.; Vukovic, B.; Horst, Y.; Destraz, M.; Smajic, J.; Leuthold, J. Dual-sideband receiver enabling 160 Gbps direct subTHz-to-optical conversion over 1400 m. In Proceedings of the Optical Fiber Communication Conference, San Diego, CA, USA, 24–28 March 2024; Optica Publishing Group: San Diego, CA, USA, 2024; p. M2F.2.
84. Sen, P.; Siles, J.V.; Thawdar, N.; Jornet, J.M. Multi-kilometre and multi-gigabit-per-second sub-terahertz communications for wireless backhaul applications. *Nat. Electron.* **2023**, *6*, 164–175. [CrossRef]
85. O'Hara, J.F.; Strecker, K.L.; Captain, G.L.; Tuschhoff, F.; Mendis, R.; D'Agostino, M.; Gates, J. Long-Distance, Multi-Gigabit-Per-Second Terahertz Wireless Communication. In Proceedings of the MILCOM 2024—2024 IEEE Military Communications Conference (MILCOM), Washington, DC, USA, 28 October–1 November 2024; IEEE: Piscataway, NJ, USA, 2024; pp. 270–275.
86. Feng, Y.; Zhang, B.; Qiao, C.; Dai, B. A 220-GHz-Band 31.2-Gbps Dual-Carrier Real-Time Wireless Link Using 64-QAM Modulation. In Proceedings of the 2021 14th UK-Europe-China Workshop on Millimetre-Waves and Terahertz Technologies (UCMMT), Lancaster, UK, 13–15 September 2021; IEEE: Piscataway, NJ, USA, 2021; pp. 1–3.
87. Wei, Y.; Yu, J.; Zhao, X.; Yang, X.; Wang, M.; Li, W.; Tian, P.; Han, Y.; Zhang, Q.; Tan, J.; et al. Demonstration of a photonics-aided 4600-m wireless transmission system in the sub-THz band. *J. Light. Technol.* **2024**, *42*, 8564–8576. [CrossRef]
88. Juan, L.; Zhou, R.; Wu, Q.; Wang, Y.; Luo, X.; Lin, C.; Deng, X. A 27 km over sea surface, 500Mbps, real time wireless communication system at 0.14 THz. In Proceedings of the 2021 46th International Conference on Infrared, Millimeter and Terahertz Waves (IRMMW-THz), Chengdu, China, 29 August–3 September 2021; IEEE: Piscataway, NJ, USA, 2021; pp. 1–2.
89. Wei, Y.; Yang, X.; Yu, J.; Zhao, X.; Wang, M.; Tan, J.; Li, W.; Zhao, F.; Bian, C.; Ma, H.; et al. Experimental demonstration of D-band signals wireless transmission over 30 km. *Sci. China Technol. Sci.* **2025**, *68*, 1280901. [CrossRef]
90. Ding, J.; Wang, M.; Li, W.; Zhu, B.; Wang, F.; Wang, C.; Wang, Y.; Zhou, W.; Zhang, J.; Zhu, M.; et al. Wireless transmission of a 200-m PS-64QAM THz-wave signal using a likelihood-based selection radius-directed equalizer. *Opt. Lett.* **2022**, *47*, 3904–3907. [CrossRef] [PubMed]
91. Ding, J.; Li, W.; Wang, Y.; Zhang, J.; Zhu, M.; Zhou, W.; Yu, J. 124.8-Gbit/s PS-256QAM signal wireless delivery over 104 m in a photonics-aided terahertz-wave system. *IEEE Trans. Terahertz Sci. Technol.* **2022**, *12*, 409–414. [CrossRef]
92. Li, W.; Yu, J.; Zhu, B.; Wang, F.; Ding, J.; Zhang, J.; Zhu, M.; Zhao, F.; Xie, T.; Wang, K.; et al. Photonic terahertz wireless communication: Towards the goal of high-speed kilometer-level transmission. *J. Light. Technol.* **2023**, *42*, 1159–1172. [CrossRef]
93. Song, L.Z.; Zhang, T.; Lai, J.X.; Yang, Y.; Du, J. A 180-GHz to 220-GHz wideband transmitarray with wide-angle beam steering for intersatellite communications. *IEEE Trans. Antennas Propag.* **2023**, *72*, 950–955. [CrossRef]
94. Balasiano, O.; Wohlgemuth, E.; Attia, I.; Roizman, A.; Falk, T.; Vered, R.; Sadot, D. Demonstration of coherent beam combining for atmospheric free space optical communication over 10 km. *J. Light. Technol.* **2024**, *42*, 7085–7094. [CrossRef]
95. Poliak, J.; Calvo, R.M.; Rein, F. Demonstration of 1.72 Tbit/s optical data transmission under worst-case turbulence conditions for ground-to-geostationary satellite communications. *IEEE Commun. Lett.* **2018**, *22*, 1818–1821. [CrossRef]
96. Dochhan, A.; Poliak, J.; Surof, J.; Richerzhagen, M.; Kelemu, H.F.; Calvo, R.M. 13.16 Tbit/s free-space optical transmission over 10.45 km for geostationary satellite feeder-links. In Proceedings of the Photonic Networks—20th ITG-Symposium, Leipzig, Germany, 8 May 2019; pp. 1–3.
97. Yu, X.; Zhang, L.; Zhang, Y.; Song, Y.; Tian, M.; Wang, T.; Tong, S. 2.5 Gbps free-space optical transmission between two 5G airship floating base stations at a distance of 12 km. *Opt. Lett.* **2021**, *46*, 2156–2159. [CrossRef]
98. Karpathakis, S.F.; Dix-Matthews, B.P.; Gozzard, D.R.; Schediwy, S.W. High-bandwidth coherent optical communication over 10.3 km of turbulent air. *Appl. Opt.* **2023**, *62*, G85–G89. [CrossRef]
99. Horst, Y.; Bitachon, B.I.; Kulmer, L.; Brun, J.; Blatter, T.; Conan, J.M.; Montmerle-Bonnefois, A.; Montri, J.; Sorrente, B.; Lim, C.B.; et al. Tbit/s line-rate satellite feeder links enabled by coherent modulation and full-adaptive optics. *Light Sci. Appl.* **2023**, *12*, 153. [CrossRef] [PubMed]
100. Mai, V.V.; Pham, A.T. Performance analysis of parallel FSO/MMW systems with adaptive rate under weather effects. In Proceedings of the 2015 21st Asia-Pacific Conference on Communications (APCC), Kyoto, Japan, 14–16 October 2015; IEEE: Piscataway, NJ, USA, 2015; pp. 193–198.
101. Zhang, J.; Wang, J.; Xu, Y.; Xu, M.; Lu, F.; Cheng, L.; Yu, J.; Chang, G.k. Fiber–wireless integrated mobile backhaul network based on a hybrid millimeter-wave and free-space-optics architecture with an adaptive diversity combining technique. *Opt. Lett.* **2016**, *41*, 1909–1912. [CrossRef] [PubMed]
102. Zhang, R.; Lu, F.; Xu, M.; Liu, S.; Peng, P.C.; Shen, S.; He, J.; Cho, H.J.; Zhou, Q.; Yao, S.; et al. An ultra-reliable MMW/FSO A-RoF system based on coordinated mapping and combining technique for 5G and beyond mobile fronthaul. *J. Light. Technol.* **2018**, *36*, 4952–4959. [CrossRef]
103. Wang, K.; Yu, J.; Wei, Y.; Chi, N.; Li, X.; Xiao, J.; Hu, F.; Zhao, M.; Zhu, B.; Zhang, J.; et al. Delivery of 1.196-Tb/s signal over 800 M based on RF/FSO convergence. In Proceedings of the 45th European Conference on Optical Communication (ECOC 2019), Dublin, Ireland, 22–26 September 2019; pp. 1–4.
104. Song, S.; Liu, Y.; Wu, J.; Wu, T.; Zhao, L.; Guo, L. Demonstration of intelligent hybrid FSO/RF system based on enhanced GRU prediction and real-world meteorological dataset. *J. Light. Technol.* **2022**, *40*, 7048–7059. [CrossRef]

105. Liu, Y.; Tan, X.; Jia, J.; Dong, B.; Huang, C.; Luo, P.; Shi, J.; Chi, N.; Zhang, J. A hybrid millimeter-wave and free-space-optics communication architecture with adaptive diversity combining and HARQ techniques. *Photonics* **2023**, *10*, 1320. [CrossRef]
106. Brandao, B.T.; Fernandes, M.A.; Loureiro, P.A.; Guiomar, F.P.; Monteiro, P.P. Cooperative FSO and mmWave system for reliable 200G wireless transmission. *IEEE Photonics Technol. Lett.* **2022**, *34*, 1333–1336. [CrossRef]
107. Gao, W.; Liu, K.; Han, C.; Chen, Z. Seamless Gbps Hybrid FSO/THz Communication With Intelligent Switching. *J. Light. Technol.* **2025**, *43*, 9079–9089. [CrossRef]
108. Nadeem, F.; Kvicera, V.; Awan, M.S.; Leitgeb, E.; Muhammad, S.S.; Kandus, G. Weather effects on hybrid FSO/RF communication link. *IEEE J. Sel. Areas Commun.* **2009**, *27*, 1687–1697. [CrossRef]
109. Touati, A.; Abdaoui, A.; Touati, F.; Uysal, M.; Bouallegue, A. On the effects of combined atmospheric fading and misalignment on the hybrid FSO/RF transmission. *J. Opt. Commun. Netw.* **2016**, *8*, 715–725. [CrossRef]
110. Sharma, S.; Madhukumar, A.; R, S. Switching-based cooperative decode-and-forward relaying for hybrid FSO/RF networks. *J. Opt. Commun. Netw.* **2019**, *11*, 267–281. [CrossRef]
111. Shah, S.; Siddharth, M.; Vishwakarma, N.; Swaminathan, R.; Madhukumar, A. Adaptive-combining-based hybrid FSO/RF satellite communication with and without HAPS. *IEEE Access* **2021**, *9*, 81492–81511. [CrossRef]
112. Singya, P.K.; Makki, B.; D'Errico, A.; Alouini, M.S. Hybrid FSO/THz-based backhaul network for mmWave terrestrial communication. *IEEE Trans. Wirel. Commun.* **2022**, *22*, 4342–4359. [CrossRef]
113. Nock, K.; Font, C.; Rupar, M. Adaptive transmission algorithms for a hard-switched FSO/RF link. In Proceedings of the MILCOM 2016—2016 IEEE Military Communications Conference, Baltimore, MD, USA, 1–3 November 2016; IEEE: Piscataway, NJ, USA, 2016; pp. 877–881.
114. Yu, L.; Chen, J.; Ding, G. Spectrum prediction via long short term memory. In Proceedings of the 2017 3rd IEEE International Conference on Computer and Communications (ICCC), Chengdu, China, 13–16 December 2017; IEEE: Piscataway, NJ, USA, 2017; pp. 643–647.
115. Amirabadi, M.A.; Kahaei, M.H.; Nezamalhosseini, S.A.; Vakili, V.T. Deep learning for channel estimation in FSO communication system. *Opt. Commun.* **2020**, *459*, 124989. [CrossRef]
116. Meng, Y.; Liu, Y.; Song, S.; Yang, Y.; Guo, L. Predictive link switching for energy efficient FSO/RF communication system. In Proceedings of the Asia Communications and Photonics Conference. Optica Publishing Group, Chengdu, China, 2–5 November 2019; p. T4C.1.
117. Zhang, W.; Hranilovic, S.; Shi, C. Soft-switching hybrid FSO/RF links using short-length raptor codes: Design and implementation. *IEEE J. Sel. Areas Commun.* **2009**, *27*, 1698–1708. [CrossRef]
118. Tang, Y.; Brandt-Pearce, M.; Wilson, S.G. Link adaptation for throughput optimization of parallel channels with application to hybrid FSO/RF systems. *IEEE Trans. Commun.* **2012**, *60*, 2723–2732. [CrossRef]
119. Amay, M.; Bas, J. On hybrid free-space optic-radio systems as enablers of 6G services over non-terrestrial networks. In Proceedings of the 2023 IEEE International Conference on Acoustics, Speech, and Signal Processing Workshops (ICASSPW), Rhodes Island, Greece, 4–10 June 2023; IEEE: Piscataway, NJ, USA, 2023; pp. 1–5.
120. He, B.; Schober, R. Bit-interleaved coded modulation for hybrid RF/FSO systems. *IEEE Trans. Commun.* **2009**, *57*, 3753–3763. [CrossRef]
121. Tang, C.; Jiang, M.; Zhao, C. Analysis and optimization design of P-LDPC coded hybrid FSO/RF systems. In Proceedings of the 2015 International Conference on Wireless Communications & Signal Processing (WCSP), Nanjing, China, 15–17 October 2015; IEEE: Piscataway, NJ, USA, 2015; pp. 1–5.
122. Merrouche, A.; Aissaoui, A. Enhancing performance in hybrid FSO/RF communication systems under atmospheric turbulence and pointing errors. *J. Opt.* **2024**, *53*, 4238–4248. [CrossRef]
123. Mai, V.V.; Pham, A.T. Adaptive multi-rate designs for hybrid FSO/RF systems over fading channels. In Proceedings of the 2014 IEEE Globecom Workshops (GC Wkshps), Austin, TX, USA, 8–12 December 2014; IEEE: Piscataway, NJ, USA, 2014; pp. 469–474.
124. Son, I.K.; Mao, S. A survey of free space optical networks. *Digit. Commun. Netw.* **2017**, *3*, 67–77. [CrossRef]
125. Wang, X.; Li, H.; Wu, Q. Optimizing adaptive coding and modulation for satellite network with ML-based CSI prediction. In Proceedings of the 2019 IEEE Wireless Communications and Networking Conference (WCNC), Marrakesh, Morocco, 15–18 April 2019; IEEE: Piscataway, NJ, USA, 2019; pp. 1–6.
126. Liu, J.; Awwad, É.; Jaouën, Y. Multi-dimensional energy limitation in sphere shaping for nonlinear interference noise mitigation. In Proceedings of the 2023 Asia Communications and Photonics Conference/2023 International Photonics and Optoelectronics Meetings (ACP/POEM), Wuhan, China, 4–7 November 2023; IEEE: Piscataway, NJ, USA, 2023; pp. 1–5.
127. Liu, J.; Awwad, É.; Hafermann, H.; Jaouën, Y. Sequence selection with dispersion-aware metric for long-haul transmission systems. *J. Light. Technol.* **2024**, *42*, 4818–4828. [CrossRef]

128. Elzanaty, A.; Alouini, M.S. Adaptive coded modulation for IM/DD free-space optical backhauling: A probabilistic shaping approach. *IEEE Trans. Commun.* **2020**, *68*, 6388–6402. [CrossRef]
129. Wang, X.; Zhao, X.; Ding, X.; Shu, C. Adaptive probabilistic shaping of free-space optical systems based on an APD detector. *Opt. Express* **2025**, *33*, 10687–10701. [CrossRef]

MDPI AG
Grosspeteranlage 5
4052 Basel
Switzerland
Tel.: +41 61 683 77 34

Micromachines Editorial Office
E-mail: micromachines@mdpi.com
www.mdpi.com/journal/micromachines

www.ingramcontent.com/pod-product-compliance
Lightning Source LLC
LaVergne TN
LVHW070453061126
851858LV00011B/227
9783725883097